Percolation

Geoffrey Grimmett

Percolation

With 76 Illustrations

Springer-Verlag
New York Berlin Heidelberg
London Paris Tokyo

Geoffrey Grimmett
School of Mathematics
University of Bristol
Bristol BS8 1TW
England

Mathematics Subject Classification (1980): 60K35

Library of Congress Cataloging-in-Publication Data
Grimmett, Geoffrey.
Percolation / Geoffrey Grimmett.
p. cm.
Bibliography: p.
Includes index.
ISBN 0-387-96843-1. ISBN 3-540-96843-1
1. Percolation (Statistical physics) I. Title.
QC174.85.P45G75 1989
530.1′3—dc19 88-24850
CIP

Typeset by Asco Trade Typesetting Ltd., Hong Kong.
Printed and bound by R.R. Donnelley & Sons, Harrisonburg, Virginia.
Printed in the United States of America.

9 8 7 6 5 4 3 2 1

ISBN 0-387-96843-1 Springer-Verlag New York Berlin Heidelberg
ISBN 3-540-96843-1 Springer-Verlag Berlin Heidelberg New York

Preface

Quite apart from the fact that percolation theory had its origin in an honest applied problem, it is a source of fascinating problems of the best kind for which a mathematician can wish: problems which are easy to state with a minimum of preparation, but whose solutions are apparently difficult and require new methods. At the same time, many of the problems are of interest to or proposed by statistical physicists and not dreamed up merely to demonstrate ingenuity.

Much progress has been made in recent years, and many of the open problems of ten years ago have been solved. With such solutions we have seen the evolution of new techniques and questions; the consequent knowledge has shifted the ground under percolation, and it is time to examine afresh the mathematics of the subject.

The quantity of literature related to percolation seems to grow hour by hour, mostly in the physics journals. It is becoming increasingly difficult to get to know the subject from scratch, and one of the principal purposes of this book is to remedy this.

This book is about the mathematics of percolation theory, with the emphasis upon presenting the shortest rigorous proofs of the main facts. I have made certain sacrifices in order to maximize the accessibility of the theory, and the major one has been to restrict myself to the special case of bond percolation on the cubic lattice $\mathbb{Z}^d$. Thus there is only little discussion of such processes as mixed, anisotropic, long-range, first-passage, and oriented percolation, some of which may soon need a book in their own right. Neither have I spent much time or space on the relationship of percolation to statistical physics, infinite particle systems, disordered media, reliability theory, and so on. With the exception of the final chapter, I have tried to stay reasonably close to core material of the sort which most graduate students in the area might aspire to know. No critical reader will agree entirely with my selection, and physicists may sometimes feel that my intuition is crooked; I have tried to straighten it out to my own satisfaction, but have had only partial success.

Almost all the results and arguments of this book are valid for all bond and site percolation models, subject to minor changes only; the principal exceptions are those results of Chapter 9 which make use of the self-duality of bond percolation on the square lattice. I have no especially convincing reason for my decision to study bond percolation rather than the more general case of site percolation, but was swayed in this direction by historical reasons as well as the consequential easy access to the famous exact calculation of the critical probability of bond percolation on the square lattice. In addition, unlike with site models, it is easy to formulate a bond model having interactions which are long-range rather than merely nearest-neighbour. Such arguments indicate the scanty importance I associate with this decision.

Here are a few words about the contents of this book. In the introductory Chapter 1 we prove the existence of a critical value p_c for the edge-probability p, marking the arrival on the scene of an infinite open cluster. The next chapter contains a general account of the three basic techniques—the FKG and BK inequalities, and Russo's formula—together with certain other useful inequalities, some drawn from reliability theory. This is followed by two proofs that the mean cluster size is finite whenever $p < p_c$; time will tell something about the relative powers of these two proofs, and it may be somewhat premature to judge this so soon after their discoveries. The next chapter contains material concerning the number of open clusters per vertex. We study subcritical percolation (that is, when $p < p_c$) in Chapter 5, and supercritical percolation (that is, when $p > p_c$) in Chapter 6. Chapter 7 contains a sketch of the physical approach to the critical phenomenon (when $p = p_c$), and includes an attempt to communicate to mathematicians the spirit of scaling theory and renormalization. Rigorous results are currently rather limited, as is witnessed by the brevity of Chapter 8. Chapter 9 is devoted to percolation in two dimensions, where the technique of planar duality leads to the famous exact calculation that $p_c = \frac{1}{2}$ for bond percolation on $\mathbb{Z}^2$. The book terminates with ten pencil sketches of related random processes, including 'continuous' percolation, first-passage percolation, and random electrical networks. The choice of the contents of Chapter 10 is rather idiosyncratic, one of the purposes of the chapter being to make available certain interesting but currently somewhat inaccessible results.

Most of this book was written in draft form while I was visiting Cornell University for the spring semester of 1987, a visit assisted by a grant from the Fulbright Commission. It is a pleasure to acknowledge the passive and active assistance of Rick Durrett, Michael Fisher, Roberto Schonmann, and Frank Spitzer during this period. Later, Pete Shiarly helped me to do simulations of percolation, and particularly with the testing matter of getting the Bristol University computer to draw round dots. The manuscript was cleaned and mended during the spring semester of 1988, which I spent at the University of Arizona at Tucson, with financial support from the Center for the Study of Complex Systems and AFOSR contract no. F49620-86-C-0130. One of the principal benefits of this visit was the opportunity for unbounded conversations with David Barsky and Chuck Newman. Rosine Bonay is responsible for the cover design and index, as well as for various other unspecifiable acts.

I make special acknowledgement to John Hammersley; not only did he oversee the early life of percolation, but also his unashamed love of a good problem has been an inspiration to many.

Unstinting in his help has been Harry Kesten. He read and commented in detail on much of the manuscript, his suggestions for improvements being so numerous to render individual acknowledgements difficult. Without his support the job would have taken much longer and been done rather worse, if at all.

G. R. G.
May 1988

Contents

1

What Is Percolation?

1.1 Modelling a Random Medium

Suppose we immerse a large porous stone in a bucket of water. What is the probability that the centre of the stone is wetted? In formulating a simple stochastic model for such a situation, Broadbent and Hammersley (1957) gave birth to the 'percolation model'. In two dimensions their model amounts to the following. Let $\mathbb{Z}^2$ be the plane square lattice and let p be a number satisfying $0 \leq p \leq 1$. We examine each edge of $\mathbb{Z}^2$ in turn, and declare this edge to be *open* with probability p and *closed* otherwise, independently of all other edges. The edges of $\mathbb{Z}^2$ represent the inner passageways of the stone, and the parameter p is the proportion of passages which are broad enough to allow water to pass along them. We think of the stone as being modelled by a large, finite subsection of $\mathbb{Z}^2$ (see Figure 1.1), perhaps those vertices and edges of $\mathbb{Z}^2$ contained in some specified connected subgraph of $\mathbb{Z}^2$. On immersion of the stone in water, a vertex x inside the stone is wetted if and only if there is a path in $\mathbb{Z}^2$ from x to a vertex on the boundary of the stone, using open edges only. Percolation theory is concerned primarily with the existence of such 'open paths'.

If we delete the closed edges, we are left with a random subgraph of $\mathbb{Z}^2$; we shall study the structure of this subgraph, particularly with regard to the way in which this structure depends on the numerical value of p. It is not unreasonable to postulate that the fine structure of the interior passageways of the stone is on a scale which is negligible when compared with the overall size of the stone. In such circumstances, the probability that a vertex buried near the centre of the stone is wetted by water permeating into the stone from its surface will behave rather similarly to the probability that this vertex is the endvertex of an infinite path of open edges in $\mathbb{Z}^2$. That is to say, the large-scale penetration

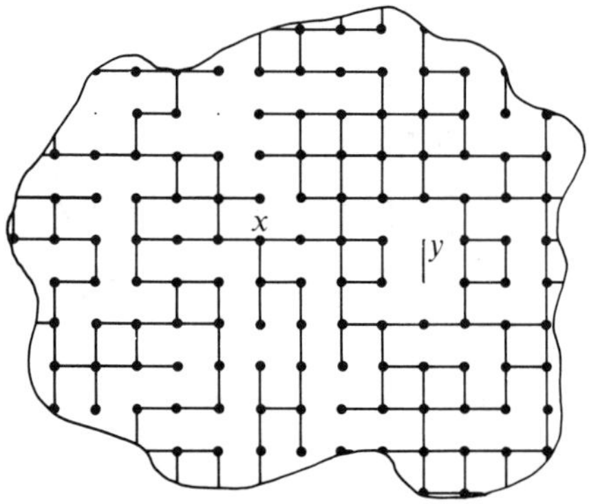

Figure 1.1. A sketch of the structure of a two-dimensional porous stone. The lines indicate the open edges; closed edges have been omitted. On immersion of the stone in water, vertex x will be wetted by the invasion of water, but vertex y will remain dry.

of the stone by water is related to the existence of infinite connected clusters of open edges. When can such infinite clusters exist? Simulations are handy indicators of the likely structure of the lattice, and Figure 1.2 contains such pictures for four different values of p. When $p = 0.25$, the connected clusters of open edges are isolated and rather small. As p increases, the sizes of clusters increase also, and there is a critical value of p at which there forms a cluster which pervades the entire picture. In loose terms, as we throw in more and more open edges, there comes a moment when large-scale connections are formed across the lattice. The pictures in Figure 1.2 are of course finite. If we were able to observe the whole of the infinite lattice $\mathbb{Z}^2$, then we would see that all open clusters are finite when p is small, but that there exists an infinite open cluster for large values of p. In other words, there exists a critical value p_c for the edge-density p such that all open clusters are finite when $p < p_c$, but there exists an infinite open cluster when $p > p_c$ (such remarks should be interpreted 'with probability 1'). Drinkers of Pernod are familiar with this type of phenomenon—the transparence of a glass of Pernod is undisturbed by the addition of a small amount of water, but in the process of adding the water drop by drop, there arrives an instant at which the mixture becomes opaque.

The occurrence of a 'critical phenomenon' is central to the appeal of percolation. In physical terms, we might say that the wetting of the stone is a 'surface effect' when the proportion p of open edges is small, and a 'volume effect' when p is large.

The above process is called 'bond percolation on the square lattice', and it is the most studied to date of all percolation processes. It is a very special process, largely because the square lattice has a certain property of self-duality which turns out to be extremely valuable. More generally, we begin with some periodic lattice in, say, d dimensions together with a number p satisfying $0 \leq p \leq 1$, and we declare each edge of the lattice to be open with probability p and closed otherwise. The resulting process is called a 'bond' model since the random blockages in the lattice are associated with the *edges*. Another type of percolation process is the 'site' model, in which the *vertices* rather than the edges are declared to be open or closed at random, the closed vertices being thought of as junctions which are blocked to the passage of fluid. It is well known that every bond model may be reformulated as a site model on a different lattice, but that the converse is false (see Section 10.1). Thus site models are more general than bond models.

We may continue to generalize in several directions such as (i) 'mixed' models, in which both edges and vertices may be blocked, (ii) anisotropic models, in which different edges may have different probabilities of being open, (iii) long-range models, in which direct flow is possible between pairs of vertices which are very distant (in the above formulation, this may require a graph with infinite vertex degrees), (iv) dependent percolation, in which the states of different edges are not independent, and so on. Mathematicians have a considerable talent in the art of generalization, and this has not been wasted on percolation theory. Such generalizations are often of considerable mathematical and physical interest; we shall however take the opposite route in this book. With few exceptions, we shall consider only bond percolation on the d-dimensional cubic lattice $\mathbb{Z}^d$ where $d \geq 1$, and the main reasons for this are as follows. As the level of geometrical generality rises, the accessibility of results in percolation theory is often diminished. Arguments which are relatively simple to explain in a special case can become concealed in morasses of geometrical and analytical detail when applied to some general model. This is not always the case, as illustrated by the recent proofs of the uniqueness of critical probabilities (see Chapter 3) and the uniqueness of the infinite open cluster when it exists (see Chapter 6). It is of course important to understand the limitations of an argument, but there may also be virtue in trying to describe something of the theory when stripped of peripheral detail. Bond percolation on $\mathbb{Z}^d$ is indeed a special case, but probably it exhibits the majority of properties expected of more general percolation-type models.

(a) $p = 0.25$

(b) $p = 0.49$

Figure 1.2. Realizations of bond percolation on a 50×60 section of the square lattice for four different values of p. The pictures have been created using the same sequence of pseudorandom numbers, with the result that each graph is a subgraph of the next. Readers with good eyesight may care to check that open

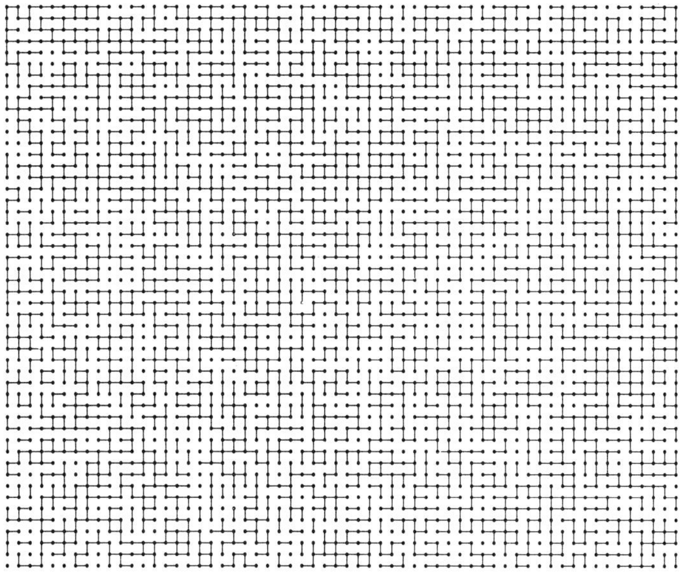

(c) $p = 0.51$

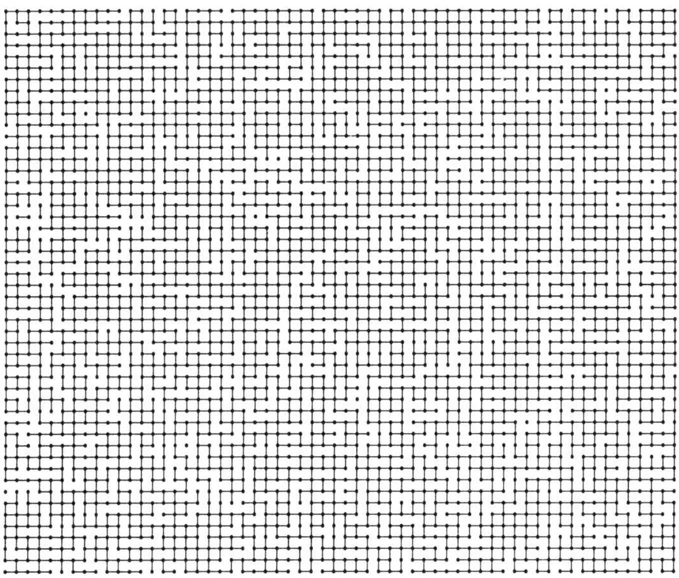

(d) $p = 0.75$

Figure 1.2 (*continued*)

paths exist joining the left to the right side when $p = 0.51$ but not when $p = 0.49$. The (random) value of p at which such paths appear for this realization is 0.5059....

1.2 Why Percolation?

As a model for a disordered medium, percolation is one of the simplest, incorporating as it does a minimum of statistical dependence. Its attractions are manyfold. First, it is easy to formulate but not unrealistic in its qualitative predictions for random media. Secondly, for those with a greater interest in more complicated processes, it is a playground for developing mathematical techniques and insight. Thirdly, it is well endowed with beautiful conjectures which are easy to state but apparently rather hard to settle.

It has been claimed that percolation theory is a cornerstone of the theory of disordered media. As evidence to support this claim, we make brief reference to four types of disordered physical systems, emphasizing the role of percolation for each.

A. *Disordered electrical networks.* It may not be too difficult to calculate the effective electrical resistance of a block of either material A or of material B, but what is the effective resistance of a mixture of these two materials? If the mixture is disordered, then it may be reasonable to assume that each component of the block is chosen at random to be of type A or of type B, independently of the types of all other components. The resulting effective resistance is a random variable whose distribution depends on the proportion p of components of type A. It seems to be difficult to say much of interest about the way in which this distribution depends on the numerical value of p. An extreme example arises when material B is a perfect insulator, and it is in this case that percolation comes to the fore. We illustrate this in a special case.

Let U_n be the square section $\{0, 1, \ldots, n\} \times \{0, 1, \ldots, n\}$ of the square lattice, and let S_n and T_n be the bottom and top sides of U_n,

$$S_n = \{(m, 0): 0 \le m \le n\}, \qquad T_n = \{(m, n): 0 \le m \le n\}.$$

We turn U_n into an electrical network as follows. We examine each edge of U_n in turn, and replace it by a wire of resistance 1 ohm with probability p, otherwise removing the connection entirely; this is done independently of all other edges. We now replace S_n and T_n by silver bars and we apply a potential difference between these bars; see Figure 1.3. What is the effective resistance R_n of the network? The value of R_n depends on the density and geometry of the set of edges having unit resistance, and such matters lie in the domain of percolation theory. We shall see in Section 10.9 that $R_n = \infty$ for all large n (almost surely) if $p < \frac{1}{2}$, whereas R_n is bounded uniformly away from 0 and ∞ if $p > \frac{1}{2}$.

B. *Ferromagnetism.* One of the most studied critical phenomena of theoretical physics is that of the ferromagnet. We position a lump of an appropriate metal in a magnetic field and we observe the way in which

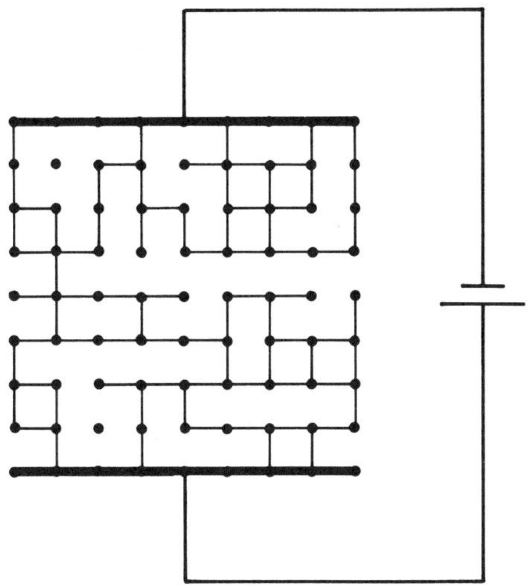

Figure 1.3. A realization of a random electrical network. Each remaining edge has unit resistance.

the magnetization of the metal varies according to imposed oscillations in the external magnetic field. Suppose that we increase the external field from 0 to some given value, and then decrease it back to 0. If the temperature is sufficiently large, the metal retains no residual magnetization, whereas at low temperatures the metal keeps some of its induced magnetization. There exists a critical value T_c of the temperature, called the Curie point, marking the borderline between the existence and non-existence of so called 'spontaneous magnetization'. The standard mathematical model for this phenomenon is the 'Ising model'. We give no definition of the Ising model here, but make instead some general remarks. In the Ising model on the lattice $\mathbb{Z}^d$, each vertex of $\mathbb{Z}^d$ may be in either of two states labelled 0 and 1. A configuration is an assignment $\omega = (\omega(x)\colon x \in \mathbb{Z}^d)$ of 0 or 1 to each vertex of the lattice. We consider probability measures on the set Ω of configurations taken in conjunction with some suitable σ-field of subsets of Ω; in particular, we are concerned with a class of measures having a type of 'spatial Markov' property: conditional on the states of all vertices outside any finite connected subgraph G of $\mathbb{Z}^d$, the states of vertices of G depend only on those of vertices in its 'external boundary'. Taken in conjunction with certain other conditions (positivity and translation invariance of the conditional probabilities, and positive correlation of increasing events), this property characterizes the class of measures of interest for the model. It turns out that there are two parameters which specify the conditional probabilities: the 'external magnetic field' h, and the strength J of interaction between neighbours. If $J = 0$ then the states of different vertices are independent, and the process is equivalent to site percolation. Whilst this particular case is of relatively little physical interest, it is worth noting that techniques and insight developed for percolation are often valuable for the Ising model in greater generality.

For more information about the relationship between percolation and the Ising model, see Essam (1971), Fortuin and Kasteleyn (1972), and Fortuin (1972); more recent references include Aizenman, Chayes, Chayes, and Newman (1988) and Imbrie and Newman (1988).

C. *Epidemics and fires in orchards.* In an early review of percolation and related topics, Frisch and Hammersley (1963) proposed the use of percolation in modelling the spread of blight in a large orchard. The problem is as follows. Hypothetical trees are grown at the vertices of a square lattice. We suppose that there is probability p that a healthy tree will be infected by a neighbouring blighted tree, where p is a known function of the distance between neighbouring trees. To prevent a single blighted tree from endangering a significant proportion of the whole orchard, it is necessary to choose the lattice spacing to be large enough that p is smaller than the critical probability of bond percolation on $\mathbb{Z}^2$.

In a forest fire, trees which are completely destroyed by fire cannot threaten their neighbours. Similarly, trees which have recovered from measles presumably gain protection from recurrence of the disease. Such observations may be incorporated into a more complicated model which takes into account the passage of time. Suppose that each tree may be in any of three states: 1 (live and not on fire), 0 (burning), and -1 (burned). We suppose that the tree at vertex x burns for a random time T_x after catching fire, where $(T_x: x \in \mathbb{Z}^2)$ is a family of independent, identically distributed random variables. A burning tree emits sparks in the manner of a Poisson process with rate α, and each spark hits one of the neighbouring trees chosen at random; the spark sets fire to that tree so long as it is neither burned nor already on fire. At time 0, an arsonist sets light to the tree at the origin. It turns out that the set C of trees which are ultimately burned in the ensuing conflagration may be identified as the set of vertices reachable from the origin by open paths of a certain percolation-type process; this process differs from ordinary bond percolation in that the states of two different edges may be dependent if the edges have a vertex in common. See Cox and Durrett (1988) and the references therein.

D. *Wafer-scale integration.* In the manufacture of microchips, silicon wafers are engraved with copies of the required circuitry, these copies being laid out in a square grid. The wafer is then broken up into the individual chips, many of which are usually found to be faulty. After elimination of the faulty chips, the remaining non-defective chips are used to build processors. There are sound engineering and computing reasons for preferring to leave the wafer intact, making use of the non-defective chips in the positions in which they occurred in the wafer. Interconnections may be made between functioning units by using channels built between the rows and columns of the grid of chips. Such questions arise as the following:

(i) How long is the longest linear chain of functioning units which may be created using interconnections each of length not exceeding δ lattice units and laid in such a way that each channel of the wafer contains no more than two such interconnections?
(ii) Find the minimal interconnection length in a wiring pattern which creates a square grid of size $k \times k$ of functioning chips out of a wafer containing $n \times n$ units in all.

Greene and El Gamal (1984) answer such questions under the hypothesis that each chip is non-defective with probability p, independently of all other chips. Under this assumption, the set of functioning chips may be identified as the set of open vertices in a site percolation process on $\mathbb{Z}^2$, and thus the theory of percolation is important.

1.3 Bond Percolation

In this section we shall establish the basic definitions and notation of bond percolation on $\mathbb{Z}^d$. We begin with some graph theory. Throughout most of this book, the letter d stands for the dimension of the process; generally $d \geq 2$, but we assume for the moment only that $d \geq 1$. We write $\mathbb{Z} = \{\ldots, -1, 0, 1, \ldots\}$ for the set of all integers, and $\mathbb{Z}^d$ for the set of all vectors $x = (x_1, x_2, \ldots, x_d)$ with integral coordinates. For $x \in \mathbb{Z}^d$ we generally write x_i for the ith coordinate of x. The *distance* $\delta(x, y)$ from x to y is defined by

(1.1) $$\delta(x, y) = \sum_{i=1}^{d} |x_i - y_i|,$$

and we write $|x|$ for the distance $\delta(0, x)$ from the origin to x. We shall sometimes have use for another distance function on $\mathbb{Z}^d$, and shall write

(1.2) $$\|x\| = \max\{|x_i|: 1 \leq i \leq d\},$$

noting that $\|x\| \leq |x| \leq d\|x\|$. We may turn $\mathbb{Z}^d$ into a graph, called the *d-dimensional cubic lattice*, by adding edges between all pairs x, y of points of $\mathbb{Z}^d$ with $\delta(x, y) = 1$. We denote this lattice by $\mathbb{L}^d$, and we write $\mathbb{Z}^d$ for the set of vertices of $\mathbb{L}^d$, and $\mathbb{E}^d$ for the set of its edges. In graph-theoretic terms, we write $\mathbb{L}^d = (\mathbb{Z}^d, \mathbb{E}^d)$. We shall often think of $\mathbb{L}^d$ as a graph embedded in $\mathbb{R}^d$, the edges being straight line segments between their endvertices. If $\delta(x, y) = 1$, then we say that x and y are *adjacent*; in this case, we write $x \sim y$ and we represent the edge from x

to y as $\langle x, y\rangle$. The edge e is *incident* to the vertex x if x is an endvertex of e. Letters such as u, v, w, x, y usually represent vertices, and letters such as e, f usually represent edges. We denote the origin of $\mathbb{Z}^d$ by 0.

Next we introduce probability. Let p and q satisfy $0 \le p \le 1$ and $p + q = 1$. We declare each edge of $\mathbb{L}^d$ to be *open* with probability p and *closed* otherwise, independently of all other edges. More formally, we consider the following probability space. As sample space we take $\Omega = \prod_{e \in \mathbb{E}^d}\{0, 1\}$, points of which are represented as $\omega = (\omega(e): e \in \mathbb{E}^d)$ and called *configurations*; the value $\omega(e) = 0$ corresponds to e being closed, and $\omega(e) = 1$ corresponds to e being open. We take $\mathscr{F}$ to be the σ-field of subsets of Ω generated by the finite-dimensional cylinders. Finally, we take product measure with density p on $(\Omega, \mathscr{F})$; this is the measure

$$P_p = \prod_{e \in \mathbb{E}^d} \mu_e$$

where μ_e is Bernoulli measure on $\{0, 1\}$, given by $\mu_e(\omega(e) = 0) = q$ and $\mu_e(\omega(e) = 1) = p$. We write P_p for this product measure, and E_p for the corresponding expectation operator. We shall occasionally need a more general construction in which different edges may have different probabilities of being open. Such a construction begins with a family $\mathbf{p} = (p(e): e \in \mathbb{E}^d)$ with $0 \le p(e) \le 1$ for all e. The appropriate probability space is now $(\Omega, \mathscr{F}, P_{\mathbf{p}})$ where $P_{\mathbf{p}} = \prod_{e \in \mathbb{E}^d}\mu_e$ and $\mu_e(\omega(e) = 0) = 1 - p(e)$, $\mu_e(\omega(e) = 1) = p(e)$ for each e.

We write A^c for the complement of an event A, and I_A for the indicator function of A:

$$I_A(\omega) = \begin{cases} 1 & \text{if } \omega \in A, \\ 0 & \text{if } \omega \notin A. \end{cases}$$

The expression $E_p(X; A)$ denotes the mean of X on the event A; that is to say, $E_p(X; A) = E_p(XI_A)$.

The following notation will be of value later. Let f be an edge of $\mathbb{L}^d$. We write P_p^f for Bernoulli product measure on $\prod_{e \ne f}\{0, 1\}$, the set of configurations of all edges of the lattice other than f. We think of P_p^f as being the measure associated with percolation on $\mathbb{L}^d$ with the edge f deleted.

There is a natural partial order on the set Ω of configurations, given by $\omega_1 \le \omega_2$ if and only if $\omega_1(e) \le \omega_2(e)$ for all $e \in \mathbb{E}^d$.

There is a one–one correspondence between Ω and the set of subsets of $\mathbb{E}^d$. For $\omega \in \Omega$, we define

$$K(\omega) = \{e \in \mathbb{E}^d: \omega(e) = 1\}; \tag{1.3}$$

thus $K(\omega)$ is the set of open edges of the lattice when the configuration is ω. Clearly, $\omega_1 \le \omega_2$ if and only if $K(\omega_1) \subseteq K(\omega_2)$.

The following device can be useful. Suppose that $(X(e): e \in \mathbb{E}^d)$ is a family of independent random variables indexed by the edge set $\mathbb{E}^d$, where each $X(e)$ is uniformly distributed on $[0, 1]$. We may couple together *all* bond percolation processes on $\mathbb{L}^d$ as p ranges over the interval $[0, 1]$ in the following way. Let p satisfy $0 \le p \le 1$ and define η_p $(\in \Omega)$ by

$$\eta_p(e) = \begin{cases} 1 & \text{if } X(e) < p, \\ 0 & \text{if } X(e) \ge p. \end{cases} \tag{1.4}$$

We call the edge e *p-open* if $\eta_p(e) = 1$. The random vector η_p has independent components and marginal distributions given by $P(\eta_p(e) = 0) = 1 - p$, $P(\eta_p(e) = 1) = p$. We may think of η_p as being the random outcome of the bond percolation process on $\mathbb{L}^d$ with edge-probability p. It is clear that $\eta_{p_1} \le \eta_{p_2}$ whenever $p_1 \le p_2$, which is to say that we may couple together the two percolation processes with edge-probabilities p_1 and p_2 in such a way that the set of open edges of the first process is a subset of the set of open edges of the second. More generally, as p increases from 0 to 1, the configuration η_p runs through typical configurations of percolation processes with all edge-probabilities.

A *path* of $\mathbb{L}^d$ is an alternating sequence $x_0, e_0, x_1, e_1, \ldots, e_{n-1}, x_n$ of distinct vertices x_i and edges $e_i = \langle x_i, x_{i+1} \rangle$; such a path has *length* n and is said to connect x_0 to x_n. A *circuit* of $\mathbb{L}^d$ is an alternating sequence $x_0, e_0, x_1, e_1, \ldots, e_{n-1}, x_n, e_n, x_0$ of vertices and edges such that $x_0, e_0, \ldots, e_{n-1}, x_n$ is a path and $e_n = \langle x_n, x_0 \rangle$; such a circuit has length $n + 1$. We call a path or circuit *open* if all of its edges are open, and *closed* if all of its edges are closed. Two subgraphs of $\mathbb{L}^d$ are called *edge-disjoint* if they have no edges in common, and *disjoint* if they have neither edges nor vertices in common.

Consider the random subgraph of $\mathbb{L}^d$ containing the vertex set $\mathbb{Z}^d$ and the open edges only. The connected components of this graph are called *open clusters*. We write $C(x)$ for the open cluster containing the vertex x, and we call $C(x)$ the *open cluster at* x. The vertex set of $C(x)$ is the set of all vertices of the lattice which are connected to x by open paths, and the edges of $C(x)$ are the open edges of $\mathbb{L}^d$ which join pairs of such vertices. By the translation invariance of the lattice and the probability measure P_p, the distribution of $C(x)$ is independent of the choice of x. The open cluster $C(0)$ at the origin is therefore typical of such clusters, and we represent this cluster by the single letter C. Occasionally we shall use the term $C(x)$ to represent the *set of vertices* joined to x by open paths, rather than the graph of this open cluster. We shall be interested in the size of $C(x)$, and we denote by $|C(x)|$ the number of vertices in $C(x)$. We note that $C(x) = \{x\}$ whenever x is incident to no open edge.

If A and B are sets of vertices of $\mathbb{L}^d$, we shall write '$A \leftrightarrow B$' if there is an open path joining some vertex in A to some vertex in B; if $A \cap B \ne \varnothing$ then $A \leftrightarrow B$ trivially. Thus, for example, $|C(x)| = |\{y \in \mathbb{Z}^d : x \leftrightarrow y\}|$. We shall write '$A \not\leftrightarrow B$' if there is no open path from any vertex of A to any vertex of B, and '$A \leftrightarrow B$ off D' if there exists an open path joining some vertex in A to some vertex in B which uses no vertex in the set D.

If A is a set of vertices of the lattice, we write ∂A for the *surface* of A, being the set of vertices in A which are adjacent to some vertex not in A.

Our notation for boxes is the following. A *box* is a subset of $\mathbb{Z}^d$ of the form $B(a, b) = \{x \in \mathbb{Z}^d: a_i \leq x_i \leq b_i \text{ for all } i\}$, where a and b lie in $\mathbb{Z}^d$; we sometimes write

$$B(a, b) = \prod_{i=1}^{d} [a_i, b_i]$$

as a convenient shorthand. We often think of $B(a, b)$ as a subgraph of the lattice $\mathbb{L}^d$ suitably endowed with the edges which it inherits from the lattice.

We shall sometimes need to 'approximate' the infinite lattice $\mathbb{L}^d$ by large finite subsections of $\mathbb{L}^d$, and we introduce the following notation. We denote by $B(n)$ the box with side-length $2n$ and centre at the origin:

$$\begin{aligned} B(n) &= [-n, n]^d \\ &= \{x \in \mathbb{Z}^d: \|x\| \leq n\}. \end{aligned} \tag{1.5}$$

We may turn $B(n)$ into a graph by adding the edges which it inherits from $\mathbb{L}^d$. If x is a vertex of the lattice, we write $B(n, x)$ for the box $x + B(n)$ with side-length $2n$ and centre at x.

We write $\lfloor a \rfloor$ and $\lceil a \rceil$ for the integer part of the real number a, and the least integer not less than a, respectively. If $(a_n: n \geq 1)$ and $(b_n: n \geq 1)$ are sequences of real numbers, we write $a_n \sim b_n$ if $a_n/b_n \to 1$ as $n \to \infty$, and $a_n \approx b_n$ if $\log a_n/\log b_n \to 1$ as $n \to \infty$. Similarly, we write $f(p) \sim g(p)$ (respectively $f(p) \approx g(p)$) as $p \to \pi$ if $f(p)/g(p) \to 1$ (respectively $\log f(p)/\log g(p) \to 1$) as $p \to \pi$.

1.4 The Critical Phenomenon

A principal quantity of interest is the *percolation probability* $\theta(p)$, being the probability that a given vertex belongs to an infinite open cluster. By the translation invariance of the lattice and probability measure, we lose no generality by taking this vertex to be the origin, and thus we define

$$\theta(p) = P_p(|C| = \infty). \tag{1.6}$$

Alternatively, we may write

$$\theta(p) = 1 - \sum_{n=1}^{\infty} P_p(|C| = n). \tag{1.7}$$

It is easy to see that $|C| = \infty$ if and only if there is an infinite sequence $x_0, x_1, x_2, \ldots$ of distinct vertices such that $x_0 = 0$, $x_i \sim x_{i+1}$, and $\langle x_i, x_{i+1} \rangle$ is open for all i. Clearly θ is a non-decreasing function of p with $\theta(0) = 0$ and $\theta(1) = 1$.

It is fundamental to percolation theory that there exists a critical value $p_c = p_c(d)$ of p such that

$$\theta(p)\begin{cases} = 0 & \text{if} \quad p < p_c, \\ > 0 & \text{if} \quad p > p_c; \end{cases}$$

$p_c(d)$ is called the *critical probability* and is defined formally by

$$p_c(d) = \sup\{p: \theta(p) = 0\}. \tag{1.8}$$

The case of one dimension is of no interest since, if $p < 1$, there are infinitely many closed edges of $\mathbb{L}^1$ to the left and to the right of the origin almost surely, so that $\theta(p) = 0$ if $p < 1$; thus $p_c(1) = 1$. The situation is quite different in two and more dimensions, and we shall see in Theorem (1.10) that $0 < p_c(d) < 1$ if $d \geq 2$. *We shall assume henceforth that, in the absence of an indication to the contrary, d is at least* 2. See Figure 1.4 for a sketch of the function θ.

The d-dimensional lattice $\mathbb{L}^d$ may be embedded in $\mathbb{L}^{d+1}$ in a natural way as the projection of $\mathbb{L}^{d+1}$ onto the subspace generated by the first d coordinates; with this embedding, the origin of $\mathbb{L}^{d+1}$ is in an infinite open cluster for a particular value of p whenever it is in an infinite open cluster of the sublattice $\mathbb{L}^d$. Thus $\theta(p) = \theta_d(p)$ is non-decreasing in d, and we have shown that

$$p_c(d+1) \leq p_c(d) \qquad \text{for} \quad d \geq 1. \tag{1.9}$$

It is not very difficult (see Section 1.6) to show that strict inequality is valid here, in that $p_c(d+1) < p_c(d)$ for all $d \geq 1$.

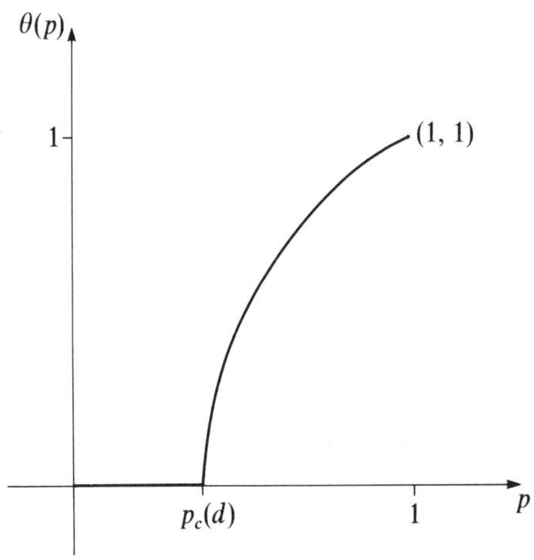

Figure 1.4. It is believed that the percolation probability $\theta(p)$ behaves roughly as indicated. It is known, for example, that θ is a continuous function of p except possibly at the critical probability $p_c(d)$. The possibility of a jump discontinuity at $p_c(d)$ has not been ruled out when $d \geq 3$.

The following theorem amounts to the statement that there is a non-trivial critical phenomenon in dimensions two and more.

(1.10) Theorem. *If* $d \geq 2$ *then* $0 < p_c(d) < 1$.

The nub of this theorem is that in two or more dimensions there are two phases of the process. In the *subcritical* phase when $p < p_c(d)$, every vertex is almost surely in a finite open cluster, so that all open clusters are almost surely finite. In the *supercritical* phase when $p > p_c(d)$, each vertex has a strictly positive probability of being in an infinite open cluster, so that there exists almost surely at least one infinite open cluster. These phases are now reasonably well understood, which is more than can be said about the intermediate *critical* percolation process with $p = p_c(d)$, to which we shall return in more detail in Chapter 7. We make more concrete the above remarks about the subcritical and supercritical phases.

(1.11) Theorem. *The probability* $\psi(p)$ *that there exists an infinite open cluster satisfies*

$$\psi(p) = \begin{cases} 0 & \text{if } \; p < p_c(d), \\ 1 & \text{if } \; p > p_c(d). \end{cases}$$

This theorem says nothing about the existence or non-existence of infinite open clusters when $p = p_c(d)$. It turns out that no infinite open cluster exists when $d = 2$ and $p = p_c(2)$, but it is an open question to determine whether or not there exists such a cluster when $d \geq 3$ and $p = p_c(d)$; it is expected that no such cluster exists. Theorem (1.11) is proved by an application of the zero–one law, and this tells us nothing about the actual *number* of infinite open clusters when $p > p_c(d)$; we shall however see in Section 6.2 that the infinite open cluster is (almost surely) unique whenever it exists.

Before proving these two theorems, we mention some associated results and open problems. First, what is the numerical value of $p_c(d)$? We know only the values $p_c(1) = 1$ and $p_c(2) = \frac{1}{2}$. The latter value is far from trivial to show, and this was the prize which attracted many people to the field in the 1970's. It is highly unlikely that there exists a useful representation of any other critical probability $p_c(d)$, although such values may be calculated with increasing degrees of accuracy with the aid of larger and faster computers. Exact values are known for the critical probabilities of certain other two-dimensional lattices (for example, $p_c = 2\sin(\pi/18)$ for bond percolation on the triangular lattice). It is the case that the value of the critical probability depends on both the dimension and the lattice structure, in contrast to certain asymptotic properties of $\theta(p)$ when p is near p_c; it is thought that, when $p - p_c$ is small and positive, then $\theta(p)$ behaves in a way which depends, to a degree, on the dimension d alone and is independent of the particular lattice structure. We return to this point in the next section and in Chapter 7.

Secondly, it is not difficult to find non-trivial upper and lower bounds for $p_c(d)$ when $d \geq 2$. We shall see in the proof of Theorem (1.10) that

$$\frac{1}{\lambda(2)} \leq p_c(2) \leq 1 - \frac{1}{\lambda(2)}, \tag{1.12}$$

and more generally

$$\frac{1}{\lambda(d)} \leq p_c(d) \qquad \text{for} \quad d \geq 3; \tag{1.13}$$

here, $\lambda(d)$ is the *connective constant* of $\mathbb{L}^d$, given by

$$\lambda(d) = \lim_{n\to\infty} \{\sigma(n)^{1/n}\}, \tag{1.14}$$

where $\sigma(n)$ is the number of paths (or 'self-avoiding walks') of $\mathbb{L}^d$ having length n and beginning at the origin. The exact value of $\lambda(d)$ is unknown for $d \geq 2$, but it is obvious that $\lambda(d) \leq 2d - 1$; to see this, note that each new step in a self-avoiding walk has at most $2d - 1$ choices since it must avoid the current position, and so $\sigma(n) \leq 2d(2d - 1)^{n-1}$.

Thirdly, how does $p_c(d)$ behave when d is large? Inequality (1.13) implies that $(2d - 1)p_c(d) \geq 1$, and Kesten (1988b) has shown that $p_c(d) \sim (2d)^{-1}$ as $d \to \infty$. This amounts to saying that, for large d, bond percolation on $\mathbb{L}^d$ behaves similarly to bond percolation on a regular branching tree in which each vertex has $2d(1 + o(1))$ descendants.

Proof of Theorem (1.10). The existence of a critical phenomenon was shown by Broadbent and Hammersley (1957) and Hammersley (1957a, 1959).

We saw in (1.9) that $p_c(d + 1) \leq p_c(d)$, and it suffices therefore to show that $p_c(d) > 0$ for $d \geq 2$, and $p_c(2) < 1$. We prove first that $p_c(d) > 0$ for $d \geq 2$. Consider bond percolation on $\mathbb{L}^d$ when $d \geq 2$. We shall show that $\theta(p) = 0$ whenever p is sufficiently close to 0. Let $\sigma(n)$ be the number of paths of $\mathbb{L}^d$ which have length n and which begin at the origin, and let $N(n)$ be the number of such paths which are open. Any such path is open with probability p^n, so that

$$E_p(N(n)) = p^n\sigma(n).$$

Now, if the origin is in an infinite open cluster then there exist open paths of all lengths beginning at the origin, so that

$$\begin{aligned} \theta(p) &\leq P_p(N(n) \geq 1) \\ &\leq E_p(N(n)) \\ &= p^n\sigma(n) \qquad \text{for all } n. \end{aligned} \tag{1.15}$$

By the definition of the connective constant $\lambda(d)$, we have that $\sigma(n) = \{\lambda(d) + o(1)\}^n$ as $n \to \infty$; we substitute this into (1.15) to obtain

$$\theta(p) \le \{p\lambda(d) + o(1)\}^n \to 0 \quad \text{if} \quad p\lambda(d) < 1 \tag{1.16}$$

as $n \to \infty$. Thus we have shown that $p_c(d) \ge \lambda(d)^{-1}$ where $\lambda(d) \ge (2d-1)^{-1} > 0$.

Finally, we show that $p_c(2) < 1$, and we use an approach which is commonly called a 'Peierls argument' in honour of Sir Rudolph Peierls and his work on the Ising model. Consider bond percolation on $\mathbb{L}^2$; we shall show that $\theta(p) > 0$ if p is sufficiently close to 1. It is here that planar duality is useful. Let G be a planar graph, drawn in the plane in such a way that edges intersect at vertices only. The *planar dual* of G is the graph obtained from G in the following way. We place a vertex in each face of G (including the infinite face if it exists) and join two such vertices by an edge whenever the corresponding faces of G share a boundary edge of G. It is easy to see (especially with the aid of Figure 1.5) that the dual of $\mathbb{L}^2$ is isomorphic to $\mathbb{L}^2$; this self-duality is not in itself important at this stage, but will be crucial to our forthcoming proof in Chapter 9 that $p_c(2) = \frac{1}{2}$. For the sake of definiteness, we take as vertices of this dual lattice the set $\{x + (\frac{1}{2}, \frac{1}{2}): x \in \mathbb{Z}^2\}$ and we join two such neighbouring vertices by a straight line segment of $\mathbb{R}^2$. There is a one–one correspondence between the edges of $\mathbb{L}^2$ and the edges of the dual, since each edge of $\mathbb{L}^2$ is crossed by a unique edge of the dual. We declare an edge of the dual to be open or closed depending respectively on whether it crosses an open or closed edge of $\mathbb{L}^2$. This assignment

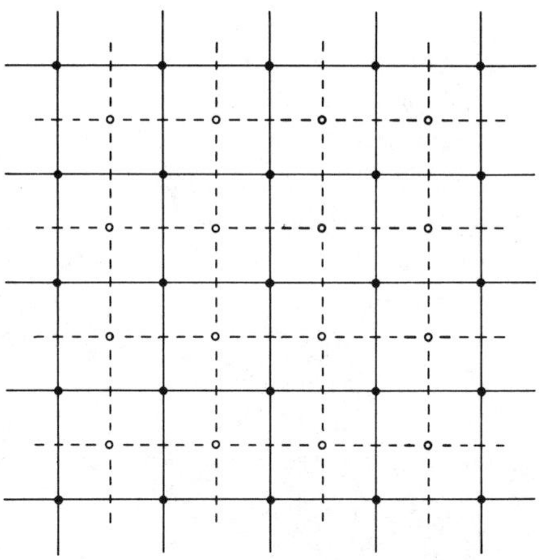

Figure 1.5. Part of the square lattice $\mathbb{L}^2$ together with its dual.

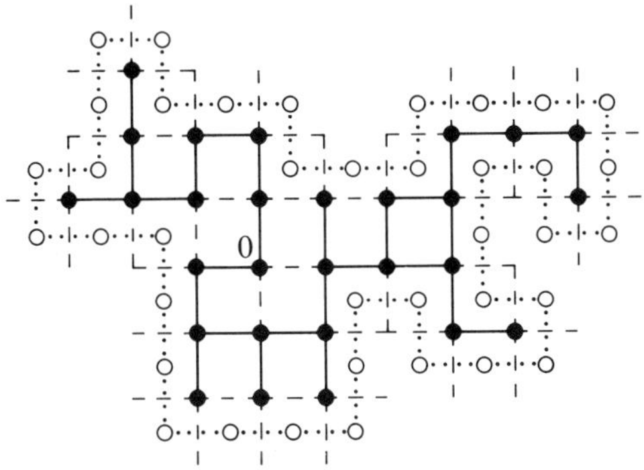

Figure 1.6. A finite open cluster at the origin, surrounded by a closed circuit in the dual lattice.

gives rise to a bond percolation process on the dual lattice with the same edge-probability p. We shall return to such matters in Chapter 9.

Suppose now that the open cluster at the origin of $\mathbb{L}^2$ is finite, and see Figure 1.6 for a sketch of the situation. We see that the origin is surrounded by a necklace of closed edges which are blocking off all possible routes from the origin to infinity. We may satisfy ourselves that the corresponding edges of the dual form a closed circuit in the dual which contains the origin of $\mathbb{L}^2$ in its interior. This is best seen by inspecting Figure 1.6 again. It is somewhat tedious to formulate and prove such a statement with complete rigour, and we shall not do so here; see Kesten (1982, p. 386) for a more careful treatment. The converse holds similarly: if the origin is in the interior of a closed circuit of the dual lattice, then the open cluster at the origin is finite. We summarize these remarks by saying that $|C| < \infty$ if and only if the origin of $\mathbb{L}^2$ is in the interior of a closed circuit of the dual.

We now proceed as in the first part of the proof, by counting the number of such closed circuits in the dual. Let $\rho(n)$ be the number of circuits in the dual which have length n and which contain in their interiors the origin of $\mathbb{L}^2$. We may estimate $\rho(n)$ as follows. Each such circuit passes through some vertex of the form $(k + \frac{1}{2}, \frac{1}{2})$ for $0 \leq k < n$ since, first, it surrounds the origin and therefore passes through $(k + \frac{1}{2}, \frac{1}{2})$ for some $k \geq 0$ and, secondly, it cannot pass through $(k + \frac{1}{2}, \frac{1}{2})$ where $k \geq n$ since then it would have length at least $2n$. Thus such a circuit contains a self-avoiding walk of length $n - 1$ starting from a vertex of the form $(k + \frac{1}{2}, \frac{1}{2})$ where $0 \leq k < n$. The number of such self-avoiding walks is at most $n\sigma(n - 1)$, giving that

$$\rho(n) \leq n\sigma(n - 1). \tag{1.17}$$

Let γ be a circuit of the dual containing the origin of $\mathbb{L}^2$ in its interior. By (1.17),

$$\text{(1.18)}\quad \sum_{\gamma} P_p(\gamma \text{ is closed}) \le \sum_{n=1}^{\infty} q^n n\sigma(n-1)$$

$$= \sum_{n=1}^{\infty} qn\{q\lambda(2) + o(1)\}^{n-1} \qquad \text{as in (1.16)}$$

$$< \infty \qquad \text{if} \quad q\lambda(2) < 1,$$

where $q = 1 - p$ and the summation is over all such γ. Furthermore,

$$\sum_{\gamma} P_p(\gamma \text{ is closed}) \to 0 \qquad \text{as} \quad q = 1 - p \downarrow 0,$$

so that we may find π satisfying $0 < \pi < 1$ such that

$$\sum_{\gamma} P_p(\gamma \text{ is closed}) \le \tfrac{1}{2} \qquad \text{if} \quad p > \pi.$$

It follows from the previous remarks that

$$P_p(|C| = \infty) = P_p(\text{no such circuit } \gamma \text{ is closed})$$

$$\ge 1 - \sum_{\gamma} P_p(\gamma \text{ is closed})$$

$$\ge \tfrac{1}{2} \qquad \text{if} \quad p > \pi,$$

giving that $p_c(2) \le \pi$. We need to work slightly harder in order to deduce that $p_c(2) \le 1 - \lambda(2)^{-1}$. Let N be a positive integer; let F_N be the event that there exists no closed circuit of the dual containing the origin in its interior and having length N or less, and let G_N be the corresponding event involving circuits with length exceeding N. If both F_N and G_N occur then $|C| = \infty$, by the earlier remarks, so that

$$\theta(p) \ge P_p(F_N \cap G_N)$$

$$= P_p(F_N | G_N) P_p(G_N).$$

We now use a correlation inequality. The events F_N and G_N are 'increasing' on the lattice of subsets of $\mathbb{E}^2$ in the sense that, if $\omega \in F_N$ (respectively G_N) and $\omega \le \omega'$ then $\omega' \in F_N$ (respectively G_N). This suggests that, if we are given that G_N occurs, then the (conditional) probability of F_N is at least as large as its unconditional probability, which is to say that $P_p(F_N | G_N) \ge P_p(F_N)$. This fact requires careful proof, and is an example of the so called 'FKG inequality' to which we shall return in Section 2.2. Assuming the validity of the above correlation inequality, we deduce that

$$\theta(p) \ge P_p(F_N) P_p(G_N).$$

Suppose now that $(1 - p)\lambda(2) < 1$. We have as in (1.18) that

$$P_p(G_N^c) \le \sum_{n=N}^{\infty} (1-p)^n n\sigma(n-1),$$

the tail of a convergent series, and therefore we may choose N such that

$P_p(G_N) \geq \frac{1}{2}$. The event F_N depends only on the states of edges within distance N of the origin; $P_p(F_N)$ is at least as large as the probability that all such edges are open, so that $P_p(F_N) > 0$. It follows that $\theta(p) > 0$ if $(1-p)\lambda(2) < 1$, which is to say that $p_c(2) \leq 1 - \lambda(2)^{-1}$ as asserted in (1.12). ■

Proof of Theorem (1.11). This is easy. First, we note that the event $\{\mathbb{L}^d$ contains an infinite open cluster$\}$ does not depend upon the states of any finite collection of edges. By the usual zero–one law (see, for example, Grimmett and Stirzaker (1982, p. 190)), ψ takes the values 0 and 1 only. If $p < p_c(d)$ then

$$\psi(p) \leq \sum_{x \in \mathbb{Z}^d} P_p(|C(x)| = \infty) = 0.$$

On the other hand, if $p > p_c(d)$ then

$$\psi(p) \geq P_p(|C| = \infty) > 0$$

so that $\psi(p) = 1$ by the zero–one law, as required. ■

1.5 The Main Questions

Consider bond percolation on $\mathbb{L}^d$ where $d \geq 2$. We are interested in the sizes and shapes of typical open clusters as the edge-probability p varies from 0 to 1, and we are particularly interested in large-scale phenomena such as the existence of infinite open clusters. We saw in the last section that 'macroscopic' quantities such as $\theta(p)$ and $\psi(p)$ have qualitatively different behaviour for small p than they have for large p. In addition to the probability that an open cluster is infinite, we may be interested in the mean size of an open cluster, and we write

(1.19) $$\chi(p) = E_p|C|$$

for the mean number of vertices in the open cluster at the origin. By the translation invariance of the process, we have that $\chi(p) = E_p|C(x)|$ for all vertices x. The functions θ and χ are two of the principal characters in percolation theory. We may express χ in terms of the distribution of $|C|$, just as we did for θ in (1.7):

(1.20) $$\begin{aligned}\chi(p) &= \infty \cdot P_p(|C| = \infty) + \sum_{n=1}^{\infty} n P_p(|C| = n)\\ &= \infty \cdot \theta(p) + \sum_{n=1}^{\infty} n P_p(|C| = n),\end{aligned}$$

so that

(1.21) $$\chi(p) = \infty \qquad \text{if} \quad p > p_c.$$

The converse is not at all obvious: is it the case that $\chi(p) < \infty$ if $p < p_c$? We answer this question affirmatively in Chapter 3 (a sketch of the function χ appears in Figure 1.7). This indicates that the 'macroscopic' quantities θ and χ manifest critical behaviour at the same value of p, and this is partial justification for the belief that there is a unique 'phase transition' in the percolation model. More precisely, it is believed that there are exactly two phases in the model—the subcritical phase ($p < p_c$) and the supercritical phase ($p > p_c$)—together with the process *at* the critical point (when $p = p_c$). We shall study these phases in some detail in Chapters 5–8, but we present here a brief preview of some of the main results and open problems.

Subcritical phase. When $p < p_c$, all open clusters are finite almost surely. We shall see in Chapter 5 that $|C|$ has a tail which decreases exponentially, which is to say that there exists $\alpha(p)$ such that

$$(1.22) \qquad P_p(|C| = n) \approx e^{-n\alpha(p)} \quad \text{as} \quad n \to \infty,$$

and $\alpha(p) > 0$ when $p < p_c$. It follows that $|C|$ has finite moments of all orders when $p < p_c$.

Supercritical phase. When $p > p_c$, there exist infinite open clusters almost surely, but how many? We shall see that the infinite open cluster is unique almost surely. If $|C| < \infty$ then how fast does the tail of $|C|$ decay? It is known that there exists $\beta(p) > 0$ such that

$$(1.23) \qquad P_p(|C| = n) \geq \exp(-\beta(p)n^{(d-1)/d}) \quad \text{for all } n,$$

and it is conjectured that the limit

$$(1.24) \qquad \delta(p) = \lim_{n\to\infty} \{-n^{-(d-1)/d} \log P_p(|C| = n)\}$$

exists and is strictly positive when $p > p_c$. The basic reason for the power $n^{(d-1)/d}$ is that this is the order of the surface area of the sphere in $\mathbb{R}^d$ with volume n. Lastly, a weaker form of the conjecture (1.24) is that finite open clusters have finite mean size; more specifically, we conjecture that

$$(1.25) \qquad \chi^f(p) < \infty \quad \text{if} \quad p > p_c$$

where

$$(1.26) \qquad \chi^f(p) = E_p(|C|; |C| < \infty),$$

the mean of $|C|$ on the event that $|C| < \infty$. The function χ^f is probably not dissimilar in general form to the sketch in Figure 1.7. The superscript 'f' refers to the condition that C be finite.

At the critical point. It is hereabouts that we find major open problems. First, is there an infinite open cluster when $p = p_c$? The answer is nega-

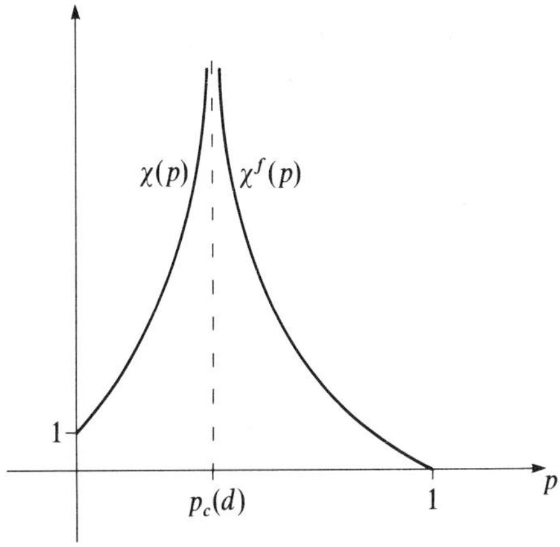

Figure 1.7. The left-hand curve is a sketch of the mean cluster size $\chi(p)$. The right-hand curve is a sketch of the mean size $\chi^f(p)$ of a finite open cluster. Note that $\chi(p) = \chi^f(p)$ if $p < p_c(d)$.

tive when $d = 2$, and is generally believed to be negative for $d \geq 3$ also. Assuming that $\theta(p_c) = 0$, so that there is no infinite open cluster at p_c, at what rate does $P_{p_c}(|C| = n)$ decay? It is believed that

(1.27) $$P_{p_c}(|C| \geq n) \approx n^{-1/\delta} \quad \text{as} \quad n \to \infty$$

for some $\delta = \delta(d) > 0$. Lower bounds for $P_p(|C| \geq n)$ of this general 'power' form are known for all dimensions $d \geq 2$, and also upper bounds when $d = 2$. Some have asked the question 'is it true that $\delta = 91/5$ when $d = 2$, and $\delta = 2$ when $d \geq 6$?'; see Newman (1987a), for example.

Near the critical point. As p approaches p_c from above (or beneath), how do such quantities as $\theta(p)$ (or $\chi(p)$) behave? It is commonly believed that such quantities behave as powers of $|p - p_c|$, and a major open problem of percolation is to prove this. That is to say, we conjecture that the limits

(1.28) $$\gamma = -\lim_{p \uparrow p_c} \frac{\log \chi(p)}{\log|p - p_c|},$$

(1.29) $$\beta = \lim_{p \downarrow p_c} \frac{\log \theta(p)}{\log|p - p_c|}$$

exist, and that the limit

(1.30) $$\delta^{-1} = -\lim_{n \to \infty} \frac{\log P_{p_c}(|C| \geq n)}{\log n}$$

exists, in agreement with (1.27). The quantities γ, β, δ are called 'critical exponents'. There are physical reasons for believing the hypothesis of 'universality': the numerical values of critical exponents may depend only on the dimension d and not on the structure of the particular lattice. We return to such questions in Chapters 7 and 8.

We close this section with a review of the principal characters. According to one method of counting, there are four such characters:

(a) the *percolation probability*

$$\theta(p) = P_p(|C| = \infty);$$

(b) the *mean size of the open cluster at the origin*

$$\chi(p) = E_p|C|;$$

(c) the *mean size of the finite open cluster at the origin*

$$\chi^f(p) = E_p(|C|; |C| < \infty).$$

(d) The fourth such principal character is the *number of open clusters per vertex*, defined by

$$\kappa(p) = E_p(|C|^{-1}), \tag{1.31}$$

with the convention that $1/\infty = 0$. That is to say,

$$\kappa(p) = \sum_{n=1}^{\infty} \frac{1}{n} P_p(|C| = n).$$

We study the function κ in more detail in Chapter 4.

We note that

$$\chi^f(p) = \chi(p) \qquad \text{whenever} \quad \theta(p) = 0. \tag{1.32}$$

There are many useful analogies between the percolation model and the Ising model. We shall not explore such analogies here, but note that θ corresponds to magnetization, χ^f to susceptibility, and κ to free energy per vertex.

The quantities χ, χ^f, and κ are moments of the number of *vertices* in C. There are good reasons to define these quantities instead in terms of the number of *edges* of C, principally since such a definition would enable a unified approach to both bond and site percolation. For bond percolation on $\mathbb{L}^d$ it matters little which route we adopt, and we have chosen that which leads to fewest technical complications later.

1.6 Notes

Section 1.1. The origins of the mathematical theory of percolation may be found in the work of Flory (1941), Broadbent (1954), and Broadbent and Hammersley (1957). Hammersley (1983) has described something of the history of the subject. We know of three books to date, the serious mathematical text of Kesten (1982), the gentle account by Efros (1986), and the book by Stauffer (1985). Of the many recent reviews, we mention Chayes and Chayes (1986a), Menshikov, Molchanov, and Sidorenko (1986), Aizenman (1987), Grimmett (1987b), Kesten (1987e), and Newman (1987a).

For general discussions of periodic lattices in d dimensions see Grimmett (1978), Godsil and McKay (1980), and Kesten (1982, Chapters 2, 3).

Introductions to mixed and anisotropic percolation models may be found in Kesten (1982, Section 3.2), Halley (1983), and Wierman (1984). For long-range models see Aizenman, Kesten, and Newman (1987) and the references therein. Oriented percolation has been reviewed by Durrett (1984); see also Durrett and Schonmann (1988b). The relationship between percolation and other models of statistical physics has been explored by Essam (1972, 1980); see also Fortuin (1972) and the references therein. Kesten (1986a, 1987e) has reviewed first-passage percolation. Wierman (1987) has studied 'high density' percolation in two dimensions; in this percolation-type process, one is interested only in those vertices which are incident to at least k open edges, for some specified value of k. Chapter 10 of this book contains some brief essays about some of the above topics.

Section 1.3. The device for defining all percolation processes on the same probability space seems to appear for the first time in Hammersley (1963).

Section 1.4. The existence of a critical phenomenon was observed by Broadbent and Hammersley (1957) and Hammersley (1957a, 1959). Some people say that 'percolation occurs' when there exists an infinite open cluster.

Bond percolation on the line $\mathbb{Z}$ is essentially the problem of 'runs'; see Feller (1968). The proof that $p_c(2) = \frac{1}{2}$ was performed by Kesten (1980a), who built upon earlier arguments of Harris (1960), Russo (1978), and Seymour and Welsh (1978). Wierman (1981) adapted Kesten's proof to calculate exact values for critical probabilities for bond percolation on the triangular and hexagonal lattices.

Kesten (1982) proved that $p_c(3) < p_c(2)$; see also the related work of Menshikov, Molchanov, and Sidorenko (1986, Section 4). J. van den Berg (unpublished) and A. Frieze (unpublished) have pointed out a simple argument which leads to the strict inequality $p_c(d+1) < p_c(d)$ for $d \geq 1$. The argument amounts to the following for $d = 2$. Each edge of $\mathbb{L}^2$ may be thought of as the bottom of two infinite ladders of $\mathbb{L}^3$. By subdividing the 'vertical' edges in such ladders, we may construct disjoint ladders above different edges. By choosing the edge-probabilities for the subdivided edges with care, we arrive at the conclusion that $p_c(3) \leq 0.4798$, whereas $p_c(2) = \frac{1}{2}$. Menshikov (1987a) has provided a more general argument (see also Zuev (1987)), showing that $p_c(\mathscr{L}_1) < p_c(\mathscr{L}_2)$ whenever $\mathscr{L}_2$ is a strict sublattice of $\mathscr{L}_1$ satisfying certain conditions; he is able to find non-trivial lower bounds for the difference $p_c(\mathscr{L}_2) - p_c(\mathscr{L}_1)$. His argument may be adapted to show strict inequality whenever $\mathscr{L}_2$ is obtained from $\mathscr{L}_1$ by the deletion of a periodic class of edges with the property that, for every such edge $e = \langle x, y \rangle$, there exist two disjoint infinite paths of $\mathscr{L}_1$ with initial vertices at x and at y. The corresponding condition for site percolation requires that each deleted vertex x be the endvertex of two infinite paths of $\mathscr{L}_1$ with the property that, apart from x, no vertex of one is a neighbour of any vertex of the second.

Kesten (1988b) has proved that $p_c(d) \sim (2d)^{-1}$ as $d \to \infty$; see also Gordon (1988), Kesten (1988c), and Kesten and Schonmann (1988).

The duality of planar lattices was explored by Hammersley (1959) and Harris (1960), and later by Fisher (1961). For an account of self-avoiding walks and the connective constant, see Hammersley (1957b).

Section 1.5. We defer the list of references until the appropriate forthcoming chapters.

2

Some Basic Techniques

2.1 Increasing Events

The probability $\theta(p)$, that the origin is in an infinite open cluster, is a non-decreasing function of p. This is intuitively obvious since an increase in the value of p leads generally to an increase in the number of open edges of $\mathbb{L}^d$, thereby increasing the number and length of open paths from the origin. Another way of putting this is to say that $\{|C| = \infty\}$ is an increasing event, in the sense that if $\omega \in \{|C| = \infty\}$ then $\omega' \in \{|C| = \infty\}$ whenever $\omega \leq \omega'$. With such an example in mind we make the following definition. The event A in $\mathscr{F}$ is called *increasing* if $I_A(\omega) \leq I_A(\omega')$ whenever $\omega \leq \omega'$, where I_A is the indicator function of A. We call A *decreasing* if its complement A^c is increasing.

More generally, a random variable N on $(\Omega, \mathscr{F}, P)$ is called *increasing* if $N(\omega) \leq N(\omega')$ whenever $\omega \leq \omega'$; N is called *decreasing* if $-N$ is increasing. Thus, an event A is increasing if and only if its indicator function is increasing.

As simple (and canonical) examples of increasing events and random variables, consider the event $A(x, y)$ that there is an open path joining x to y, and the number $N(x, y)$ of different open paths from x to y.

The following result is intuitively clear.

(2.1) Theorem. *If N is an increasing random variable on $(\Omega, \mathscr{F})$ then*

$$E_{p_1}(N) \leq E_{p_2}(N) \qquad \textit{whenever} \quad p_1 \leq p_2, \tag{2.2}$$

so long as these mean values exist. If A is an increasing event in $\mathscr{F}$ then

$$P_{p_1}(A) \leq P_{p_2}(A) \qquad \textit{whenever} \quad p_1 \leq p_2. \tag{2.3}$$

There are good reasons for working with monotonic (that is, increasing or decreasing) events whenever possible. In doing calculations, we are often required to estimate such quantities as (i) the probability $P_p(A \cap B)$ that both A and B occur, in terms of the individual probabilities $P_p(A)$ and $P_p(B)$, or (ii) the rate of change of the probability $P_p(A)$ as p increases. For *increasing* events A and B, there are general techniques for performing such estimates. We describe such techniques in the next three sections. The principal results of this chapter are the FKG and BK inequalities and Russo's formula. The reader is counselled to read at least Sections 2.2–2.4 before moving to the next chapter; some will choose to omit the formal proof of the BK inequality (Theorem (2.12)).

Before we prove Theorem (2.1), here is a general remark. The concept of monotonic events and the consequent techniques are rather general, and rely in no way upon the lattice structure of the percolation process. In the simple form in which we describe such techniques here, we require merely a probability space $(\Omega, \mathscr{F}, P)$ of the form $\Omega = \prod_{s \in S} \{0, 1\}$ where S is finite or countably infinite, $\mathscr{F}$ is the σ-field generated by the finite-dimensional cylinders of Ω, and P is a product measure on $(\Omega, \mathscr{F})$:

$$P = \prod_{s \in S} \mu_s$$

where μ_s is given by $\mu_s(\omega(s) = 0) = 1 - p(s)$, $\mu_s(\omega(s) = 1) = p(s)$ for vectors $\omega = (\omega(s): s \in S) \in \Omega$ and some specified collection $(p(s): s \in S)$ of numbers from the interval $[0, 1]$. We have such a probability space to hand already, and thus we describe the basic techniques for dealing with increasing events in terms of this space. We may not always make explicit references to the underlying probability space, and so we adopt the following convention. Unless otherwise stated, the word 'event' shall refer to a member of $\mathscr{F}$, and the term 'random variable' shall refer to a measurable real-valued function on $(\Omega, \mathscr{F})$, where $(\Omega, \mathscr{F}, P_p)$ is the basic probability space of bond percolation on $\mathbb{L}^d$.

Proof. Let the random variables $(X(e): e \in \mathbb{E}^d)$ be independent and uniformly distributed on $[0, 1]$. As in Section 1.3, we write $\eta_p(e) = 1$ if $X(e) < p$ and $\eta_p(e) = 0$ otherwise. If $p_1 \leq p_2$ then $\eta_{p_1} \leq \eta_{p_2}$, giving that $N(\eta_{p_1}) \leq N(\eta_{p_2})$ if N is an increasing random variable on $(\Omega, \mathscr{F})$. We take expectations of this inequality to obtain (2.2). The second part of the theorem follows immediately by applying the first part to the increasing random variable $N = I_A$. ∎

2.2 The FKG Inequality

It is highly plausible that increasing events A and B are positively correlated, in the sense that $P_p(A \cap B) \geq P_p(A)P_p(B)$. For example, if we know that there is an open path joining vertex u to vertex v, then it becomes more likely that there is an open path joining vertex x to vertex y. Such a correlation inequality was first proved by Harris (1960). Since then, such inequalities have been reworked in a more general context than product measure on $(\Omega, \mathscr{F})$; by current convention, they are named after Fortuin, Kasteleyn, and Ginibre (1971).

(2.4) Theorem. FKG inequality.

(a) *If X and Y are increasing random variables such that $E_p(X^2) < \infty$ and $E_p(Y^2) < \infty$, then*

$$E_p(XY) \geq E_p(X)E_p(Y). \tag{2.5}$$

(b) *If A and B are increasing events then*

$$P_p(A \cap B) \geq P_p(A)P_p(B). \tag{2.6}$$

Similar inequalities are valid for decreasing random variables and events. For example, if X and Y are both decreasing then $-X$ and $-Y$ are increasing, giving that

$$E_p(XY) \geq E_p(X)E_p(Y),$$

so long as X and Y have finite second moments. Similarly, if X is increasing and Y is decreasing, then we may apply the FKG inequality to X and $-Y$ to find that

$$E_p(XY) \leq E_p(X)E_p(Y).$$

As an example of the FKG inequality in action, we describe the way in which it is normally applied. Let $\Pi_1, \Pi_2, \ldots, \Pi_k$ be families of paths in $\mathbb{L}^d$, and let A_i be the event that some path in Π_i is open. The A_i's are increasing events, and therefore

$$P_p\left(\bigcap_{i=1}^{k} A_i\right) \geq P_p(A_1)P_p\left(\bigcap_{i=2}^{k} A_i\right),$$

since the intersection of increasing events is increasing. We iterate this to obtain

$$P_p\left(\bigcap_{i=1}^{k} A_i\right) \geq \prod_{i=1}^{k} P_p(A_i). \tag{2.7}$$

Proof of FKG inequality. We need only prove part (a), since part (b) follows by applying the first part to the indicator functions of A and B. First we prove part (a) for random variables X and Y which depend on the states of only finitely many edges; later we shall remove this restriction.

Suppose then that X and Y are increasing random variables which depend only on the states of the edges $e_1, e_2, \ldots, e_n$. We proceed by induction on n. First, suppose $n = 1$. Then X and Y are functions of the state $\omega(e_1)$ of e_1, which takes the values 0 and 1 with probabilities $1 - p$ and p, respectively. Now,

$$\{X(\omega_1) - X(\omega_2)\}\{Y(\omega_1) - Y(\omega_2)\} \geq 0$$

for all pairs ω_1, ω_2, each taking the value 0 or 1; this is trivial if $\omega_1 = \omega_2$, and follows from the monotonicity of X and Y otherwise. Thus

$$\begin{aligned} 0 &\leq \sum_{\omega_1, \omega_2} \{X(\omega_1)-X(\omega_2)\}\{Y(\omega_1)-Y(\omega_2)\}P_p(\omega(e_1)=\omega_1)P_p(\omega(e_1)=\omega_2) \\ &= 2\{E_p(XY)-E_p(X)E_p(Y)\} \end{aligned}$$

as required. Suppose now that the result is valid for values of n satisfying $n < k$, and that X and Y are increasing functions of the states $\omega(e_1)$, $\omega(e_2)$, ..., $\omega(e_k)$ of the edges e_1, e_2, ..., e_k. Then

$$\begin{aligned} E_p(XY) &= E_p(E_p\{XY|\omega(e_1), \ldots, \omega(e_{k-1})\}) \\ &\geq E_p(E_p\{X|\omega(e_1), \ldots, \omega(e_{k-1})\}E_p\{Y|\omega(e_1), \ldots, \omega(e_{k-1})\}), \end{aligned} \tag{2.8}$$

since, for given $\omega(e_1)$, ..., $\omega(e_{k-1})$, it is the case that X and Y are increasing in the single variable $\omega(e_k)$. Now, $E_p\{X|\omega(e_1), \ldots, \omega(e_{k-1})\}$ is an increasing function of the states of $k - 1$ edges, as is the corresponding function of Y. It follows from the induction hypothesis that the last mean value in (2.8) is no smaller than the product of the means, so that

$$\begin{aligned} E_p(XY) &\geq E_p(E_p\{X|\omega(e_1), \ldots, \omega(e_{k-1})\})E_p(E_p\{Y|\omega(e_1), \ldots, \omega(e_{k-1})\}) \\ &= E_p(X)E_p(Y). \end{aligned}$$

We now lift the condition that X and Y depend on the states of only finitely many edges. Suppose that X and Y are increasing random variables with finite second moments. Let e_1, e_2, ... be a (fixed) ordering of the edges of $\mathbb{L}^d$, and define

$$X_n = E_p(X|\omega(e_1), \ldots, \omega(e_n)), \qquad Y_n = E_p(Y|\omega(e_1), \ldots, \omega(e_n)). \tag{2.9}$$

Now X_n and Y_n are increasing functions of the states of e_1, e_2, ..., e_n, and therefore

$$E_p(X_n Y_n) \geq E_p(X_n)E_p(Y_n) \tag{2.10}$$

by the discussion above. As a consequence of the martingale convergence theorem (see Grimmett and Stirzaker (1982, p. 204)) we have that, as $n \to \infty$,

$$X_n \to X \quad \text{and} \quad Y_n \to Y \quad \text{a.s. } (P_p) \text{ and in } L^2(P_p),$$

so that

$$E_p(X_n) \to E_p(X) \quad \text{and} \quad E_p(Y_n) \to E_p(Y) \quad \text{as} \quad n \to \infty. \tag{2.11}$$

Also, by the triangle and Cauchy–Schwarz inequalities,

$$\begin{aligned} E_p|X_nY_n - XY| &\le E_p(|(X_n - X)Y_n| + |X(Y_n - Y)|) \\ &\le \sqrt{E_p((X_n - X)^2)E_p(Y_n^2)} + \sqrt{E_p(X^2)E_p((Y_n - Y)^2)} \\ &\to 0 \quad \text{as} \quad n \to \infty, \end{aligned}$$

so that $E_p(X_nY_n) \to E_p(XY)$. We take the limit as $n \to \infty$ in (2.10) to obtain the result. ∎

2.3 The BK Inequality

The positive correlation of increasing events is only half of the story; in many situations the direction of the FKG inequality is useless for the problem, and so we seek some complementary correlation-type inequality. It is, of course, hopeless to expect some useful general upper bound on $P_p(A \cap B)$ for increasing events A and B. It turns out that the right way to approach the question is to replace the event $A \cap B$, that both A and B occur, by the event $A \circ B$, that A and B occur on disjoint edge sets of $\mathbb{L}^d$. We make precise this idea of 'disjoint occurrence' as follows.

Let $e_1, e_2, \ldots, e_n$ be n distinct edges of $\mathbb{L}^d$, and let A and B be increasing events which depend on the vector $\omega = (\omega(e_1), \omega(e_2), \ldots, \omega(e_n))$ of the states of these edges only. Each such ω is specified uniquely by the set $K(\omega) = \{e_i : \omega(e_i) = 1\}$ of edges with state 1. We define the event $A \circ B$ to be the set of all ω for which there exists a subset H of $K(\omega)$ such that ω', determined by $K(\omega') = H$, belongs to A, and ω'', determined by $K(\omega'') = K(\omega) \backslash H$, belongs to B. We say that $A \circ B$ is the event that A and B 'occur disjointly'. Thus $A \circ B$ is the set of configurations ω for which there exist disjoint sets of open edges with the property that the first such set guarantees the occurrence of A and the second guarantees the occurrence of B.

The following is the canonical example of disjoint occurrence. Let G be a finite subgraph of $\mathbb{L}^d$, and let $A_G(x, y)$ be the event that there is an open path joining vertex x of G to vertex y of G, lying entirely within G. Then $A_G(u, v) \circ A_G(x, y)$ is the event that there exist two edge-disjoint open paths in G, the first joining u to v and the second joining x to y. Suppose now that we are given that $A_G(u, v)$ occurs, and we ask for the (conditional) probability of $A_G(u, v) \circ A_G(x, y)$. The conditioning on $A_G(u, v)$ amounts to knowing some information about the occurrence of open edges, but we are not allowed to use such open edges in finding an open path from x to y disjoint from one of the open paths from u to v. Thus it is plausible that

$$P_p\big(A_G(u, v) \circ A_G(x, y) | A_G(u, v)\big) \le P_p(A_G(x, y)).$$

This is essentially the assertion of the BK inequality, proved by van den Berg and Kesten (1985).

We shall state the BK inequality in a slightly more general context than that described above. Let m be a positive integer, $\Gamma = \prod_{i=1}^{m} \{0, 1\}$, $\mathscr{G}$ be the set of all subsets of Γ, and P be product measure on $(\Gamma, \mathscr{G})$ with density $p(i)$ on the ith coordinate of vectors in Γ; thus

$$P = \prod_{i=1}^{m} \mu_i,$$

where $\mu_i(0) = 1 - p(i)$ and $\mu_i(1) = p(i)$.

(2.12) Theorem. BK inequality. *If A and B are increasing events in $\mathscr{G}$ then*

$$P(A \circ B) \leq P(A)P(B). \tag{2.13}$$

We note that $A \circ B \subseteq A \cap B$ and that $A \circ B$ is increasing whenever A and B are increasing. The BK inequality may be generalized in the following direction. Let $A_1, A_2, \ldots, A_k$ be increasing events in $\mathscr{G}$. It is easily checked that the operation $\circ$ is associative, in that $A_i \circ (A_j \circ A_k) = (A_i \circ A_j) \circ A_k$, and repeated application of (2.13) yields

$$P(A_1 \circ A_2 \circ \cdots \circ A_k) \leq \prod_{i=1}^{k} P(A_i). \tag{2.14}$$

In the majority of applications of the BK inequality, we shall use the following version, obtained immediately from Theorem (2.12) by specializing to the case of the probability space $(\Omega, \mathscr{F}, P_p)$ of bond percolation on $\mathbb{L}^d$.

(2.15) Theorem. *Consider bond percolation on $\mathbb{L}^d$ and let A and B be increasing events which depend on only finitely many edges. Then*

$$P_p(A \circ B) \leq P_p(A)P_p(B). \tag{2.16}$$

The condition that the events depend on only finitely many edges is easily dropped in most potential applications, and we describe how to do this in a particular important context. Let $\Pi_1, \Pi_2, \ldots, \Pi_k$ be collections of paths in $\mathbb{L}^d$; remember that paths have finite length. Let $A_i(n)$ be the event that some path in Π_i, entirely contained in the box $B(n)$, is open; we write $A_i = \lim_{n\to\infty} A_i(n)$ for the event that some path in Π_i is open. It is not difficult to see that, as $n \to \infty$,

$$A_1(n) \circ A_2(n) \circ \cdots \circ A_k(n) \to A_1 \circ A_2 \circ \cdots \circ A_k.$$

We apply (2.14) and let $n \to \infty$ to deduce that

(2.17)

$$P_p(\text{there exist disjoint open paths } \pi_1 \in \Pi_1, \ldots, \pi_k \in \Pi_k) \leq P_p(A_1) \ldots P_p(A_k),$$

in contrast to the conclusion of the FKG inequality that

(2.18) $P_p(\text{there exist open paths } \pi_1 \in \Pi_1, \ldots, \pi_k \in \Pi_k) \geq P_p(A_1) \ldots P_p(A_k)$.

Van den Berg and Kesten have conjectured that a version of the BK inequality is valid for all pairs A, B of events, regardless of whether or not they are increasing. We perform a preliminary skirmish in advance of stating the conjecture formally. Let A and B be events of the probability space $(\Gamma, \mathscr{G}, P)$. For $\omega = (\omega_1, \omega_2, \ldots, \omega_m) \in \Gamma$ and $K \subseteq \{1, 2, \ldots, m\}$ we define the cylinder event $C(\omega, K)$ generated by ω on K by

$$C(\omega, K) = \{\omega' \in \Gamma: \omega'_i = \omega_i \text{ for } i \in K\}.$$

We now define $A \square B$ to be the set of all $\omega \in \Gamma$ for which there exist $K \subseteq \{1, 2, \ldots, m\}$ such that $C(\omega, K) \subseteq A$ and $C(\omega, K^c) \subseteq B$, where $K^c = \{1, 2, \ldots, m\} \backslash K$.

It is immediate that $A \square B = A \circ B$ if A and B are increasing events, and a little thought leads to the observation that $A \square B = A \cap B$ if A is increasing and B is decreasing. The conjecture of van den Berg and Kesten is that

$$P(A \square B) \leq P(A)P(B) \tag{2.19}$$

for *all* events A and B. We note that an affirmation of this conjecture would include the FKG inequality: for increasing events A and B, apply (2.19) to A and B^c, and use the fact that B^c is decreasing so that $A \square B^c = A \cap B^c$. It is currently an open problem to prove or disprove the general inequality (2.19). Certain special cases have been dealt with by van den Berg and Fiebig (1987), and we mention one of these. They show that (2.19) is valid whenever each of A and B is the intersection of some increasing event with some decreasing event.

There are various proofs of the BK inequality, of which we shall present possibly the easiest. We motivate this proof by describing the intuitive approach of van den Berg (1985). Let G be a finite graph and let $0 \leq p \leq 1$. We declare each edge of G to be open with probability p and closed otherwise, independently of all other edges. Let u, v, x, y be vertices of G, and let A and B be the events that there are open paths from u to v and from x to y, respectively. We concentrate on the event $A \circ B$ that there exist disjoint open paths joining u to v and x to y. Let e be an edge of G. We replace e by two parallel edges e' and e'', each of which is open with probability p, independently of all other edges. Suppose that, in this new graph, we look for open paths from u to v which do not use e'', and we look for open paths from x to y which do not use e'. Some thought suggests that this 'splitting' of e cannot decrease the probability of disjoint open paths from u to v and from x to y. We now 'split' each edge of G in turn, replacing each edge f by two parallel edges f' and f''. At each stage we seek two disjoint paths, the first being from u to v not using edges marked $''$, and the second being from x to y not using edges marked $'$. The probability of two such disjoint paths either increases or remains unchanged at each stage. When all the edges of G have been split, we are left with two independent copies of G, in the first of which we look for an open path from u to v and in the second of which we look for an open path from x to y; such paths occur independently, so that the probability that they both occur is $P_p(A)P_p(B)$. Thus $P_p(A \circ B) \leq P_p(A)P_p(B)$ as required. We now write this out formally.

Proof of Theorem (2.12). We first produce two copies of the probability space $(\Gamma, \mathscr{G}, P)$, and we denote these copies by $(\Gamma_1, \mathscr{G}_1, P_1)$ and $(\Gamma_2, \mathscr{G}_2, P_2)$. We are concerned with the product space $(\Gamma_1 \times \Gamma_2, \mathscr{G}_1 \times \mathscr{G}_2, P_{12})$ where $P_{12} = P_1 \times P_2$, and we write $x \times y$ for a typical point in $\Gamma_1 \times \Gamma_2$; thus $x = (x_1, x_2, \ldots, x_m)$ and $y = (y_1, y_2, \ldots, y_m)$ where each x_i and y_j equals either 0 or 1.

Let A and B be increasing events in $\mathscr{G}$. We define the following events in the product space: we write A' for the set of all points $x \times y \in \Gamma_1 \times \Gamma_2$ for which $x \in A$, and, for $0 \leq k \leq m$, we write B'_k for the set of points $x \times y$ for which the composite vector $(y_1, y_2, \ldots, y_k, x_{k+1}, \ldots, x_m)$ belongs to B. We note that A' and B'_k are increasing events in the product space. For each point $x \times y \in \Gamma_1 \times \Gamma_2$, we say that the subset I of $\{1, 2, \ldots, m\}$ *forces* A' if $u \times v \in A'$ whenever $u \times v \in \Gamma_1 \times \Gamma_2$ and $u_i = x_i$ for each $i \in I$; similarly, we say that I forces B'_k if $u \times v \in B'_k$ whenever $u_i = y_i$ for each $i \in I$ satisfying $i \leq k$ and $v_i = x_i$ for each $i \in I$ satisfying $i > k$.

The left-hand side of (2.13) is

$$P(A \circ B) = P_{12}(A' \circ B'_0). \tag{2.20}$$

On the other hand A' and B'_m are defined in terms of disjoint sets of coordinates of $\Gamma_1 \times \Gamma_2$, so that $A' \circ B'_m = A' \cap B'_m$. Thus

$$\begin{aligned} P_{12}(A' \circ B'_m) &= P_{12}(A' \cap B'_m) \\ &= P_{12}(A')P_{12}(B'_m) \qquad \text{since } P_{12} \text{ is a product measure} \\ &= P(A)P(B). \end{aligned}$$

We shall therefore be finished once we have shown that $P_{12}(A' \circ B'_0) \leq P_{12}(A' \circ B'_m)$, and to this end we claim that

$$P_{12}(A' \circ B'_{k-1}) \leq P_{12}(A' \circ B'_k) \qquad \text{for} \quad 1 \leq k \leq m. \tag{2.21}$$

The remainder of the proof is devoted to proving this inequality. It is somewhat tedious and some readers may prefer to leave the matter to their intuition.

We partition $A' \circ B'_{k-1}$ into two events: $A' \circ B'_{k-1} = C_1 \cup C_2$ where

$$C_1 = \{x \times y\colon A' \circ B'_{k-1} \text{ occurs regardless of the value of } x_k\},$$

$$C_2 = \{x \times y\colon x_k = 1 \text{ and } A' \circ B'_{k-1} \text{ occurs, but } A' \circ B'_{k-1} \text{ would not occur if the value of } x_k \text{ were changed to } 0\}.$$

Thus C_2 is the event $\{A' \circ B'_{k-1}$ occurs if and only if $x_k = 1\} \cap \{x_k = 1\}$, and C_1 is the remainder of $A' \circ B'_{k-1}$. We further partition C_2 into two events: $C_2 = C'_2 \cup C''_2$ where

$$C'_2 = C_2 \cap \{x \times y\text{: there exists } I \subseteq \{1, 2, \ldots, m\} \text{ such that } k \in I,\ I \text{ forces } A', \text{ and } I^c \text{ forces } B'_{k-1}\},$$

$$C''_2 = C_2 \setminus C'_2,$$

where $I^c = \{1, 2, \ldots, m\} \setminus I$. Thus C'_2 is the sub-event of C_2 on which we may think of x_k as contributing essentially to A'. We now construct a measure-preserving injection φ mapping $A' \circ B'_{k-1}$ into $A' \circ B'_k$, and we do this as follows. First, for $x \times y \in \Gamma_1 \times \Gamma_2$ we define $x' \times y'$ to be the point in $\Gamma_1 \times \Gamma_2$ obtained from $x \times y$ by interchanging x_k and y_k, so that

$$x'_i = x_i \quad \text{for} \quad i \neq k, \quad \text{and} \quad x'_k = y_k,$$

$$y'_i = y_i \quad \text{for} \quad i \neq k, \quad \text{and} \quad y'_k = x_k.$$

Let φ be defined on $A' \circ B'_{k-1}$ by

$$\varphi(x \times y) = \begin{cases} x \times y & \text{if} \quad x \times y \in C_1 \cup C'_2, \\ x' \times y' & \text{if} \quad x \times y \in C''_2. \end{cases} \tag{2.22}$$

We note first that $C_1 \subseteq A' \circ B'_k$, since if $x \times y \in C_1$ then A' and B'_{k-1} occur disjointly regardless of the values of x_k and y_k. Secondly, $C'_2 \subseteq A' \circ B'_k$, since if $x \times y \in C'_2$ then A' and B'_{k-1} occur disjointly and there exists such a disjoint occurrence in which x_k contributes essentially to A' and not therefore to B'_{k-1}; it follows from the definitions of A' and B'_k that $x \times y \in A' \circ B'_k$ also. Thirdly, if $x \times y \in C''_2$ then $\varphi(x \times y) = x' \times y' \in A' \circ B'_k$, since in this case $x_k = 1$ and there exists $I \subseteq \{1, 2, \ldots, m\} \setminus \{k\}$ such that I forces A' and I^c forces B'_{k-1}; from the point of view of the configuration $x' \times y'$, I forces A' and I^c forces B'_k, giving that $x' \times y' \in A' \circ B'_k$ as required. We have shown that φ maps $A' \circ B'_{k-1}$ into $A' \circ B'_k$. Certainly, φ is one–one on $C_1 \cup C'_2$, since φ coincides with the identity map here. Also, φ is one–one on C''_2, and therefore φ fails to be an injection on $A' \circ B'_{k-1}$ only if there exists $x \times y \in C''_2$ with $\varphi(x \times y) = x' \times y' \in C_1 \cup C'_2$. Suppose that $x \times y$ is such a configuration in C''_2. We cannot have $x' \times y' \in C_1$, since this would imply that the values of x_k and y_k are irrelevant to the occurrence of $A' \circ B'_{k-1}$, in contradiction of the assumption that $x \times y \in C''_2$, a subset of C_2. It is impossible also that $x' \times y' \in C'_2$, since in this case $x_k = 1$ and $y_k = 1$ by virtue of the assumptions that $x \times y \in C''_2$ and $x' \times y' \in C'_2$, respectively; thus $x' \times y' = x \times y$, which contradicts the assumption that one lies in C'_2 and the other in C''_2. Therefore φ is an injection from $A' \circ B'_{k-1}$ into $A' \circ B'_k$ as claimed. Finally, φ preserves measure since $P_{12}(\varphi(x \times y)) = P_{12}(x \times y)$ whenever $x \times y \in A' \circ B'_{k-1}$. Putting this together, we deduce that $P_{12}(A' \circ B'_{k-1}) \leq P_{12}(A' \circ B'_k)$, and (2.21) is proved. ■

2.4 Russo's Formula

We saw in Theorem (2.1) that, for increasing events A, the probability $P_p(A)$ is a non-decreasing function of p. Our third basic technique is a method for estimating the rate of change of $P_p(A)$ as a function of p. We may motivate the argument as follows. First, in comparing $P_p(A)$ and $P_{p+\delta}(A)$, it is useful to construct the two processes having densities p and $p + \delta$ on the same probability space in the usual way. Let $(X(e)\colon e \in \mathbb{E}^d)$ be independent random variables with the uniform distribution on $[0, 1]$, and define $\eta_p(e) = 1$ if $X(e) < p$ and $\eta_p(e) = 0$ otherwise. Then, for increasing events A,

$$P_{p+\delta}(A) - P_p(A) = P(\eta_p \notin A, \eta_{p+\delta} \in A). \tag{2.23}$$

If $\eta_p \notin A$ but $\eta_{p+\delta} \in A$, there must exist edges e satisfying $\eta_p(e) = 0$, $\eta_{p+\delta}(e) = 1$, which is to say that $p \leq X(e) < p + \delta$. Let $E_{p,\delta}$ be the set of such edges, and assume that A depends on the states of only finitely many edges. Clearly $P(|E_{p,\delta}| \geq 2) = o(\delta)$ as $\delta \downarrow 0$, and so we may neglect the possibility that there are two or more edges in $E_{p,\delta}$. If e is the unique edge satisfying $p \leq X(e) < p + \delta$, then e must be 'essential' for A in the sense that $\eta_p \notin A$ but $\eta'_p \in A$ where η'_p is obtained from η_p by changing the state of e from 0 to 1. The latter property of e does not depend on the state of e, so that each edge e contributes roughly an amount

$$P(p \leq X(e) < p+\delta) P_p(e \text{ is 'essential' for } A) = \delta P_p(e \text{ is 'essential' for } A) \tag{2.24}$$

towards the quantity in (2.23). We divide by δ and take the limit as $\delta \downarrow 0$ to obtain without rigour the formula

$$\frac{d}{dp} P_p(A) = \sum_e P_p(e \text{ is 'essential' for } A).$$

We shall now derive a rigorous version of this formula, valid for all increasing events A which depend on only finitely many edges.

Let A be an event, not necessarily increasing, and let ω be a configuration. We call the edge e *pivotal* for the pair (A, ω) if $I_A(\omega) \neq I_A(\omega')$, where ω' is the configuration which agrees with ω on all edges except e, and $\omega'(e) = 1 - \omega(e)$. Thus e is pivotal for (A, ω) if the occurrence or non-occurrence of A depends crucially on the state of e. The event that 'e is pivotal for A' is the set of configurations ω for which e is pivotal for (A, ω). We note that this event depends only on the states of edges other than e; it is independent of the state of e itself. We shall be interested particularly in *increasing* events A; for such an event A, an edge e is pivotal if and only if A does not occur when e is closed but A does occur when e is open.

Here are two examples:

(a) Let A be the event that the origin is in an infinite open cluster. The edge e is pivotal for A if, when e is removed from the lattice, one

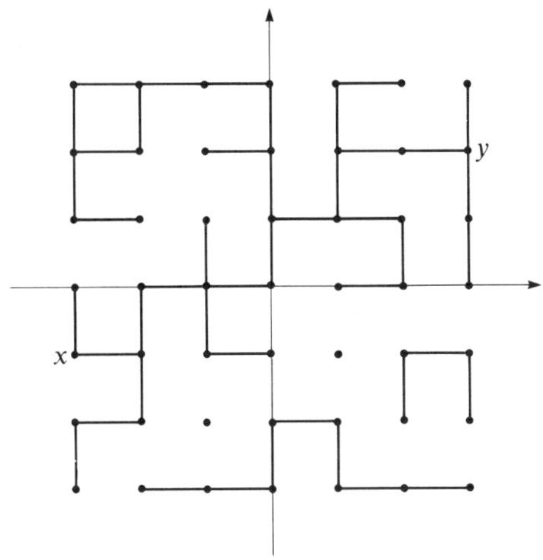

Figure 2.1. The box $B(3)$, with an open path joining a vertex x on the left side to a vertex y on the right side. This configuration has four edges which are pivotal for the existence of an open path joining the left side to the right side.

endvertex of e is in a finite open cluster which includes the origin, and the other endvertex of e is in an infinite open cluster.

(b) Let A be the event that there is an open path within the box $B(n)$ in two dimensions, joining a vertex on the left-hand side of $B(n)$ to a vertex on the right-hand side (see Figure 2.1). The edge e of $B(n)$ is pivotal for A if, when e is removed from the lattice, no such open crossing of $B(n)$ exists, but one endvertex of e is joined by an open path to the left side of $B(n)$ and the other endvertex of e is joined similarly to the right side of $B(n)$.

(2.25) Theorem. Russo's formula. *Let A be an increasing event which depends on only finitely many edges of $\mathbb{L}^d$. Then*

$$\frac{d}{dp} P_p(A) = E_p(N(A)), \tag{2.26}$$

where $N(A)$ is the number of edges which are pivotal for A.

Formula (2.26) may be written as

$$\frac{d}{dp} P_p(A) = \sum_{e \in \mathbb{E}^d} P_p(e \text{ is pivotal for } A). \tag{2.27}$$

Such formulae are not generally valid for events which depend on more than finitely many edges; indeed $P_p(A)$ need not be differentiable for all values of p (for example, we shall remark in Section 8.3 that, for percolation in two dimensions, $(\theta(p) - \theta(p_c))/(p - p_c)$ equals 0 if $p < p_c$ but tends to ∞ as $p \downarrow p_c$). For general increasing events A, the best that the method allows is a lower bound on the right-hand derivative of $P_p(A)$:

$$\liminf_{\delta \downarrow 0} \frac{1}{\delta}\big(P_{p+\delta}(A) - P_p(A)\big) \ge E_p(N(A)). \tag{2.28}$$

The event $\{e \text{ is pivotal for } A\}$ is independent of the state of e, and thus

$$P_p(e \text{ is pivotal for } A) = \frac{1}{p} P_p(e \text{ is open and pivotal for } A).$$

Hence, if A is increasing and depends on only finitely many edges,

$$\begin{aligned} \frac{d}{dp} P_p(A) &= \frac{1}{p} \sum_e P_p(e \text{ is open and pivotal for } A) \\ &= \frac{1}{p} \sum_e P_p(A \cap \{e \text{ is pivotal for } A\}) \\ &= \frac{1}{p} \sum_e P_p(e \text{ is pivotal for } A \mid A) P_p(A) \\ &= \frac{1}{p} E_p(N(A) \mid A) P_p(A). \end{aligned} \tag{2.29}$$

We may make use of this expression in the following way. Divide throughout by $P_p(A)$ and integrate over the interval $[p_1, p_2]$ to obtain

$$P_{p_2}(A) = P_{p_1}(A) \exp\left(\int_{p_1}^{p_2} \frac{1}{p} E_p(N(A) \mid A)\, dp \right) \quad \text{if } 0 \le p_1 < p_2 \le 1. \tag{2.30}$$

Applications of this expression require estimates for $E_p(N(A) \mid A)$. Accurate estimates for the probability that a given edge e is pivotal are often rather hard to derive, and it is more common that we resort to an inequality of the FKG or BK type in this context. If we use the trivial bound

$$P_p(e \text{ is pivotal for } A \mid A) \le 1,$$

then (2.30) yields

$$P_{p_2}(A) \le (p_2/p_1)^m P_{p_1}(A) \qquad \text{if} \quad 0 \le p_1 < p_2 \le 1, \tag{2.31}$$

where m is the (finite) number of edges on which A depends. This inequality is a quantified version of the statement that $P_p(A)$ cannot increase 'too fast' as p increases. We shall make use of it later.

There is a trivial example of Russo's formula in which the calculations may be done exactly, and we describe this in the hope that it illuminates the principles of the method. Let $E = \{e_1, e_2, \dots, e_m\}$ be a collection of m distinct edges, and let A be the event that k or more of these edges are open, where $1 \le k \le m$. The edge $e \in E$ is pivotal for A if and only if $A \backslash \{e\}$ has exactly $k - 1$ open edges. The probability of this event is

$\binom{m-1}{k-1}p^{k-1}(1-p)^{m-k}$, so that

$$\frac{d}{dp}P_p(A) = \sum_{e \in E}\binom{m-1}{k-1}p^{k-1}(1-p)^{m-k} = m\binom{m-1}{k-1}p^{k-1}(1-p)^{m-k},$$

which is easily integrated subject to the boundary condition $P_0(A) = 0$, to obtain

$$P_p(A) = \sum_{l=k}^{m}\binom{m}{l}p^l(1-p)^{m-l}$$

in agreement with the binomial distribution.

Proof of Russo's formula. We shall make rigorous the discussion at the beginning of the section. Let A be an increasing event and let $\mathbf{p} = (p(e): e \in \mathbb{E}^d)$ be a collection of numbers satisfying $0 \le p(e) \le 1$ for all e. Let $(X(e): e \in \mathbb{E}^d)$ be independent random variables each having the uniform distribution on $[0, 1]$. We construct the configuration $\eta_{\mathbf{p}}$ on $\mathbb{E}^d$ by defining $\eta_{\mathbf{p}}(e) = 1$ if $X(e) < p(e)$ and $\eta_{\mathbf{p}}(e) = 0$ otherwise. Writing $P_{\mathbf{p}}$ for the probability measure on Ω in which the state $\omega(e)$ of the edge e equals 1 with probability $p(e)$, we have that

$$P_{\mathbf{p}}(A) = P(\eta_{\mathbf{p}} \in A)$$

as usual. Now, choose an edge f and define $\mathbf{p}' = (p'(e): e \in \mathbb{E}^d)$ by

$$p'(e) = \begin{cases} p(e) & \text{if } \ e \ne f, \\ p'(f) & \text{if } \ e = f; \end{cases}$$

$\mathbf{p}$ and $\mathbf{p}'$ differ only at the edge f. Now, if $p(f) \le p'(f)$ then

$$\begin{aligned} P_{\mathbf{p}'}(A) - P_{\mathbf{p}}(A) &= P(\eta_{\mathbf{p}} \notin A, \eta_{\mathbf{p}'} \in A) \\ &= \{p'(f) - p(f)\}P_{\mathbf{p}}(f \text{ is pivotal for } A), \end{aligned}$$

by the discussion leading to (2.24). We divide by $p'(f) - p(f)$ and take the limit as $p'(f) - p(f) \to 0$ to obtain

$$\frac{\partial}{\partial p(f)}P_{\mathbf{p}}(A) = P_{\mathbf{p}}(f \text{ is pivotal for } A). \tag{2.32}$$

So far we have assumed nothing about A save that it be increasing. If A depends on finitely many edges only, then $P_{\mathbf{p}}(A)$ is a function of a finite collection $(p(f_i): 1 \le i \le m)$ of edge-probabilities, and the chain rule gives that

$$\begin{aligned} \frac{d}{dp}P_p(A) &= \sum_{i=1}^{m}\frac{\partial}{\partial p(f_i)}P_{\mathbf{p}}(A)\bigg|_{\mathbf{p}=(p,p,\dots,p)} \\ &= \sum_{i=1}^{m}P_p(f_i \text{ is pivotal for } A) \\ &= E_p(N(A)) \end{aligned}$$

as required. If, on the other hand, A depends on infinitely many edges, we let E be a finite collection of edges and define

$$\mathbf{p}_E(e) = \begin{cases} p & \text{if} \quad e \notin E, \\ p + \delta & \text{if} \quad e \in E, \end{cases}$$

where $0 \leq p \leq p + \delta \leq 1$. Now A is increasing, so that

$$P_{p+\delta}(A) \geq P_{\mathbf{p}_E}(A),$$

and hence

$$\begin{aligned} \frac{1}{\delta}\big(P_{p+\delta}(A) - P_p(A)\big) &\geq \frac{1}{\delta}\big(P_{\mathbf{p}_E}(A) - P_p(A)\big) \\ &\to \sum_{e \in E} P_p(e \text{ is pivotal for } A) \end{aligned}$$

as $\delta \downarrow 0$. We let $E \uparrow \mathbb{E}^d$ to obtain (2.28). ∎

2.5 Inequalities of Reliability Theory

Reliability theory studies, amongst other things, the probability of a route through a network whose individual links are unreliable. More precisely, let G be a finite graph, and declare each edge of G to be open with probability p, independently of all other edges. Let x and y be specified vertices of G and let $h(p)$ be the probability that there is an open path of G from x to y. In studying the behaviour of h as a function of p, reliability theorists have developed various tools which are relevant to percolation. These include versions of the FKG inequality and Russo's formula; see Moore and Shannon (1956), Esary and Proschan (1963), and Barlow and Proschan (1965, 1975). We present some of their results here, in forms suitable for application to percolation.

Let E be a finite set of edges of $\mathbb{L}^d$. We shall confine ourselves to events which depend only on the edges in E. We write $\text{cov}_p(X, Y)$ for the covariance of two random variables X and Y under the measure P_p.

(2.33) Theorem. *Let A be an event which depends only on the edges in E, and let N be the (random) number of edges of E which are open. Then*

(2.34) $$\frac{d}{dp} P_p(A) = \frac{1}{p(1-p)} \text{cov}_p(N, I_A) \qquad \textit{for} \quad 0 < p < 1.$$

This theorem is valid for all events A regardless of whether or not they are monotone. This version of Russo's formula amounts to saying that

(2.35) $$\frac{d}{dp} P_p(A) = \frac{1}{p(1-p)} \big(E_p(N; A) - E_p(N) P_p(A)\big)$$

for all suitable events A. The following is an immediate corollary, of which the second part is the main step in the celebrated 'S-shape' theorem of reliability theory.

(2.36) Theorem. *Let A be an event which depends only on the edges in E, and suppose that $0 < p < 1$. Then*

(a) $$\frac{d}{dp} P_p(A) \leq \sqrt{\frac{mP_p(A)(1 - P_p(A))}{p(1-p)}} \qquad \text{where} \quad m = |E|,$$

(b) *if A is increasing, we have that*

$$\frac{d}{dp} P_p(A) \geq \frac{P_p(A)(1 - P_p(A))}{p(1-p)}. \tag{2.37}$$

Before proving these results, we mention a related inequality which we shall use later.

(2.38) Theorem. *Let A be an increasing event which depends on only finitely many edges of $\mathbb{L}^d$, and suppose that $0 < p < 1$. Then $\log P_p(A)/\log p$ is a non-increasing function of p.*

This last theorem amounts to saying that $h(p) = P_p(A)$ satisfies

$$h(p^\gamma) \leq h(p)^\gamma \qquad \text{if} \quad 0 < p < 1 \quad \text{and} \quad \gamma \geq 1, \tag{2.39}$$

whenever A is increasing. Rewriting the conclusion of the theorem in terms of the derivative of $\log P_p(A)/\log p$, we obtain

$$\frac{d}{dp} P_p(A) \geq \frac{P_p(A) \log P_p(A)}{p \log p}. \tag{2.40}$$

This is an improvement over inequality (2.37) whenever $P_p(A) < p$.

Proof of Theorem (2.33). We write ω for a typical configuration of the edges in E, and $N(\omega)$ for the number of open edges of ω. Clearly

$$P_p(A) = \sum_\omega I_A(\omega) p^{N(\omega)} q^{m-N(\omega)},$$

where $p + q = 1$ and $m = |E|$. Thus

$$\begin{aligned} \frac{d}{dp} P_p(A) &= \sum_\omega I_A(\omega) p^{N(\omega)} q^{m-N(\omega)} \left(\frac{N(\omega)}{p} - \frac{m - N(\omega)}{q} \right) \\ &= \frac{1}{pq} \sum_\omega I_A(\omega) p^{N(\omega)} q^{m-N(\omega)} (N(\omega) - mp) \\ &= \frac{1}{pq} \operatorname{cov}_p(N, I_A) \end{aligned}$$

since $E_p(N) = mp$. ∎

Proof of Theorem (2.36). We apply the Cauchy–Schwarz inequality to the right-hand side of (2.34) to obtain

$$\frac{d}{dp} P_p(A) \le \frac{1}{p(1-p)} \sqrt{\operatorname{var}_p(N) \operatorname{var}_p(I_A)}$$

and part (a) follows from the observation that N has the binomial distribution with parameters m and p, and I_A has the Bernoulli distribution.

Suppose now that A is increasing. Then

$$\begin{aligned} \text{(2.41)} \qquad \operatorname{cov}_p(N, I_A) &= \operatorname{cov}_p(I_A, I_A) + \operatorname{cov}_p(N - I_A, I_A) \\ &\ge \operatorname{var}_p(I_A) \end{aligned}$$

by the FKG inequality, since $N - I_A$ is an increasing random variable. We substitute this into (2.34) to obtain (2.37). ■

Proof of Theorem (2.38). We shall show that $h(p) = P_p(A)$ satisfies

$$\text{(2.42)} \qquad h(p^\gamma) \le h(p)^\gamma \quad \text{if} \quad 0 < p < 1 \quad \text{and} \quad \gamma \ge 1,$$

whenever A is increasing and depends on only finitely many edges of $\mathbb{L}^d$. To see that this implies the result, substitute $p_1 = p^\gamma$ in (2.42) and take logarithms to find that

$$\frac{\log h(p_1)}{\log p_1} \ge \frac{\log h(p)}{\log p}$$

as required.

We shall prove (2.42) by induction on the number of edges which are relevant to A. Suppose that A is increasing and depends only on the edges in the finite set E, and write $m = |E|$. If $m = 1$ then A is either $\{0, 1\}$ or $\{1\}$, and so $h(p) = P_p(A)$ equals either 1 or p; in either case (2.42) is valid. Suppose therefore that k (≥ 1) is such that (2.42) is valid whenever $m \le k$, and consider the case when $m = k + 1$. Let $0 < p < 1$, $\gamma \ge 1$, and let e be an edge in E. Then

$$\begin{aligned} \text{(2.43)} \qquad h(p^\gamma) &= P_{p^\gamma}(A \mid \omega(e) = 1)p^\gamma + P_{p^\gamma}(A \mid \omega(e) = 0)(1 - p^\gamma) \\ &\le P_p(A \mid \omega(e) = 1)^\gamma p^\gamma + P_p(A \mid \omega(e) = 0)^\gamma (1 - p^\gamma) \end{aligned}$$

by the induction hypothesis. It is not difficult to check that

$$\text{(2.44)} \qquad x^\gamma p^\gamma + y^\gamma(1 - p^\gamma) \le \{xp + y(1-p)\}^\gamma \qquad \text{when} \quad x \ge y \ge 0;$$

to see this, check that equality holds when $x = y \ge 0$ and that the derivative of the left-hand side with respect to x is at most the corresponding derivative of the right-hand side when $x, y \ge 0$. We note that $P_p(A \mid \omega(e) = 1) \ge P_p(A \mid \omega(e) = 0)$ since A is increasing, and (2.43) now yields

$$\begin{aligned} h(p^\gamma) &\le \big(P_p(A \mid \omega(e) = 1)p + P_p(A \mid \omega(e) = 0)(1 - p)\big)^\gamma \\ &= h(p)^\gamma \end{aligned}$$

as required. ■

2.6 Another Inequality

There is a case for delaying the presentation of our final inequality until it is needed, but we include it here with the recommendation that the reader note its existence and move on.

Let r be a positive integer. For any configuration ω of edge-states, we define the *sphere* with radius r and centre at ω by

$$S_r(\omega) = \left\{\omega' \in \Omega: \sum_{e \in \mathbb{E}^d} |\omega'(e) - \omega(e)| \leq r\right\};$$

$S_r(\omega)$ is the collection of configurations which differ from ω on at most r edges. For any event A, we define the *interior* of A with depth r by

$$I_r(A) = \{\omega \in \Omega: S_r(\omega) \subseteq A\};$$

thus $I_r(A)$ is the set of configurations in A which are still in A even if we perturb the states of up to r edges.

(2.45) Theorem. *Let A be an increasing event and let r be a positive integer. Then*

(2.46) $$1 - P_{p_2}(I_r(A)) \leq \left(\frac{p_2}{p_2 - p_1}\right)^r \{1 - P_{p_1}(A)\}$$

whenever $0 \leq p_1 < p_2 \leq 1$.

We may think of $I_r(A)$ as the event that A occurs and is 'stable' with respect to changes in the states of r or fewer edges. The theorem amounts to the assertion that, if A is likely to occur when the edge-probability is p_1, then $I_r(A)$ is likely to occur when the edge-probability exceeds p_1. Such an inequality is valuable in the context of the following type of construction. Let A_n be the event that there exists an open path between the left and right sides of the rectangle $[0, n]^2$ in two dimensions; such a path is called a 'left–right crossing'. In this case, $I_r(A_n)$ is the event that such a path exists irrespective of the states of any collection of r edges. Those familiar with the max–flow min–cut theorem of network flows (see Section 10.8) and Menger's theorem (see Wilson (1979, p. 126)) will recognize $I_r(A_n)$ as the event that there are at least $r + 1$ edge-disjoint left–right crossings of $[0, n]^2$. We shall see in Chapter 9 that $P_p(A_n) \geq 1 - e^{-n\alpha(p)}$ for some $\alpha(p) > 0$, so long as $p > \frac{1}{2}$. We apply (2.46) with $r = \beta n$ to find that

(2.47) $$P_{p_2}(K_n \leq \beta n) \leq \left(\frac{p_2}{p_2 - p_1}\right)^{\beta n} e^{-n\alpha(p_1)},$$

where K_n is the maximal number of edge-disjoint left–right crossings of the rectangle, and $\frac{1}{2} < p_1 < p_2 \leq 1$. If

$$\gamma(p_1, p_2, \beta) = \beta \log\left(\frac{p_2}{p_2 - p_1}\right) - \alpha(p_1)$$

satisfies $\gamma(p_1, p_2, \beta) < 0$, then the left-hand side of (2.47) decays exponentially as $n \to \infty$. We have proved that, if for some value of p there is an exponentially decaying probability that there is no left–right crossing of $[0, n]^2$, then for larger values of p there is an exponentially decaying probability that there are $o(n)$ such crossings.

Proof of Theorem (2.45). We follow Aizenman, Chayes, Chayes, Fröhlich, and Russo (1983). Let $(X(e): e \in \mathbb{E}^d)$ and $(\eta_p: 0 \leq p \leq 1)$ be defined in the usual way. Suppose that $0 \leq p_1 < p_2 \leq 1$ and that A is an increasing event. If $\eta_{p_2} \notin I_r(A)$ then there exists a (random) collection $B = B(\eta_{p_2})$ of edges such that:

(a) $|B| \leq r$;
(b) $\eta_{p_2}(e) = 1$ for all $e \in B$; and
(c) the configuration η, obtained from η_{p_2} be declaring all edges in B to be closed, satisfies $\eta \notin A$.

There may exist many such sets B, in which case we choose the earliest in some fixed ordering of all possible such sets. Suppose now that $\eta_{p_2} \notin I_r(A)$, and that every edge e in the set B satisfies $p_1 \leq X(e) < p_2$; it follows from (c) above that $\eta_{p_1} \notin A$. Conditional on B, there is probability $\{(p_2 - p_1)/p_2\}^{|B|}$ that $p_1 \leq X(e) < p_2$ for all $e \in B$, and therefore

$$P(\eta_{p_1} \notin A \mid \eta_{p_2} \notin I_r(A)) \geq \left(\frac{p_2 - p_1}{p_2}\right)^r$$

since $|B| \leq r$. Inequality (2.46) follows easily. ∎

2.7 Notes

Section 2.2. The second inequality of Theorem (2.4) might well be dubbed Harris's inequality after Harris (1960); a version of this inequality is standard in reliability theory, and may be found in Barlow and Proschan (1965, p. 207) for example. There is a more general version due to Fortuin, Kasteleyn, and Ginibre (1971) which is of great value in statistical physics. For a general mathematical approach to such correlation inequalities, see Ahlswede and Daykin (1979), Batty and Bollman (1980), den Hollander and Keane (1986), and the references therein. The proof given here is derived from material in Kesten (1982) and Durrett (1988, Chapter 6).

Section 2.3. The BK inequality was first stated and proved in a more general form by van den Berg and Kesten (1985). There are many refinements in van den Berg and Fiebig (1987). The proof given here has its origins in the work of McDiarmid (1980, 1981, 1983). Van den Berg (1985) has pointed out that a similar proof may be used to show the

FKG inequality. Similar ideas occur in papers of Hammersley (1961), Rüschendorf (1982), and Campanino and Russo (1985).

Section 2.4. Russo's formula was proved by Russo (1981). Such techniques are well known in reliability theory. For example, Russo's formula is essentially equation (4.4) of Barlow and Proschan (1965, p. 210). There are several interesting approaches to its proof; see Chayes and Chayes (1986a) for example. Equation (3.15) of Kesten (1981) is a special case of inequality (2.31).

Section 2.5. Theorem (2.33) may be found in Barlow and Proschan (1965, p. 210). The inequalities of Theorem (2.36) appear in Moore and Shannon (1956). Part (a) has been rediscovered by Chayes, Chayes, Fisher, and Spencer (1986), and our proof resembles theirs. Theorem (2.38) was suggested by J. van den Berg.

Section 2.6. The inequality of Theorem (2.45) was proved by Aizenman, Chayes, Chayes, Fröhlich, and Russo (1983), and has been put to use by Chayes and Chayes (1986a, d). It refers to the effect of adding a low density of extra open edges, a technique known to some as 'sprinkling'.

3

The Uniqueness of the Critical Point

3.1 The Problem

Let us consider bond percolation on $\mathbb{L}^d$ with edge-density p, where $d \geq 2$. As we have remarked, the following two functions are two of the principal characters in the action:

$$\theta(p) = P_p(|C| = \infty), \qquad \chi(p) = E_p|C|,$$

where C is the open cluster containing the origin and $|C|$ is the number of vertices in C. We have seen that there exists $p_c = p_c(d)$ in (0, 1) such that

$$\theta(p) \begin{cases} = 0 & \text{if} \quad p < p_c, \\ > 0 & \text{if} \quad p > p_c. \end{cases} \tag{3.1}$$

The critical value p_c is the value of p above which infinite open clusters appear. Much of the past romance of percolation theory has been associated with the question of whether or not p_c is the unique value of p at which a critical phenomenon occurs; in particular, we shall see in Chapter 9 that relatively simple arguments imply that $p_c(2) = \frac{1}{2}$ once we have answered this question affirmatively.

The function χ experiences a critical phenomenon also. Clearly χ is non-decreasing in p, and $\chi(0) = 1$, $\chi(1) = \infty$. We have, as in Section 1.5, that

$$\chi(p) = \infty \cdot P_p(|C| = \infty) + \sum_{n=1}^{\infty} nP_p(|C| = n),$$

so that $\chi(p) = \infty$ whenever $\theta(p) = P_p(|C| = \infty)$ satisfies $\theta(p) > 0$. Thus $\chi(p) = \infty$ when $p > p_c$. It is not so easy to show that $\chi(p) < \infty$ when $p < p_c$, and this result is perhaps the principal target of this chapter.

(3.2) Theorem. *The mean size $\chi(p)$ of the open cluster containing the origin is finite if $p < p_c$.*

One of the greatest recent challenges in percolation theory has been to prove this result, which has customarily been reformulated as follows. We define a second 'critical probability' $p_T = p_T(d)$ by

$$p_T = \sup\{p: \chi(p) < \infty\}, \tag{3.3}$$

the value of p at which χ changes from being finite to being infinite. It follows from the remarks above that $p_T \le p_c$, and the problem was to prove that $p_T = p_c$. The special case of this conjecture for the planar lattice $\mathbb{L}^2$ is of particular historical interest, since this was the major obstacle in the way of a complete proof that the critical probability $p_c(2)$ equals $\frac{1}{2}$. With this exact calculation as a target, the conjecture attracted many by its simplicity but apparent difficulty. It is relatively easy to show that $p_T(2) + p_c(2) \le 1$, and it has been known since 1960 that $p_c(2) \ge \frac{1}{2}$, so that it remained to show the equality of $p_T(2)$ and $p_c(2)$; there is therefore a sense in which Theorem (3.2) is the natural generalization to higher dimensions of the exact calculation in two dimensions.

We present two proofs of Theorem (3.2). The first is due to Menshikov (1986) (see also Menshikov, Molchanov, and Sidorenko (1986)). Menshikov has proved the stronger result that the tail of the radius of the open cluster at the origin decays exponentially whenever $p < p_c$. As a by-product of the proof, we obtain that there exists a (>0) such that $\theta(p) - \theta(p_c) \ge a(p - p_c)$ when $p - p_c$ is small and positive; this observation will be of value in the discussion concerning critical exponents (see Chapters 7 and 8). The second proof of Theorem (3.2) is due to Aizenman and Barsky (1987) and is perhaps less basic and intuitive than the first proof. At first sight these two proofs may appear to differ fundamentally, but actually they have a lot in common. The Aizenman–Barsky proof yields also the fact that θ grows at least linearly in $p - p_c$, as well as an estimate for the decay rate of $P_{p_c}(|C| = n)$ as $n \to \infty$.

3.2 Exponential Decay of the Radius Distribution beneath p_c

We prove the uniqueness of the critical probability by the method of Menshikov (1986) in conjunction with an improvement due to H. Kesten (unpublished).

Let $S(n)$ be the ball of radius n with centre at the origin: $S(n)$ is the set of all vertices x in $\mathbb{Z}^d$ for which $\delta(0, x) \le n$. We write $\partial S(n)$ for the surface of $S(n)$, being the set of all x with $\delta(0, x) = n$. Let A_n be the event that there is an open path joining the origin to some vertex in $\partial S(n)$. The following is the main result.

(3.4) Theorem. Exponential tail decay of the radius of an open cluster. *If $p < p_c$ then there exists $\psi(p) > 0$ such that*

(3.5) $$P_p(A_n) < e^{-n\psi(p)} \qquad \textit{for all } n.$$

This theorem amounts to the assertion that the radius of the cluster at the origin has a tail which decays at least exponentially fast when $p < p_c$. The uniqueness of the critical probability follows immediately, using the following elementary argument. It is easily seen that the number of vertices in $S(n)$ is no larger than the volume of the euclidean ball in $\mathbb{R}^d$ with radius $n + 1$, and so there exists $\pi = \pi(d)$ such that

(3.6) $$|S(n)| \le \pi(n + 1)^d.$$

Let $M = \max\{n: A_n \text{ occurs}\}$. If $p < p_c$ then $P_p(M < \infty) = 1$, giving that

$$\begin{aligned} E_p|C| &\le \sum_n E_p(|C| \,|\, M = n)P_p(M = n) \\ &\le \sum_n |S(n)| P_p(A_n) \\ &\le \sum_n \pi(n + 1)^d e^{-n\psi(p)} \\ &< \infty, \end{aligned}$$

where $\psi(p)$ is the quantity given in Theorem (3.4).

The above theorem is a lot stronger than is needed to demonstrate Theorem (3.2). It implies not only that $E_p|C| < \infty$ when $p < p_c$, but also an estimate on the tail probabilities of $|C|$. To see this, suppose that $p < p_c$ and $|C| > n$. If $C \subseteq S(m)$ then $n \le \pi(m + 1)^d$ by (3.6), giving that $m \ge (n/\pi)^{1/d} - 1$. Thus, if $m < (n/\pi)^{1/d} - 1$ then C is not contained in $S(m)$, giving that C contains some vertex in $\partial S(k)$, where $k = \lfloor (n/\pi)^{1/d} \rfloor$. It follows that

(3.7) $$\begin{aligned} P_p(|C| > n) &\le P_p(A_k) \\ &\le \exp(-n^{1/d}\beta(p)) \qquad \text{when} \quad p < p_c, \end{aligned}$$

for some $\beta(p) > 0$. We shall see in Chapter 5 that this estimate is not best possible: the term $n^{1/d}$ in the exponent may be replaced by n.

The principal step in proving Theorem (3.4) involves an estimate which is valid for all $p \in (0, 1)$, and not merely for values of p less than p_c. This estimate has an interesting consequence when $p > p_c$, in that it implies that $\theta(p) - \theta(p_c)$ grows at least linearly in $p - p_c$. We note this fact here, and shall make use of it in Chapter 8 when discussing the matter of 'critical exponents'.

(3.8) Theorem. *There exist $a, b > 0$ such that*

$$\theta(p) - \theta(p_c) \ge a(p - p_c) \qquad \textit{if} \quad 0 \le p - p_c \le b.$$

We move on to the proof of Theorem (3.4). Let $S(n, x)$ be the ball of radius n with centre at the vertex x, and let $\partial S(n, x)$ be the surface of $S(n, x)$; thus $S(n, x) = x + S(n)$ and $\partial S(n, x) = x + \partial S(n)$. Similarly, let $A_n(x)$ be the event that there is an open path from the vertex x to some vertex in $\partial S(n, x)$. We are concerned with the probabilities

$$g_p(n) = P_p(A_n) = P_p(A_n(x)) \qquad \text{for any } x.$$

Now A_n is an increasing event which depends on the edges joining vertices in $S(n)$ only. We apply Russo's formula to $P_p(A_n)$ to obtain

(3.9) $$g'_p(n) = E_p(N(A_n))$$

where the prime denotes differentiation with respect to p, and $N(A_n)$ is the number of edges which are pivotal for A_n. It follows as in (2.29) that

$$\begin{aligned} g'_p(n) &= \frac{1}{p} E_p(N(A_n); A_n) \\ &= \frac{1}{p} E_p(N(A_n) \mid A_n) g_p(n) \end{aligned}$$

so that

(3.10) $$\frac{1}{g_p(n)} g'_p(n) = \frac{1}{p} E_p(N(A_n) \mid A_n).$$

Let $0 \leq \alpha < \beta \leq 1$, and integrate (3.10) from $p = \alpha$ to $p = \beta$ to obtain

(3.11) $$\begin{aligned} g_\alpha(n) &= g_\beta(n) \exp\left(- \int_\alpha^\beta \frac{1}{p} E_p(N(A_n) \mid A_n)\, dp \right) \\ &\leq g_\beta(n) \exp\left(- \int_\alpha^\beta E_p(N(A_n) \mid A_n)\, dp \right), \end{aligned}$$

as in (2.30). We need now to show that $E_p(N(A_n) \mid A_n)$ grows roughly linearly in n when $p < p_c$, and then this inequality will yield an upper bound for $g_\alpha(n)$ of the form required in (3.5). The vast majority of the work in the proof is devoted to estimating $E_p(N(A_n) \mid A_n)$, and the argument is roughly as follows. If $p < p_c$ then $P_p(A_n) \to 0$ as $n \to \infty$, so that for large n we are conditioning on an event of small probability. If A_n occurs, 'but only just', then the connections between the origin and $\partial S(n)$ must be sparse; indeed, there must exist many open edges in $S(n)$ which are crucial for the occurrence of A_n (see Figure 3.1). It is plausible that the number of such pivotal edges in paths from the origin to $\partial S(2n)$ is approximately twice the number of such edges in paths to $\partial S(n)$, since these sparse paths have to traverse twice the distance. Thus the number $N(A_n)$ of edges pivotal for A_n should grow linearly in n.

Suppose that the event A_n occurs, and denote by $e_1, e_2, \ldots, e_N$ the (random) edges which are pivotal for A_n. Since A_n is increasing, each e_j has the property that A_n occurs if and only if e_j is open; thus all open

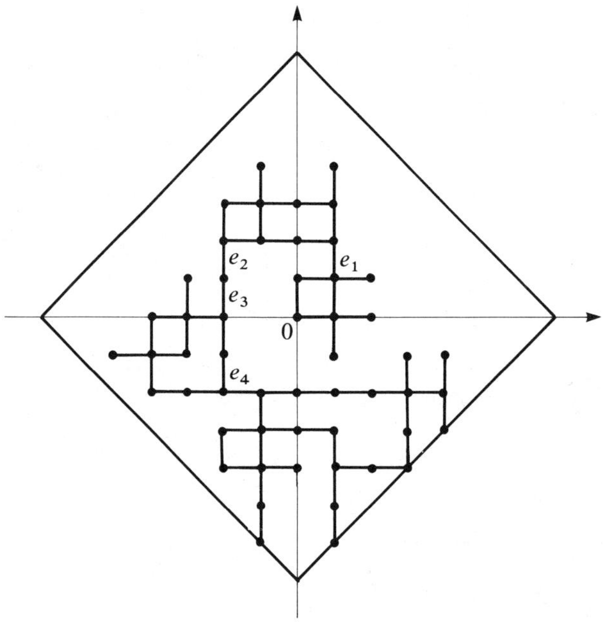

Figure 3.1. A picture of the open cluster of $S(7)$ at the origin. There are exactly four pivotal edges for A_n in this configuration, and these are labelled e_1, e_2, e_3, e_4.

paths from the origin to $\partial S(n)$ traverse e_j, for every j (see Figure 3.1). Let π be such an open path; we assume that the edges $e_1, e_2, \ldots, e_N$ have been enumerated in the order in which they are traversed by π. A glance at Figure 3.1 confirms that this ordering is independent of the choice of π. We denote by x_i the endvertex of e_i encountered first by π, and by y_i the other endvertex of e_i. We observe that there exist at least two edge-disjoint open paths joining 0 to x_1, since, if two such paths cannot be found then, by Menger's theorem (Wilson (1979, p. 126)), there exists a pivotal edge in π which is encountered prior to x_1, a contradiction. Similarly, for $1 \le i < N$, there exist at least two edge-disjoint open paths joining y_i to x_{i+1}; see Figure 3.2. In the words of the discoverer of this proof, the open cluster containing the origin resembles a chain of sausages.

As before, let $M = \max\{k: A_k \text{ occurs}\}$ be the radius of the largest ball whose surface contains a vertex which is joined to the origin by an open path. We note that, if $p < p_c$, then M has a non-defective distribution in that $P_p(M \ge k) = g_p(k) \to 0$ as $k \to \infty$. We shall show that, conditional on A_n, $N(A_n)$ is at least as large as the number of renewals up to time n of a renewal process whose inter-renewal times have approximately the same distribution as M. In order to compare $N(A_n)$ with such a renewal process, we introduce the following notation. Let $\rho_1 = \delta(0, x_1)$ and $\rho_{i+1} = \delta(y_i, x_{i+1})$ for $1 \le i < N$. The first step is to show that, roughly

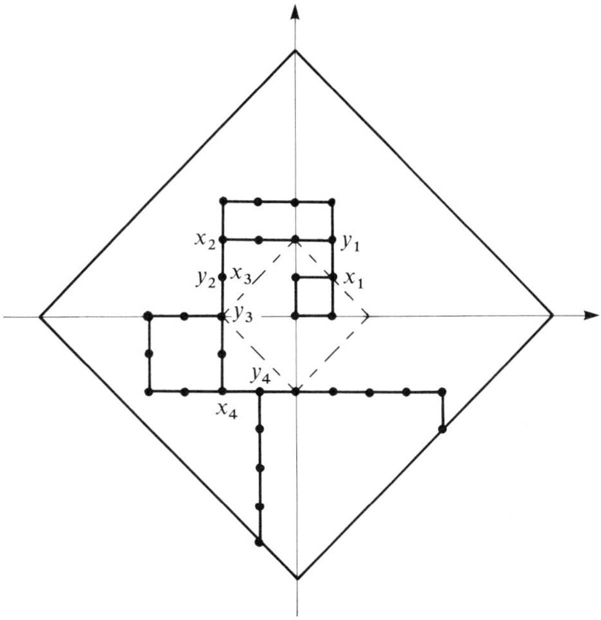

Figure 3.2. The pivotal edges are $e_i = \langle x_i, y_i \rangle$ for $i = 1, 2, 3, 4$. Note that $x_3 = y_2$ in this configuration. The dashed line is the surface $\partial S(\rho_1)$ of $S(\rho_1)$. Note the two edge-disjoint paths from the origin to $\partial S(\rho_1)$.

speaking, the random variables $\rho_1, \rho_2, \ldots$ are jointly smaller in distribution than a sequence $M_1, M_2, \ldots$ of independent random variables distributed as M.

(3.12) Lemma. *Let k be a positive integer, and let $r_1, r_2, \ldots, r_k$ be non-negative integers such that $\sum_{i=1}^{k} r_i \leq n - k$. Then, for $0 < p < 1$,*

$$(3.13) \qquad P_p(\rho_k \leq r_k,\ \rho_i = r_i \text{ for } 1 \leq i < k \mid A_n) \geq P_p(M \leq r_k) P_p(\rho_i = r_i \text{ for } 1 \leq i < k \mid A_n).$$

Proof. Suppose by way of illustration that $k = 1$ and $0 \leq r_1 < n$. Then

$$(3.14) \qquad \{\rho_1 > r_1\} \cap A_n \subseteq A_{r_1+1} \circ A_n,$$

since if $\rho_1 > r_1$ then the first endvertex of the first pivotal edge lies either outside $S(r_1 + 1)$ or on its surface $\partial S(r_1 + 1)$; see Figure 3.2. However, A_{r_1+1} and A_n are increasing events which depend on the edges within $S(n)$ only, and the BK inequality yields

$$P_p(\{\rho_1 > r_1\} \cap A_n) \leq P_p(A_{r_1+1}) P_p(A_n).$$

We divide by $P_p(A_n)$ to obtain

$$P_p(\rho_1 > r_1 \mid A_n) \leq g_p(r_1 + 1);$$

however $P_p(M \geq m) = g_p(m)$, and thus we have obtained (3.13) in the case $k = 1$.

We now prove the lemma for general values of k. Suppose that $k \geq 1$, and let $r_1, r_2, \ldots, r_k$ be non-negative integers with sum not exceeding $n - k$. Let N be the number of edges which are pivotal for A_n; we enumerate and label these edges as $e_i = \langle x_i, y_i \rangle$ as before. Let

$$B = \{\rho_i = r_i \text{ for } 1 \leq i < k\},$$

and note that B is a sub-event of the event that $N \geq k - 1$. On the event B, we consider the set of vertices and open edges attainable along open paths from the origin without using e_{k-1}; to this graph we append e_{k-1} and its other endvertex y_{k-1}, and we place a mark over y_{k-1} in order to distinguish it from the other vertices. We denote by G the resulting (marked) graph, and we write $y(G)$ for the unique marked vertex of G. We condition on G to obtain

$$P_p(A_n \cap B) = \sum_{\Gamma} P_p(B, G = \Gamma) P_p(A_n | B, G = \Gamma),$$

where the sum is over all possible values Γ of G. The final term in this summation is the probability that $y(\Gamma)$ is joined to $\partial S(n)$ by an open path which has no vertex other than $y(\Gamma)$ in common with Γ. Thus, in the obvious terminology,

$$P_p(A_n \cap B) = \sum_{\Gamma} P_p(B, G = \Gamma) P_p(y(\Gamma) \leftrightarrow \partial S(n) \text{ off } \Gamma). \tag{3.15}$$

Similarly,

$$\begin{aligned} P_p(\{\rho_k > r_k\} \cap A_n \cap B) &= \sum_{\Gamma} P_p(B, G = \Gamma) P_p(\{\rho_k > r_k\} \cap A_n | B, G = \Gamma) \\ &= \sum_{\Gamma} P_p(B, G = \Gamma) P_p(\text{there exist edge-disjoint open paths joining } y(\Gamma) \text{ to } \partial S(r_k + 1, y(\Gamma)) \text{ and joining } y(\Gamma) \text{ to } \partial S(n), \text{ off } \Gamma). \end{aligned}$$

We apply the BK inequality to the last term to obtain

(3.16)

$$\begin{aligned} &P_p(\{\rho_k > r_k\} \cap A_n \cap B) \\ &\quad \leq \sum_{\Gamma} P_p(B, G = \Gamma) P_p(y(\Gamma) \leftrightarrow \partial S(n) \text{ off } \Gamma) P_p(y(\Gamma) \leftrightarrow \partial S(r_k + 1, y(\Gamma)) \text{ off } \Gamma) \\ &\quad \leq g_p(r_k + 1) P_p(A_n \cap B) \end{aligned}$$

by (3.15) and the fact that, for each possible Γ,

$$\begin{aligned} P_p(y(\Gamma) \leftrightarrow \partial S(r_k + 1, y(\Gamma)) \text{ off } \Gamma) &\leq P_p(y(\Gamma) \leftrightarrow \partial S(r_k + 1, y(\Gamma))) \\ &= P_p(A_{r_k + 1}) \\ &= g_p(r_k + 1). \end{aligned}$$

We divide each side of (3.16) by $P_p(A_n \cap B)$ to obtain

$$P_p(\rho_k \le r_k | A_n \cap B) \ge 1 - g_p(r_k + 1),$$

throughout which we multiply by $P_p(B|A_n)$ to obtain the result. ∎

(3.17) Lemma. *For $0 < p < 1$, it is the case that*

$$E_p(N(A_n)|A_n) \ge \frac{n}{\sum_{i=0}^{n} g_p(i)} - 1. \tag{3.18}$$

Proof. It follows from Lemma (3.12) that

(3.19)

$$P_p(\rho_1 + \rho_2 + \cdots + \rho_k \le n - k | A_n) \ge P(M_1 + M_2 + \cdots + M_k \le n - k),$$

where $k \ge 1$ and $M_1, M_2, \ldots$ is a sequence of independent random variables distributed as M. We defer until the end of this proof the minor chore of deducing (3.19) from (3.13). Now $N(A_n) \ge k$ if $\rho_1 + \rho_2 + \cdots + \rho_k \le n - k$, so that

$$P_p(N(A_n) \ge k | A_n) \ge P(M_1 + M_2 + \cdots + M_k \le n - k). \tag{3.20}$$

A minor difficulty is that the M_i's may have a defective distribution. Indeed,

$$\begin{aligned} P(M \ge r) &= g_p(r) \\ &\to \theta(p) \qquad \text{as} \quad r \to \infty; \end{aligned}$$

thus we allow the M_i's to take the value ∞ with probability $\theta(p)$. On the other hand, we are not concerned with atoms at ∞, since

$$P(M_1 + M_2 + \cdots + M_k \le n - k) = P(M_1' + M_2' + \cdots + M_k' \le n),$$

where $M_i' = 1 + \min\{M_i, n\}$, and we work henceforth with these truncated random variables. Summing (3.20) over k, we obtain

$$\begin{aligned} E_p(N(A_n)|A_n) &\ge \sum_{k=1}^{\infty} P(M_1' + M_2' + \cdots + M_k' \le n) \\ &= \sum_{k=1}^{\infty} P(K \ge k + 1) \\ &= E(K) - 1, \end{aligned} \tag{3.21}$$

where $K = \min\{k: M_1' + M_2' + \cdots + M_k' > n\}$. Let $S_k = M_1' + M_2' + \cdots + M_k'$, the sum of independent, identically distributed, bounded random variables. By Wald's equation (see Chow and Teicher (1978, pp. 137, 150)),

$$n < E(S_K) = E(K)E(M_1'),$$

giving that

$$E(K) > \frac{n}{E(M_1')} = \frac{n}{1 + E(\min\{M_1, n\})} = \frac{n}{\sum_{i=0}^{n} g_p(i)}$$

since

$$E(\min\{M_1, n\}) = \sum_{i=1}^{n} P(M \geq i) = \sum_{i=1}^{n} g_p(i).$$

It remains to show that (3.19) follows from Lemma (3.12). We have that

$$\begin{aligned}
&P_p(\rho_1 + \rho_2 + \cdots + \rho_k \leq n - k \mid A_n) \\
&\quad = \sum_{i=0}^{n-k} P_p(\rho_1 + \rho_2 + \cdots + \rho_{k-1} = i,\ \rho_k \leq n - k - i \mid A_n) \\
&\quad \geq \sum_{i=0}^{n-k} P(M \leq n - k - i) P_p(\rho_1 + \rho_2 + \cdots + \rho_{k-1} = i \mid A_n) \quad \text{by (3.13)} \\
&\quad = P_p(\rho_1 + \rho_2 + \cdots + \rho_{k-1} + M_k \leq n - k \mid A_n),
\end{aligned}$$

where M_k is a random variable which is independent of all edge-states in $S(n)$ and is distributed as M. There is a mild abuse of notation here, since P_p is not the correct probability measure unless M_k is measurable on the usual σ-field of events, but we need not trouble ourselves overmuch about this. We iterate the above argument in the obvious way to deduce (3.19), thereby completing the proof of the lemma. ∎

The conclusion of Theorem (3.8) is easily obtained from this lemma, but we delay this step until the end of the section. The proof of Theorem (3.4) proceeds by substituting (3.18) into (3.11) to obtain that, for $0 \leq \alpha < \beta \leq 1$,

$$g_\alpha(n) \leq g_\beta(n) \exp\left(- \int_\alpha^\beta \left[\frac{n}{\sum_{i=0}^{n} g_p(i)} - 1 \right] dp \right).$$

It is difficult to calculate the integral in the exponent, and so we use the inequality $g_p(i) \leq g_\beta(i)$ for $p \leq \beta$ to obtain

$$g_\alpha(n) \leq g_\beta(n) \exp\left(-(\beta - \alpha) \left[\frac{n}{\sum_{i=0}^{n} g_\beta(i)} - 1 \right] \right), \tag{3.22}$$

and it is from this relation that the conclusion of Theorem (3.4) will be extracted. Before continuing, it is interesting to observe that by combining (3.10) and (3.18) we obtain a differential-difference inequality involving the function

$$G(p, n) = \sum_{i=0}^{n} g_p(i);$$

rewriting this equation rather informally as a partial differential inequality, we obtain

$$\frac{\partial^2 G}{\partial p\, dn} \geq \frac{\partial G}{\partial n}\left(\frac{n}{G} - 1\right). \tag{3.23}$$

Efforts to integrate this inequality directly have failed so far.

Once we know that

$$E_\beta(M) = \sum_{i=1}^{\infty} g_\beta(i) < \infty \qquad \text{for all} \quad \beta < p_c,$$

then (3.22) gives us that

$$g_\alpha(n) \leq e^{-n\psi(\alpha)} \qquad \text{for all} \quad \alpha < p_c,$$

for some $\psi(\alpha) > 0$, as required. At the moment we know rather less than the finite summability of the $g_p(i)$ for $p < p_c$, knowing only that $g_p(i) \to 0$ as $i \to \infty$. In order to estimate the rate at which $g_p(i) \to 0$, we shall use (3.22) as a mathematical turbocharger.

(3.24) Lemma. *For $p < p_c$, there exists $\delta(p)$ such that*

$$g_p(n) \leq \delta(p) n^{-1/2} \qquad \textit{for} \quad n \geq 1. \tag{3.25}$$

Once this lemma has been proved, the theorem follows quickly. To see this, note that (3.25) implies the existence of $\Delta(p) < \infty$ such that

$$\sum_{i=0}^{n} g_p(i) \leq \Delta(p) n^{1/2} \qquad \text{for} \quad p < p_c. \tag{3.26}$$

Let $\alpha < p_c$, and find β such that $\alpha < \beta < p_c$. Substitute (3.26) with $p = \beta$ into (3.22) to find that

$$\begin{aligned} g_\alpha(n) &\leq g_\beta(n) \exp\left\{-(\beta - \alpha)\left(\frac{n^{1/2}}{\Delta(\beta)} - 1\right)\right\} \\ &\leq \exp\left\{1 - \frac{(\beta - \alpha)}{\Delta(\beta)} n^{1/2}\right\}. \end{aligned}$$

Thus

$$\sum_{n=1}^{\infty} g_\alpha(n) < \infty \qquad \text{for} \quad \alpha < p_c,$$

and the theorem follows from the observations made prior to the statement of Lemma (3.24). We shall now prove this lemma.

Proof. First, we shall show the existence of a subsequence $n_1, n_2, \ldots$ along which $g_p(n)$ approaches 0 rather quickly; secondly, we shall fill in the gaps in this subsequence.

Fix $\beta < p_c$ and a positive integer n. Let α satisfy $0 < \alpha < \beta$ and let $n' \geq n$; later we shall choose α and n' explicitly in terms of β and n. From (3.22),

$$g_\alpha(n') \leq g_\beta(n') \exp\left(1 - \frac{n'(\beta - \alpha)}{\sum_{i=0}^{n'} g_\beta(i)}\right) \tag{3.27}$$

$$\leq g_\beta(n) \exp\left(1 - \frac{n'(\beta - \alpha)}{\sum_{i=0}^{n'} g_\beta(i)}\right)$$

since $n \leq n'$. We wish to write the exponent in terms of $g_\beta(n)$, and to this end we shall choose n' appropriately. We split the summation into two parts corresponding to $i < n$ and $i \geq n$, and we use the monotonicity of $g_\beta(i)$ to find that

$$\frac{1}{n'} \sum_{i=0}^{n'} g_\beta(i) \leq \frac{1}{n'} \{n g_\beta(0) + n' g_\beta(n)\}$$

$$\leq 3 g_\beta(n) \qquad \text{if} \quad n' \geq n \lfloor g_\beta(n)^{-1} \rfloor.$$

We now define

$$n' = n\gamma_\beta(n) \qquad \text{where} \quad \gamma_\beta(n) = \lfloor g_\beta(n)^{-1} \rfloor \tag{3.28}$$

and deduce from (3.27) that

$$g_\alpha(n') \leq g_\beta(n) \exp\left(1 - \frac{\beta - \alpha}{3 g_\beta(n)}\right). \tag{3.29}$$

Next we choose α by setting

$$\beta - \alpha = 3 g_\beta(n)\{1 - \log g_\beta(n)\}. \tag{3.30}$$

Now $g_\beta(m) \to 0$ as $m \to \infty$, so that $0 < \alpha < \beta$ if n has been picked large enough; (3.29) then yields

$$g_\alpha(n') \leq g_\beta(n)^2. \tag{3.31}$$

This conclusion is the basic recursion step which we shall use repeatedly. We have shown that, for $\beta < p_c$, there exists $n_0(\beta)$ such that (3.31) holds for all $n \geq n_0(\beta)$ whenever n' and α are given by (3.28) and (3.30), respectively.

Next, we fix $p < p_c$ and choose π such that $p < \pi < p_c$. We now construct sequences $(p_i: i \geq 0)$ of probabilities and $(n_i: i \geq 0)$ of integers as follows. We set $p_0 = \pi$ and shall pick n_0 later. Having found $p_0, p_1, \ldots, p_i$ and $n_0, n_1, \ldots, n_i$, we define

$$n_{i+1} = n_i \gamma_i \qquad \text{and} \qquad p_i - p_{i+1} = 3 g_i (1 - \log g_i) \tag{3.32}$$

where $g_i = g_{p_i}(n_i)$ and $\gamma_i = \lfloor g_i^{-1} \rfloor$. We note that $n_i \leq n_{i+1}$ and $p_i > p_{i+1}$.

The recursion (3.32) is valid so long as $p_{i+1} > 0$, and this is indeed the case so long as n_0 has been chosen to be sufficiently large. To see this we argue as follows. From the definition of $p_0, \ldots, p_i$ and $n_0, \ldots, n_i$ and the discussion leading to (3.31), we find that

(3.33) $$g_{j+1} \le g_j^2 \quad \text{for} \quad j = 0, 1, \ldots, i-1.$$

If a real sequence $(x_j\colon j \ge 0)$ satisfies $0 < x_0 < 1$, $x_{j+1} = x_j^2$ for $j \ge 0$, then it is easy to check that

$$s(x_0) = \sum_{j=0}^{\infty} 3x_j(1 - \log x_j) < \infty,$$

and furthermore that $s(x_0) \to 0$ as $x_0 \to 0$. We may pick x_0 sufficiently small such that

(3.34) $$s(x_0) \le \pi - p$$

and then we pick n_0 sufficiently large that $g_0 = g_\pi(n_0) < x_0$. Now $h(x) = 3x(1 - \log x)$ is an increasing function on $[0, x_0]$, giving from (3.32) and (3.33) that

$$\begin{aligned} p_{i+1} &= p_i - 3g_i(1 - \log g_i) \\ &= \pi - \sum_{j=0}^{i} 3g_j(1 - \log g_j) \\ &\ge \pi - \sum_{j=0}^{\infty} 3x_j(1 - \log x_j) \\ &\ge p \qquad \text{by (3.34).} \end{aligned}$$

Thus, by a suitable choice of n_0 we may guarantee not only that $p_{i+1} > 0$ for all i but also that

$$\tilde{p} = \lim_{i \to \infty} p_i$$

satisfies $\tilde{p} \ge p$. Let us suppose that n_0 has been chosen accordingly, so that the recursion (3.32) is valid and $\tilde{p} \ge p$. We have from (3.32) and (3.33) that

$$n_k = n_0 \gamma_0 \gamma_1 \cdots \gamma_{k-1} \qquad \text{for} \quad k \ge 1$$

and

(3.35) $$\begin{aligned} g_{k-1}^2 &= g_{k-1} g_{k-1} \\ &\le g_{k-1} g_{k-2}^2 \le \cdots \\ &\le g_{k-1} g_{k-2} \cdots g_1 g_0^2 \\ &\le (\gamma_{k-1}\gamma_{k-2} \cdots \gamma_0)^{-1} g_0 \\ &= \delta^2 n_k^{-1}, \end{aligned}$$

where $\delta^2 = n_0 g_0$.

We are essentially finished. Let $n > n_0$, and find an integer k such that $n_{k-1} \leq n < n_k$; this is always possible since $g_k \to 0$ as $k \to \infty$, and therefore $n_{k-1} < n_k$ for all large k. Then

$$\begin{aligned} g_p(n) &\leq g_{p_{k-1}}(n_{k-1}) && \text{since} \quad p \leq p_{k-1} \\ &= g_{k-1} \\ &\leq \delta n_k^{-1/2} && \text{by (3.35)} \\ &\leq \delta n^{-1/2} && \text{since} \quad n < n_k \end{aligned}$$

as required. This is valid for $n > n_0$, but we may adjust the constant δ so that a similar inequality is valid for all $n \geq 1$. ■

Proof of Theorem (3.8). Combining (3.10) with (3.18), we obtain the first ingredient of the proof,

$$g'_\pi(n) \geq g_\pi(n)\left(\frac{n}{\sum_{i=0}^{n} g_\pi(i)} - 1\right), \tag{3.36}$$

valid for $0 < \pi < 1$. The following argument is informative, if naive and incorrect. Suppose that $\pi > p_c$, and use the fact that $g_\pi(n) \to \theta(\pi)$ as $n \to \infty$, to deduce (somewhat rashly) from (3.36) that

$$\theta'(\pi) \geq 1 - \theta(\pi),$$

an inequality which is easily integrated over the interval (p_c, p) to obtain

$$\theta(p)e^p - \theta(p_c)e^{p_c} \geq e^p - e^{p_c} \qquad \text{if} \quad p > p_c.$$

It is now a short step to deduce that

$$\theta(p) - \theta(p_c) \geq \tfrac{1}{2}(1 - \theta(p_c))(p - p_c)$$

if $p - p_c$ is positive but sufficiently small.

A more rigorous argument proceeds as follows. The second ingredient of the proof is the fact that θ is a right-continuous function on the interval $(0, 1)$. No special technique is required to prove this; instead we use a 'soft' argument. It is clear that $\theta(p)$ is the decreasing limit of $g_p(n)$ as $n \to \infty$. However, $g_p(n)$ is a polynomial in p and is therefore continuous, so that θ is the decreasing limit of continuous functions. Hence θ is upper semi-continuous. On the other hand, θ is non-decreasing, and therefore right-continuous also. We shall return in Section 6.3 to the question of the continuity of θ; it turns out to be rather more difficult to show that θ is left-continuous, and indeed it is an open problem to prove that θ is left-continuous at p_c.

Let π be such that $p_c < \pi < 1$, and let ε satisfy $0 < \varepsilon < \frac{1}{2}\{1 - \theta(\pi)\}$. Let $\alpha \in (p_c, \pi)$. By the right-continuity of θ, we may find $\beta = \beta(\alpha) \in (\alpha, \pi]$

such that

$$\theta(\beta) \le (1 + \varepsilon)\theta(\alpha). \tag{3.37}$$

Next, we pick $N = N(\alpha, \beta)$ such that

$$\frac{1}{n}\sum_{i=0}^{n} g_\beta(i) \le \frac{1}{1-\varepsilon}\,\theta(\beta) \qquad \text{if} \quad n \ge N; \tag{3.38}$$

this is possible since the left-hand side converges to $\theta(\beta)$ as $n \to \infty$. We have that

$$\theta(\alpha) \le g_\alpha(n) \le g_\gamma(n) \le g_\beta(n) \qquad \text{for all } n \text{ and } \gamma \in [\alpha, \beta],$$

giving from (3.36) that, for $n \ge N$ and $\alpha \le \gamma \le \beta$,

$$\begin{aligned} g_\gamma'(n) &\ge \theta(\alpha)\left(\frac{n}{\sum_{i=0}^{n} g_\beta(i)} - 1\right) \\ &\ge \theta(\alpha)\left(\frac{1-\varepsilon}{\theta(\beta)} - 1\right) && \text{by (3.38)} \\ &\ge \frac{1}{1+\varepsilon}\{1 - \varepsilon - \theta(\beta)\} && \text{by (3.37)} \\ &\ge \frac{1}{1+\varepsilon}\{1 - \varepsilon - \theta(\pi)\} && \text{since} \quad \beta \le \pi \\ &\ge \frac{\varepsilon}{1+\varepsilon} && \text{since} \quad 1 - \theta(\pi) > 2\varepsilon. \end{aligned}$$

We integrate this inequality and take the limit as $n \to \infty$ to obtain

$$\theta(\gamma) - \theta(\alpha) \ge (\gamma - \alpha)\frac{\varepsilon}{1+\varepsilon} \qquad \text{for} \quad \alpha \le \gamma < \beta. \tag{3.39}$$

If β did not depend on the choice of α, then we would obtain the desired result by taking the limit as $\alpha \downarrow p_c$ and using the right-continuity of θ at p_c. The following argument shows that we may replace β by π in (3.39), thus completing the proof of the theorem. Define

$$\mu(\alpha) = \sup\left\{\beta \in (\alpha, \pi]\colon \theta(\gamma) - \theta(\alpha) \ge (\gamma - \alpha)\frac{\varepsilon}{1+\varepsilon} \text{ for all } \gamma \in [\alpha, \beta)\right\}.$$

It suffices to prove that $\mu(\alpha) = \pi$. Suppose on the contrary that $\alpha < \mu(\alpha) < \pi$. We may repeat the argument above to find that there exists $\zeta \in (\mu(\alpha), \pi]$ such that

$$\theta(\nu) - \theta(\mu(\alpha)) \ge \{\nu - \mu(\alpha)\}\frac{\varepsilon}{1+\varepsilon} \qquad \text{for} \quad \mu(\alpha) \le \nu < \zeta.$$

On the other hand, by the monotonicity of θ,

$$\begin{aligned}\theta(\mu(\alpha)) - \theta(\alpha) &\geq \lim_{\xi \uparrow \mu(\alpha)} \{\theta(\xi) - \theta(\alpha)\}\\ &\geq \lim_{\xi \uparrow \mu(\alpha)} \left\{(\xi - \alpha)\frac{\varepsilon}{1+\varepsilon}\right\}\\ &= \{\mu(\alpha) - \alpha\}\frac{\varepsilon}{1+\varepsilon},\end{aligned}$$

so that, for $\nu \in [\mu(\alpha), \zeta)$,

$$\begin{aligned}\theta(\nu) - \theta(\alpha) &\geq \{\nu - \mu(\alpha)\}\frac{\varepsilon}{1+\varepsilon} + \{\mu(\alpha) - \alpha\}\frac{\varepsilon}{1+\varepsilon}\\ &= (\nu - \alpha)\frac{\varepsilon}{1+\varepsilon}.\end{aligned}$$

This contradicts the maximality of $\mu(\alpha)$, and thus $\mu(\alpha) = \pi$ as claimed. ∎

3.3 Uniqueness via Differential Inequalities

In this section we shall prove Theorem (3.2) by the method of Aizenman and Barsky (1987). Their proof is rather different from the argument in the last section, although there are certain striking similarities. The general structure of the Aizenman–Barsky proof is as follows. We introduce a new parameter $\gamma \in (0, 1)$ in such a way that ordinary percolation is obtained by taking the limit as $\gamma \downarrow 0$; this new parameter corresponds to an external magnetic field in the Ising model. For pairs (p, γ) of parameter values with $\gamma > 0$, we find that the two quantities analogous to $\theta(p)$ and $\chi(p)$ are analytic. We then establish certain differential inequalities for these quantities using Russo's formula and the BK inequality. We integrate these differential inequalities and take the limit as $\gamma \downarrow 0$ to find, as required, that θ and χ have singularities at the same value of p.

Our reasons for including this proof are primarily twofold. First, the introduction of a new parameter is an idea with considerable physical appeal, and allows us to explore in greater depth both rigorous and intuitive links between percolation and certain models of statistical physics. Secondly, such methods yield some information about the behaviour of θ and χ for values of p at and near the critical value p_c. This is related to the problem of so called 'critical exponents' to which we shall return in Chapters 7 and 8.

We introduce the new parameter γ as follows. We create a new vertex, labelled g, and we join each vertex of $\mathbb{L}^d$ by an open edge to g with probability γ, independently of all other open edges. For each vertex x of $\mathbb{L}^d$, we write $\omega(x) = 1$ if x is thus joined to g and $\omega(x) = 0$ otherwise. Whereas previously the set of realizations was the set $\prod_{e \in \mathbb{E}^d}\{0, 1\}$ of assignments of 'closed' or 'open' to the edges of $\mathbb{L}^d$, the set of realizations is now $\prod_{e \in \mathbb{E}^d} \{0, 1\} \times \prod_{x \in \mathbb{Z}^d} \{0, 1\}$, since each edge is either closed or open, and in addition each vertex is either non-adjacent or adjacent to g. The original product measure $\prod_{e \in \mathbb{E}^d} \mu_e$ has been replaced by a product measure $\prod_{e \in \mathbb{E}^d} \mu_e \times \prod_{x \in \mathbb{Z}^d} \mu_x$, and we write $P_{p,\gamma}$ for this product measure and $E_{p,\gamma}$ for the corresponding expectation. We retrieve the original percolation process by neglecting g and its incident edges.

The purpose of this construction is as follows. We think of g as a surrogate 'point at infinity'. We shall concentrate on the existence of open paths joining the origin to g, and then we shall take the limit as $\gamma \downarrow 0$. As γ approaches 0, the density of vertices adjacent to g drops to 0, so that the open cluster at the origin is forced to be larger and larger in order that there exist an open path from the origin to g. Writing G for the set of vertices of $\mathbb{L}^d$ which are adjacent to g, we may see that there is an open path from the origin to g if and only if some vertex in C lies also in G; we write $\{C \cap G \neq \varnothing\}$ for this event. It is not difficult to see that, as $\gamma \to 0$,

$$P_{p,\gamma}(C \cap G \neq \varnothing) \to P_p(|C| = \infty) = \theta(p).$$

It is in this sense that ordinary percolation is retrieved by taking the limit as γ tends to 0. We shall not generally refer to the new vertex g, preferring to talk in terms of the random set G of vertices of $\mathbb{L}^d$ which are adjacent to g. The advantage of this is that the term 'open path' shall retain its meaning as a path of open edges in $\mathbb{L}^d$. Thus we shall speak in such terms as '0 is joined to a vertex in G by an open path' rather than '0 is joined to g by an open path'.

Suppose now that $0 < \gamma < 1$, and define

$$\theta(p, \gamma) = P_{p,\gamma}(C \cap G \neq \varnothing), \tag{3.40}$$

the probability that there is an open path from the origin to a vertex in G. Just as $\theta(p)$ may be represented as

$$\theta(p) = 1 - \sum_{n=1}^{\infty} P_p(|C| = n), \tag{3.41}$$

so may we write

$$\begin{aligned} \theta(p, \gamma) &= 1 - \sum_{n=1}^{\infty} P_{p,\gamma}(C \cap G = \varnothing \,|\, |C| = n) P_p(|C| = n) \\ &= 1 - \sum_{n=1}^{\infty} (1 - \gamma)^n P_p(|C| = n). \end{aligned} \tag{3.42}$$

We note that, for $\gamma > 0$, $C \cap G \neq \varnothing$ almost surely when $|C| = \infty$, and it is for this reason that the summation in (3.42) contains no term corresponding to this case. The sum in (3.42) contains non-negative terms and converges when $\gamma > 0$, and so we may take the limit as $\gamma \to 0$ to obtain that

$$\theta(p, \gamma) \downarrow \theta(p) \qquad \text{as} \quad \gamma \downarrow 0. \tag{3.43}$$

We define $\theta(p, 0)$ to be $\theta(p)$.

The second quantity of interest is analogous to $\chi(p)$. For $0 < \gamma < 1$, we define $\chi(p, \gamma)$ to be the mean number of vertices in C on the event that there is no open path from the origin to any vertex of G; thus

$$\chi(p, \gamma) = E_{p,\gamma}(|C|; C \cap G = \varnothing). \tag{3.44}$$

By conditioning on $|C|$ we see that

$$\begin{aligned} \chi(p, \gamma) &= \sum_{n=1}^{\infty} n P_{p,\gamma}(C \cap G = \varnothing \,|\, |C| = n) P_p(|C| = n) \\ &= \sum_{n=1}^{\infty} n(1-\gamma)^n P_p(|C| = n). \end{aligned} \tag{3.45}$$

As before, there is no term corresponding to the event that $|C| = \infty$, since $C \cap G \neq \varnothing$ almost surely in this case. Here also we may take the limit as $\gamma \downarrow 0$ to find that

$$\chi(p, \gamma) \uparrow \chi^f(p) \qquad \text{as} \quad \gamma \downarrow 0, \tag{3.46}$$

where $\chi^f(p) = \sum_n n P_p(|C| = n)$ is the mean size of the finite open cluster at the origin. We define $\chi(p, 0)$ to be equal to $\chi^f(p)$, and note that $\chi^f(p) = \chi(p)$ whenever $\chi(p) < \infty$.

We see in (3.42) that θ is a power series in $1 - \gamma$ with radius of convergence at least 1. We may differentiate θ with respect to γ within its disc of convergence to obtain that

$$\frac{\partial \theta}{\partial \gamma} = \sum_{n=1}^{\infty} n(1-\gamma)^{n-1} P_p(|C| = n) \qquad \text{if} \quad 0 < \gamma < 1,$$

which we compare with (3.45) to see that

$$\chi(p, \gamma) = (1 - \gamma) \frac{\partial \theta}{\partial \gamma}(p, \gamma) \qquad \text{if} \quad 0 < \gamma < 1. \tag{3.47}$$

It is not difficult to show that θ is continuously differentiable with respect to p also, so long as $\gamma > 0$; to see this, it is necessary to expand $P_p(|C| = n)$ as a polynomial in p and bound its derivative with respect to p. Rather than go through the details here, we defer this to Appendix I.

The above formulation is simple and appealing in probabilistic terms, but obscures the physical motivation underlying the parameter γ. Readers with a background in statistical physics may prefer to write $1 - \gamma = e^{-h}$ where $0 \leq h < \infty$, and to think of h as an external magnetic field. The quantity $\theta(p, \gamma)$ is essentially the Laplace transform of the mass function of $|C|$:

$$\theta(p, \gamma) = 1 - E\{\exp(-h|C|)\} \qquad \text{when} \quad h = -\log(1-\gamma) > 0.$$

In some articles, g is called the 'ghost' vertex, and vertices in G are said to be 'green'. In this jargon, $\theta(p, \gamma)$ is the probability of an open path from the origin to the ghost, and $\chi(p, \gamma)$ is the mean size of the 'green-free' open cluster at the origin.

We shall prove the following main result.

(3.48) Theorem. *If p is such that $\chi^f(p) = \infty$ then*

either (a) $\theta(p) > 0$,

or (b) $\theta(p) = 0$ *and* $\theta(p') \geq \dfrac{1}{2p'}(p' - p)$ *for* $p' \geq p$.

This proves that $\chi(p) < \infty$ when $p < p_c$, since suppose otherwise that $\chi(p) = \infty$ for some $p < p_c$. For such a value of p, we have that

$$\begin{aligned} \chi^f(p) &= E_p(|C|; |C| < \infty) \\ &= E_p|C| \qquad \text{since } \theta(p) = 0 \\ &= \chi(p) \\ &= \infty \end{aligned}$$

giving by the theorem that $p \geq p_c$, a contradiction.

An important step in the proof of this theorem is the following proposition, which we extract in order that we may locate it more easily later.

(3.49) Proposition. *If p is such that $\chi^f(p) = \infty$ then there exists a constant $\alpha = \alpha(p) > 0$ such that*

$$\theta(p, \gamma) \geq \alpha \gamma^{1/2} \tag{3.50}$$

for all small positive values of γ.

Our tactics will involve using Russo's formula and the BK inequality to establish certain differential inequalities for θ and its derivatives $\partial\theta/\partial p$ and $\partial\theta/\partial\gamma = (1 - \gamma)^{-1}\chi$. Unlike the differential inequality (3.23) in the proof of the last section, we shall obtain inequalities which may be integrated directly to give the result.

(3.51) Lemma. *If $0 < \gamma < 1$ and $0 < p < 1$ then*

$$(1 - p)\frac{\partial\theta}{\partial p} \leq 2d(1 - \gamma)\theta\frac{\partial\theta}{\partial\gamma}. \tag{3.52}$$

(3.53) Lemma. *If $0 < \gamma < 1$ and $0 < p < 1$, then*

$$\theta \leq \gamma\frac{\partial\theta}{\partial\gamma} + \theta^2 + p\theta\frac{\partial\theta}{\partial p}. \tag{3.54}$$

As remarked above, these inequalities are fairly straightforward applications of Russo's formula and the BK inequality. There is, however, an unfortunate technicality in their proofs, arising from the fact that these techniques apply to events which depend on only finitely many edges, whereas events such as $\{C \cap G \neq \varnothing\}$ depend on infinitely many edges. Thus it will be necessary to express such events as the limits of events depending on only finitely many edges; in physical terms, we need to perform a finite-volume approximation. Certain properties of differentiability of θ are necessary in order to take the infinite-volume limit with rigour.

In advance of proving Lemmas (3.51) and (3.53), we show how they may be used to demonstrate Theorem (3.48) and Proposition (3.49).

Proof of Proposition (3.49). Suppose that $0 < \gamma < 1$, $0 < p < 1$, and $\chi^f(p) = \infty$. If $\theta(p) = \theta(p, 0) > 0$ then (3.50) holds vacuously, and so we suppose that $\theta(p) = 0$. With this fixed value of p, we write $\theta(p, \gamma) = f(\gamma)$, where f is strictly increasing and continuously differentiable on (0, 1) with $f(0) = 0$ and $f(1) = 1$. By the mean value theorem, for each $\gamma \in (0, 1)$, there exists $\psi = \psi(\gamma)$ satisfying $\psi \in (0, \gamma)$ such that

$$\frac{1}{\gamma} f(\gamma) = f'(\psi).$$

As $\gamma \downarrow 0$, we have that $\psi \to 0$, giving by (3.47) and (3.46) that

$$\lim_{\gamma \downarrow 0} \frac{\gamma}{f(\gamma)} = 0. \tag{3.55}$$

Let g be the inverse function of f: g is strictly increasing and continuously differentiable with $g(0) = 0$ and $g(1) = 1$, and satisfies

$$\lim_{\varphi \downarrow 0} \frac{1}{\varphi} g(\varphi) = 0 \tag{3.56}$$

by (3.55).

We substitute (3.52) into (3.54) to find that

$$f \leq \gamma \frac{df}{d\gamma} + f^2 + 2d(1 - \gamma)p(1 - p)^{-1} f^2 \frac{df}{d\gamma}.$$

Next we substitute $\varphi = f(\gamma)$ (>0), $\gamma = g(\varphi)$, and

$$\frac{dg}{d\varphi} = \left(\frac{df}{d\gamma}\right)^{-1},$$

to obtain

$$\frac{1}{\varphi} \frac{dg}{d\varphi} - \frac{1}{\varphi^2} g \leq 2d(1 - g)p(1 - p)^{-1} + \frac{dg}{d\varphi}.$$

By (3.56) and the preceding remarks, $g(\varphi) \geq 0$ and $g'(\varphi)$ is bounded on $(0, \Phi)$ for some $\Phi > 0$, giving that there exists a positive constant $\beta =$

$\beta(p)$ such that

$$\frac{1}{\varphi}\frac{dg}{d\varphi} - \frac{1}{\varphi^2} g \le \beta \qquad \text{if} \quad 0 < \varphi < \Phi,$$

where $0 < \Phi < 1$. We integrate this equation from $\varphi = 0$ to $\varphi = x$ where $x \le \Phi$ to obtain

$$\left[\frac{1}{\varphi} g(\varphi)\right]_0^x \le \beta x;$$

we now use (3.56) to deduce that $g(x) \le \beta x^2$ for all $x \le \Phi$. We substitute back in terms of f to obtain $\gamma \le \beta f(\gamma)^2$ for $\gamma \le g(\Phi)$, or $\theta(p, \gamma) = f(\gamma) \ge \alpha \gamma^{1/2}$ where $\alpha = \beta^{-1/2}$. ■

Proof of Theorem (3.48). Let $0 < \gamma < 1$, $0 < a < 1$, and suppose that $\chi^f(a) = \infty$. If $\theta(a) > 0$, then there is nothing to prove. We assume henceforth that $\theta(a) = 0$, but note that $\theta(a, \gamma) > 0$. We shall show $\theta(b) > 0$ if $b > a$ by integrating inequality (3.54), which we write in the form

$$0 \le \frac{1}{\theta}\frac{\partial \theta}{\partial \gamma} + \frac{1}{\gamma}\frac{\partial}{\partial p}(p\theta - p). \tag{3.57}$$

Suppose $a < b < 1$. We integrate (3.57) over the rectangle $a \le p \le b$, $\delta \le \gamma \le \varepsilon$, where $0 < \delta < \varepsilon$; see Figure 3.3. On this rectangle, θ takes its minimum value at $(p, \gamma) = (a, \delta)$ and its maximum value at $(p, \gamma) = (b, \varepsilon)$. Thus

$$0 \le (b - a)\log\left(\frac{\theta(b, \varepsilon)}{\theta(a, \delta)}\right) + \{b\theta(b, \varepsilon) - b + a\}\log(\varepsilon/\delta),$$

where we have used the fact that $\theta(a, \delta) \ge 0$. We shall divide by $\log(\varepsilon/\delta)$ and take the limit as $\delta \downarrow 0$. From Proposition (3.49),

$$\limsup_{\delta \to 0} \frac{\log \theta(a, \delta)}{\log \delta} \le \frac{1}{2},$$

giving as $\delta \downarrow 0$ that

$$0 \le \tfrac{1}{2}(b - a) + b\theta(b, \varepsilon) - b + a.$$

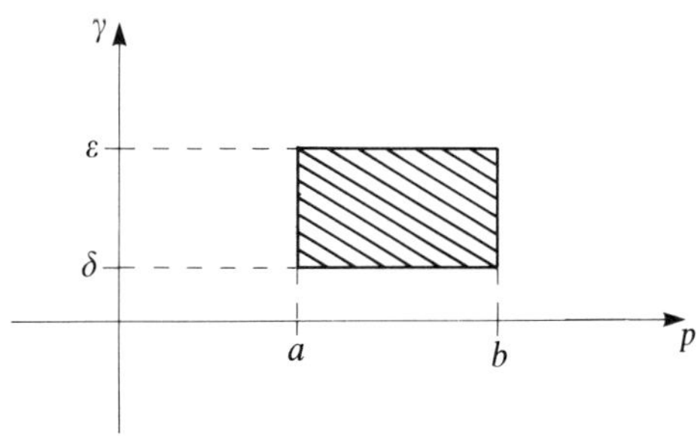

Figure 3.3. The rectangle over which we integrate the quantities in (3.57).

Finally, we let ε approach 0 to obtain

$$\theta(b) = \theta(b, 0) \geq \frac{1}{2b}(b - a) > 0. \tag{3.58}$$ ■

We turn next to the differential inequalities (3.52) and (3.54). As remarked earlier, we shall require a finite-volume approximation. Let $B(N)$ be the cube $[-N, N]^d$ as before; we shall study the percolation process restricted to $B(N)$. It is convenient for this finite-volume process to have a property of translation invariance, and to this end we assign periodic boundary conditions to $B(N)$ by identifying opposite faces of $B(N)$. Rather than describing this in detail, we remark only that we replace by single vertices all sets of vertices of the form $\{x \in B(N): x_i = \pm N \text{ if } i \in I, \text{ and } x_i = z_i \text{ if } i \notin I\}$ as I ranges over subsets of $\{1, 2, \ldots, d\}$ and z ranges over $\{-N+1, \ldots, N-1\}^d$; in this new graph we identify parallel edges, leaving all other edges untouched, so that each composite vertex has degree $2d$ exactly. Thus in two dimensions we may think of $B(N)$ as being embedded in the torus. We denote the resulting graph by $L(N)$, writing $B^*(N)$ and $E(N)$ for the sets of its vertices and edges, respectively. We examine each edge of $L(N)$ in turn and declare it to be open with probability p and closed otherwise, independently of all other edges; we then examine each vertex of $L(N)$ in turn, and join this vertex to g by an open edge with probability γ, independently of all other vertices. Writing G_N for the set of vertices of $L(N)$ which are adjacent to g, we define quantities analogous to $\theta(p, \gamma)$ and $\chi(p, \gamma)$ by

$$\theta_N(p, \gamma) = P_{p,\gamma}(C_N \cap G_N \neq \varnothing), \tag{3.59}$$

$$\chi_N(p, \gamma) = E_{p,\gamma}(|C_N|; C_N \cap G_N = \varnothing), \tag{3.60}$$

where $C_N(x)$ is the open cluster of $L(N)$ at the vertex x, and $C_N = C_N(0)$. Thus

$$\theta_N(p, \gamma) = 1 - \sum_{n=1}^{\infty} (1-\gamma)^n P_p(|C_N| = n), \tag{3.61}$$

$$\chi_N(p, \gamma) = \sum_{n=1}^{\infty} n(1-\gamma)^n P_p(|C_N| = n). \tag{3.62}$$

These two series are finite, since $|B^*(N)| < \infty$, so that θ_N and χ_N are analytic functions of p and γ. Just as in (3.47),

$$\chi_N(p, \gamma) = (1-\gamma)\frac{\partial \theta_N}{\partial \gamma}(p, \gamma). \tag{3.63}$$

In the limit as $N \to \infty$, it is the case that $P_p(|C_N| = n) \to P_p(|C| = n)$; this rather weak remark is easy to check, since if $N > n$ and $|C_N| = n$ then there is no vertex on the boundary of $B(N)$ which is in C_N also. Actually, rather more is true about the convergence of $P_p(|C_N| = n)$. By expanding this probability as the sum over animals of appropriate sizes, it may be shown without undue difficulty that there is sufficient regularity in such probabilities and their derivatives to ensure that $\theta_N(p, \gamma)$

and its derivatives converge as $N \to \infty$ to the appropriate derivatives of $\theta(p, \gamma)$ whenever $0 < \gamma < 1$. We do not prove this here, but summarize the argument in Appendix I. We shall therefore assume that the following limits are valid whenever $0 < \gamma < 1$ and $0 < p < 1$:

$$\theta_N(p, \gamma) \to \theta(p, \gamma) \qquad \text{as} \quad N \to \infty, \tag{3.64}$$

$$\frac{\partial \theta_N}{\partial p}(p, \gamma) \to \frac{\partial \theta}{\partial p}(p, \gamma) \qquad \text{as} \quad N \to \infty, \tag{3.65}$$

$$\frac{\partial \theta_N}{\partial \gamma}(p, \gamma) \to \frac{\partial \theta}{\partial \gamma}(p, \gamma) \qquad \text{as} \quad N \to \infty. \tag{3.66}$$

Having made these assumptions, it remains only to prove (3.52) and (3.54) with θ_N in place of θ.

Proof of Lemma (3.51). Suppose $0 < \gamma < 1$ and $0 < p < 1$. We shall apply Russo's formula to the event $\{C_N \cap G_N \neq \varnothing\}$. One way of doing this is to condition on G_N, say $G_N = \Gamma$, and apply Russo's formula to the increasing event $A_N(\Gamma) = \{C_N \cap \Gamma \neq \varnothing\}$. Thus

$$\frac{d}{dp} P_p(A_N(\Gamma)) = \sum_e P_p(e \text{ is pivotal for } A_N(\Gamma)),$$

where the sum is over all edges e of $L(N)$. The edge $e = \langle x, y \rangle$ is pivotal for $A_N(\Gamma)$ if and only if, in the graph $L(N)$ with e deleted, the following three events occur:

(a) no vertex in Γ is joined to the origin by an open path;
(b) exactly one of x and y is joined to the origin by an open path; and
(c) the other endvertex of e is joined to a vertex of Γ by an open path.

See Figure 3.4 for a pictorial representation of the event that e is pivotal

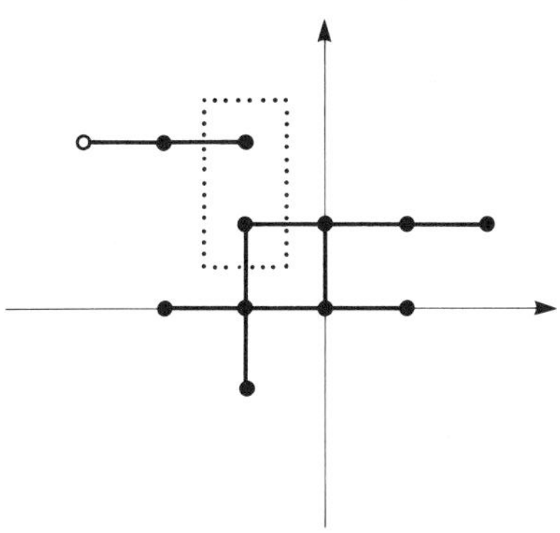

Figure 3.4. The open circle is a vertex in Γ, and the dense circles are vertices not in Γ. The origin is joined by an open path to a vertex in Γ if the two vertices in the dotted box are joined by an open edge, and not otherwise. That is to say, the edge in the dotted box is pivotal for the event $A_N(\Gamma)$.

for $A_N(\Gamma)$. We may now see that

$$(1-p)\frac{d}{dp}P_p(A_N(\Gamma)) = \sum_e P_p(e \text{ is closed})P_p(e \text{ is pivotal for } A_N(\Gamma))$$
$$= \sum_{x\sim y} P_p(x \in C_N,\ C_N \cap \Gamma = \varnothing,\ C_N(y)\cap\Gamma \neq \varnothing),$$

where the sum is over all ordered pairs x, y of adjacent vertices in $L(N)$. We now average over Γ, noting that

$$\sum_\Gamma (1-p)P_{p,\gamma}(G_N=\Gamma)\frac{d}{dp}P_p(A_n(\Gamma))$$
$$= (1-p)\frac{\partial}{\partial p}\left\{\sum_\Gamma P_{p,\gamma}(C_N\cap G_N \neq \varnothing \mid G_N=\Gamma)P_{p,\gamma}(G_N=\Gamma)\right\}$$
$$= (1-p)\frac{\partial\theta_N}{\partial p}$$

since G_N takes one of only finitely many values and its distribution depends on γ only. Thus

$$(3.67)\quad (1-p)\frac{\partial\theta_N}{\partial p} = \sum_{x\sim y} P_{p,\gamma}(x\in C_N,\ C_N\cap G_N=\varnothing,\ C_N(y)\cap G_N\neq\varnothing).$$

The last event occurs only if $C_N(y)$ has no vertex in common with C_N. In order to bound the last probability, we condition on C_N, say $C_N = \Sigma$, and rewrite (3.67) as

$$(1-p)\frac{\partial\theta_N}{\partial p} = \sum_{x\sim y}\sum_\Sigma P_p(C_N=\Sigma)P_{p,\gamma}(C_N\cap G_N=\varnothing,\ C_N(y)\cap G_N\neq\varnothing \mid C_N=\Sigma),$$

where the second sum is over all sets Σ of vertices containing 0 and x but not y. However, conditional on $C_N=\Sigma$, the events $\{C_N\cap G_N=\varnothing\}$ and $\{C_N(y)\cap G_N\neq\varnothing\}$ are independent, the first depending only on the vertices of Σ and the second depending only on the vertices outside Σ and the edges of $L(N)$ having no endvertex in Σ. Furthermore, using the translation invariance of $L(N)$,

$$P_{p,\gamma}(C_N(y)\cap G_N\neq\varnothing \mid C_N=\Sigma) \le P_{p,\gamma}(C_N(y)\cap G_N\neq\varnothing)$$
$$= \theta_N(p,\gamma),$$

since the condition that $C_N=\Sigma$ where $y\notin\Sigma$ limits the possibilities for there to be an open path from y to a vertex in G_N. Thus we obtain

$$(1-p)\frac{\partial\theta_N}{\partial p} \le \sum_{x\sim y}\sum_\Sigma P_p(C_N=\Sigma)P_{p,\gamma}(C_N\cap G_N=\varnothing \mid C_N=\Sigma)\theta_N$$
$$= \theta_N\sum_{x\sim y}P_{p,\gamma}(x\in C_N,\ y\notin C_N,\ C_N\cap G_N=\varnothing)$$

$$\leq \theta_N \sum_{x \sim y} P_{p,\gamma}(x \in C_N, C_N \cap G_N = \varnothing)$$

$$\leq 2d\theta_N \sum_{x} P_{p,\gamma}(x \in C_N, C_N \cap G_N = \varnothing) \quad \text{since } x \text{ has } 2d \text{ neighbours}$$

$$= 2d\theta_N E_{p,\gamma}(|C_N|; C_N \cap G_N = \varnothing)$$

$$= 2d\theta_N \chi_N$$

as required, by (3.63). ■

Proof of Lemma (3.53). This is rather similar to the last proof. Suppose $0 < \gamma < 1$ and $0 < p < 1$, and note that

(3.68)
$$\begin{aligned}\theta_N &= P_{p,\gamma}(C_N \cap G_N \neq \varnothing)\\ &= P_{p,\gamma}(|C_N \cap G_N| = 1) + P_{p,\gamma}(|C_N \cap G_N| \geq 2).\end{aligned}$$

The first term here is easily calculated by conditioning on $|C_N|$:

(3.69)
$$\begin{aligned}P_{p,\gamma}(|C_N \cap G_N| = 1) &= \sum_{n=1}^{\infty} n\gamma(1-\gamma)^{n-1} P_p(|C_N| = n)\\ &= \frac{\gamma}{1-\gamma}\chi_N\\ &= \gamma\frac{\partial \theta_N}{\partial \gamma} \qquad \text{by (3.63).}\end{aligned}$$

The last term in (3.68) is harder to estimate. For each vertex x of $L(N)$, we write A_x for the event that either $x \in G_N$ or x is joined by an open path to a vertex in G_N; we note that A_x is defined on the set $\prod_{e \in E(N)}\{0, 1\} \times \prod_{x \in B^*(N)}\{0, 1\}$ of realizations. Remembering the discussion leading to the BK inequality in Section 2.3, we write $A_x \circ A_x$ for the set of such realizations ω for which there exist disjoint sets V_1 and V_2 of vertices in G_N and disjoint sets E_1 and E_2 of open edges of $L(N)$, such that A_x occurs for all realizations ω' which agree with ω on (V_1, E_1) and for all realizations ω'' which agree with ω on (V_2, E_2). Thus, $A_x \circ A_x$ is the event that there exist two distinct vertices in G_N and two edge-disjoint open paths joining these two vertices to x; if $x \in G_N$, one of these paths may be the singleton x itself. Now,

(3.70)
$$\begin{aligned}&P_{p,\gamma}(|C_N \cap G_N| \geq 2)\\ &\quad = P_{p,\gamma}(A_0 \circ A_0) + P_{p,\gamma}(|C_N \cap G_N| \geq 2,\ A_0 \circ A_0 \text{ does not occur}).\end{aligned}$$

By the BK inequality,

(3.71)
$$P_{p,\gamma}(A_0 \circ A_0) \leq P_{p,\gamma}(A_0)^2 = \theta_N^2.$$

The remaining term in (3.70) is the probability that $|C_N \cap G_N| \geq 2$ but there do not exist two disjoint connections from the origin to G_N in the

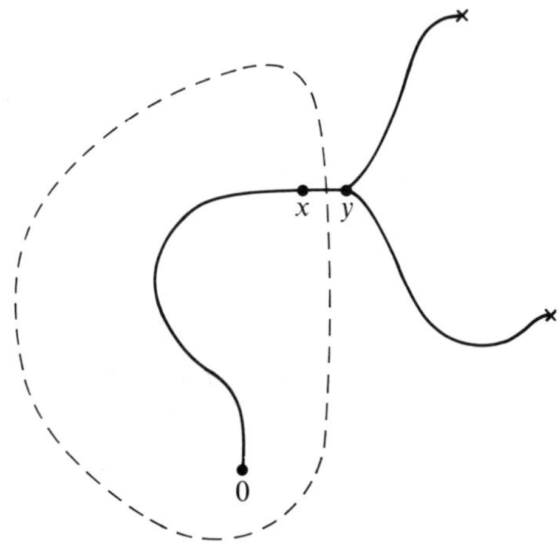

Figure 3.5. A sketch of the event that $|C_N \cap G_N| \geq 2$ but $A_0 \circ A_0$ does not occur. The crosses represent vertices in G_N, and the dashed line represents a 'hypersurface' which would cut off the origin from all vertices in G_N if $e = \langle x, y \rangle$ were closed.

sense described above. It is not hard to convince oneself (possibly with the aid of Figure 3.5) that this event occurs if and only if there exists a pair x, y of adjacent vertices of $L(N)$ such that:

(i) the edge $e = \langle x, y \rangle$ is open; and
(ii) in the subgraph of $L(N)$ obtained by deleting e, the following three events occur:
   (a) no vertex of G_N is joined to the origin by an open path,
   (b) x is joined to the origin by an open path,
   (c) the event $A_y \circ A_y$ occurs.

The event (ii) is independent of the state of e, so that the probability that (i) and (ii) occur is exactly

$$pq^{-1}P_{p,\gamma}(e \text{ is closed, } x \in C_N,\ C_N \cap G_N = \varnothing,\ A_y \circ A_y),$$

where $p + q = 1$, giving that the probability that such a pair x, y exists is no larger than

$$pq^{-1} \sum_{x \sim y} P_{p,\gamma}(x \in C_N,\ C_N \cap G_N = \varnothing,\ A_y \circ A_y),$$

since e is necessarily closed if $C_N(y)$ intersects G_N but $C_N(x)$ does not. We now condition on C_N as usual to obtain

(3.72) $$P_{p,\gamma}(|C_N \cap G_N| \geq 2,\ A_0 \circ A_0 \text{ does not occur})$$
$$\leq pq^{-1} \sum_{x \sim y} \sum_{\Sigma} P_p(C_N = \Sigma) P_{p,\gamma}(C_N \cap G_N = \varnothing,\ A_y \circ A_y | C_N = \Sigma),$$

where the second sum is over all sets Σ of vertices which contain 0 and x but not y. Conditional on $C_N = \Sigma$, the events $\{C_N \cap G_N = \varnothing\}$ and

$A_y \circ A_y$ are independent, since $A_y \circ A_y$ is defined on the vertices of $L(N)$ outside Σ together with the edges between such vertices. Conditional on $C_N = \Sigma$, the states of these vertices and edges are chosen according to the appropriate product measure, so that

(3.73)

$$\begin{aligned}
P_{p,\gamma}(C_N \cap G_N = \varnothing,\, A_y \circ A_y | C_N = \Sigma) \\
&= P_{p,\gamma}(C_N \cap G_N = \varnothing | C_N = \Sigma) P_{p,\gamma}(A_y \circ A_y | C_N = \Sigma) \quad \text{by independence} \\
&\leq P_{p,\gamma}(C_N \cap G_N = \varnothing | C_N = \Sigma) P_{p,\gamma}(A_y | C_N = \Sigma)^2 \quad \text{by the BK inequality} \\
&\leq P_{p,\gamma}(C_N \cap G_N = \varnothing | C_N = \Sigma) P_{p,\gamma}(A_y | C_N = \Sigma) P_{p,\gamma}(A_y) \\
&= P_{p,\gamma}(C_N \cap G_N = \varnothing,\, A_y | C_N = \Sigma)\theta_N \quad \text{by independence,}
\end{aligned}$$

where we have used the fact that A_y is more likely to occur on the whole of $L(N)$ than on the restricted graph obtained by removing Σ. We substitute this last equation into (3.72) to obtain

$$\begin{aligned}
(3.74)\quad P_{p,\gamma}(|C_N \cap G_N| \geq 2,\, A_0 \circ A_0 \text{ does not occur}) \\
&\leq pq^{-1}\theta_N \sum_{x \sim y} \sum_{\Sigma} P_{p,\gamma}(C_N \cap G_N = \varnothing,\, C_N(y) \cap G_N \neq \varnothing,\, C_N = \Sigma) \\
&= pq^{-1}\theta_N \sum_{x \sim y} P_{p,\gamma}(x \in C_N,\, C_N \cap G_N = \varnothing,\, C_N(y) \cap G_N \neq \varnothing) \\
&= p\theta_N \frac{\partial \theta_N}{\partial p} \qquad \text{by (3.67),}
\end{aligned}$$

and the result is proved by combining (3.68)–(3.71) and (3.74). ∎

3.4 Notes

Section 3.1. Much of the past romance of percolation theory has been associated with the problem of proving $p_c = \frac{1}{2}$ for (bond) percolation on $\mathbb{L}^2$, and this was finally proved by Kesten (1980a), who built upon arguments of Harris (1960), Russo (1978), and Seymour and Welsh (1978). Earlier, Sykes and Essam (1964) developed a beautiful intuitive argument to show $p_c(2) = \frac{1}{2}$ (amongst other things), although their argument remains somewhat in the air even at the time of this book (see Section 9.2). The main step in Kesten's proof was the demonstration that $p_T = p_c$ for the square lattice $\mathbb{L}^2$, where $p_T = \sup\{p\colon \chi(p) < \infty\}$ as in (3.3); combined with the Russo–Seymour–Welsh result that $p_T + p_c = 1$, this yielded $p_c = \frac{1}{2}$. Later, Kesten (1982) generalized his proof that $p_T = p_c$ to certain families of two-dimensional lattices; his methods were good for the plane only.

The first proof that $p_T = p_c$ in any number of dimensions was announced by Menshikov in Moscow in 1985, and appeared in Menshikov (1986) and Menshikov, Molchanov, and Sidorenko (1986). The second proof was announced in Minneapolis in early 1986 by Aizenman and Barsky (1987). It appears that Menshikov's proof first flew over the Soviet border in September 1986.

Once we know that $p_T = p_c$ in two dimensions, it is not difficult to deduce that $p_c = \frac{1}{2}$, and we shall return to this point in Chapter 9.

The following remark is of some historical relevance but little importance. Many authors have denoted the critical probability p_c by the symbol p_H; the notation p_H and p_T was introduced by Seymour and Welsh (1978) in honour of Hammersley and Temperley (see Temperley and Lieb (1971)).

Section 3.2. Menshikov (1986) reaches a marginally weaker conclusion than that of Theorem (3.4), but in a more general context. He was interested in finite-range percolation on a certain class of graphs with conditions of regularity on the automorphism groups of such graphs. When applied to the lattice $\mathbb{L}^d$, his result amounts to the statement that $P_p(A_n) < \exp(-\psi(p)n/\log n)$ for some $\psi(p) > 0$. The proof given here is an improved version. The renewal argument of Lemma (3.17) was found by Kesten, with a prompt from Gallavotti. Much of Menshikov's work is based upon a weaker hypothesis about the underlying graph than that it be a lattice in the conventional sense. Crucial is his assumption that the number of vertices within distance n of any given vertex is at most $\exp(n^\gamma)$ for some $\gamma < 1$; Grigorchuk (1983) has shown the existence of graphs which satisfy Menshikov's hypotheses but for which the number of such vertices grows faster than any power of n.

The linear lower bound for $\theta(p)$ was obtained first by Chayes and Chayes (1986c, 1987). Our argument is a variation of that of Menshikov (1987b).

Hammersley (1957a) proved a weaker form of Theorem (3.4), valid for values of p satisfying $\chi(p) < \infty$ rather than $p < p_c$; see Theorem (5.1).

Section 3.3. Ghost sites, green sites, or their equivalent, appear in Suzuki (1965), Griffiths (1967), Kasteleyn and Fortuin (1969), Fortuin and Kasteleyn (1972), and Fortuin (1972). Differential inequalities have been used by Russo (1981, 1982), and an interesting example is in Aizenman and Newman (1984).

The Aizenman–Barsky proof is rather general; the authors show uniqueness for all finite- and long-range bond models, with or without orientations of edges, and their method applies also to finite-range site models. One of the main ideas behind Lemma (3.53) is due to J. T. Chayes and L. Chayes. Aizenman and Barsky note that their technology has close relations with technology developed for the Ising model.

When $\gamma > 0$, the quantity analogous to the number of open clusters per vertex is

$$\kappa(p, \gamma) = \sum_{n=1}^{\infty} \frac{1}{n} (1 - \gamma)^n P_p(|C| = n).$$

We note that κ plays the role of the partition function of statistical mechanics in the sense that $\theta(p, \gamma)$ and $\chi(p, \gamma)$ are expressible in terms of the derivatives of κ:

$$\theta(p, \gamma) = 1 + (1 - \gamma) \frac{\partial \kappa}{\partial \gamma}, \qquad \chi(p, \gamma) = (1 - \gamma) \frac{\partial \theta}{\partial \gamma}.$$

4

The Number of Open Clusters per Vertex

4.1 Definition

In addition to the percolation probability $\theta(p)$ and the mean cluster size $\chi(p)$, there is a third principal character in percolation theory. The *number of open clusters per vertex* is defined as in (1.31) by

$$\kappa(p) = E_p(|C|^{-1}) = \sum_{n=1}^{\infty} \frac{1}{n} P_p(|C| = n), \tag{4.1}$$

where $|C|$ is the number of vertices in the open cluster at the origin. The function κ crops up frequently but usually peripherally in calculations: for example, it was a central figure in the first proof that the infinite open cluster is (almost surely) unique when $p > p_c$ (see Section 6.2), not to mention the famous Sykes–Essam 'proof' that $p_c = \frac{1}{2}$ for bond percolation on $\mathbb{L}^2$ (see Section 9.2). We begin by exploring the origins of κ and its rather unwieldy title.

Let $B(n)$ be the box with side-length $2n$ and centre at the origin, as usual. We think of $B(n)$ as being a graph, by adding the edges inherited from $\mathbb{L}^d$. The principal object of current attention is the number K_n of open clusters of $B(n)$. More precisely, K_n is the number of connected components obtained in $B(n)$ by deleting all closed edges. It turns out that K_n is approximately a linear function of the volume of $B(n)$, in that $K_n |B(n)|^{-1}$ converges to a non-trivial limit as $n \to \infty$.

(4.2) Theorem. *Suppose $0 \le p \le 1$. The number K_n of open clusters of $B(n)$ satisfies*

$$\frac{1}{|B(n)|} K_n \to \kappa(p) \qquad a.s.\ (P_p) \quad and\ in \quad L^1(P_p) \tag{4.3}$$

as $n \to \infty$, *where*

$$\kappa(p) = E_p(|C|^{-1}). \tag{4.4}$$

This explains the title of the function κ. The idea of the proof is simple. For each $x \in \mathbb{Z}^d$, we define

$$\Gamma(x) = \begin{cases} |C(x)|^{-1} & \text{if } |C(x)| < \infty, \\ 0 & \text{if } |C(x)| = \infty. \end{cases} \tag{4.5}$$

Then $\sum_{x \in B(n)} \Gamma(x)$ is approximately equal to K_n, since any open cluster of $\mathbb{L}^d$ which is contained entirely in $B(n)$ contributes exactly 1 to this sum. There are certain 'boundary effects' which arise from points x in open clusters not contained entirely in $B(n)$, but these turn out to be negligible when we take the limit:

$$\lim_{n \to \infty} \left\{ \frac{1}{|B(n)|} \sum_{x \in B(n)} \Gamma(x) \right\} = \lim_{n \to \infty} \left\{ \frac{1}{|B(n)|} K_n \right\}.$$

Finally, we use the ergodic theorem to see that the first limit equals $E_p(\Gamma(0)) = E_p(|C|^{-1})$. Here is a formal proof.

Proof of Theorem (4.2). For $x \in B(n)$ we define $C_n(x)$ to be the open cluster of $B(n)$ containing x; that is to say, we delete all closed edges from $B(n)$, and we write $C_n(x)$ for the connected component of the ensuing graph which contains the vertex x. As usual, we write $|C_n(x)|$ for the number of vertices of $C_n(x)$.

For each $x \in \mathbb{Z}^d$, we define $\Gamma(x) = |C(x)|^{-1}$. For $x \in B(n)$, we define $\Gamma_n(x) = |C_n(x)|^{-1}$, noting that

$$\Gamma(x) \le \Gamma_n(x) \qquad \text{for all} \quad x \in B(n). \tag{4.6}$$

The basic formula of this proof is

$$\sum_{x \in B(n)} \Gamma_n(x) = K_n. \tag{4.7}$$

This is valid for the following reason. Let Σ be an open cluster of $B(n)$. Each vertex x of Σ contributes $|\Sigma|^{-1}$ to the sum in (4.7), so that the aggregate contribution from the vertices of Σ equals 1. Thus each open cluster of $B(n)$ contributes 1 to the left-hand side of (4.7).

We have from (4.6) and (4.7) that

$$\frac{1}{|B(n)|} K_n \ge \frac{1}{|B(n)|} \sum_{x \in B(n)} \Gamma(x). \tag{4.8}$$

However, $(\Gamma(x): x \in \mathbb{Z}^d)$ is a collection of bounded functions of independent edge-states, and is stationary under translations of the lattice $\mathbb{L}^d$. It follows from an appropriate ergodic theorem (see Dunford and Schwartz (1958, Theorem VIII.6.9) or Tempel'man (1972, Theorem 6.1 and Corollary 6.2)) that

$$\frac{1}{|B(n)|} \sum_{x \in B(n)} \Gamma(x) \to E_p(\Gamma(0)) \quad \text{a.s.} \tag{4.9}$$

as $n \to \infty$, giving by (4.8) that

$$\liminf_{n\to\infty} \frac{1}{|B(n)|} K_n \geq E_p(\Gamma(0)) = \kappa(p) \quad \text{a.s.} \tag{4.10}$$

It is not difficult to construct a useful upper bound for K_n. Clearly

$$\begin{aligned} \sum_{x\in B(n)} \Gamma_n(x) &= \sum_{x\in B(n)} \Gamma(x) + \sum_{x\in B(n)} \{\Gamma_n(x) - \Gamma(x)\} \\ &\leq \sum_{x\in B(n)} \Gamma(x) + \sum_{x:\, x\leftrightarrow \partial B(n)} \Gamma_n(x) \end{aligned} \tag{4.11}$$

by (4.6), where the final summation is over all vertices x of $B(n)$ which are joined to the surface $\partial B(n) = \{y \in B(n): \|y\| = n\}$ of $B(n)$ by open paths. The inequality is valid since $C_n(x) = C(x)$, and therefore $\Gamma_n(x) = \Gamma(x)$, whenever there is no open path joining x to any vertex of $\partial B(n)$. However, the final sum of (4.11) equals the number of open clusters of $B(n)$ which contain vertices in $\partial B(n)$, and this is no larger than the cardinality $|\partial B(n)|$ of $\partial B(n)$. Therefore

$$\begin{aligned} \frac{1}{|B(n)|} \sum_{x\in B(n)} \Gamma_n(x) &\leq \frac{1}{|B(n)|} \left\{ \sum_{x\in B(n)} \Gamma(x) \right\} + \frac{|\partial B(n)|}{|B(n)|} \\ &\to \kappa(p) \quad \text{a.s.} \end{aligned} \tag{4.12}$$

as $n \to \infty$, by (4.9). We combine this with (4.7) and (4.10) to deduce that

$$\frac{1}{|B(n)|} K_n \to \kappa(p) \quad \text{a.s. as} \quad n \to \infty. \tag{4.13}$$

Convergence in L^1 is immediate since $0 \leq K_n |B(n)|^{-1} \leq 1$. ■

4.2 Lattice Animals and Large Deviations

Each of the principal characters θ, χ, and κ of percolation theory may be expressed in terms of the distribution of the number $|C|$ of vertices in the open cluster at the origin. To calculate or estimate this distribution is a question involving combinatorics.

We begin by introducing some notation. An *animal* is defined to be a finite connected subgraph of $\mathbb{L}^d$ containing the origin; see Figure 4.1

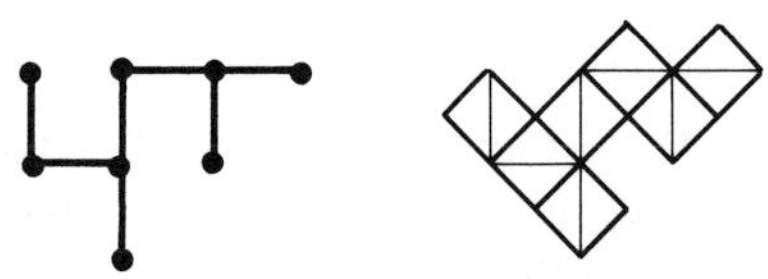

Figure 4.1. The left-hand picture is a two-dimensional animal with eight vertices and seven edges. If we replace each edge by a square having the edge as diagonal, then we obtain a picture which emphasizes the cellular structure of the animal.

for two representations of the same two-dimensional animal. If A is an animal, we write A_v and A_e for the sets of vertices and edges of A, respectively, and $\triangle A$ for the set of edges of $\mathbb{L}^d$ which do not belong to A but which have at least one endvertex in A; $\triangle A$ is called the *edge boundary* of A. We write $\mathscr{A}_{nmb}$ for the set of animals A with $|A_v| = n$, $|A_e| = m$, and $|\triangle A| = b$, and we let $a_{nmb} = |\mathscr{A}_{nmb}|$ be the number of such animals. It is easy to see that $a_{nmb} = 0$ unless n, m, and b satisfy certain inequalities. For example, we may assume that

$$(4.14) \qquad 1 \le b \le 2dn$$

since each vertex of A is incident to no more than $2d$ edges which are not in A; similarly

$$(4.15) \qquad n - 1 \le m \le dn,$$

where the first inequality follows from the fact that every connected graph on n vertices has at least $n - 1$ edges, and the second inequality is a consequence of the observation that each vertex of A has at most $2d$ neighbours in A.

The open cluster C is a random animal:

$$(4.16) \qquad P_p(C = A) = p^m(1-p)^b \qquad \text{where} \quad m = |A_e| \quad \text{and} \quad b = |\triangle A|,$$

for all $A \in \mathscr{A}$, the set of all animals. Thus, the probability that C contains exactly n vertices is

$$(4.17) \qquad P_p(|C| = n) = \sum_{m,b} a_{nmb} p^m (1-p)^b,$$

showing that calculating the distribution of $|C|$ amounts to finding the generating function of the number of animals. The above series expansion for $P_p(|C| = n)$ is finite, so that $P_p(|C| = n)$ is a finite polynomial in p and is thus analytic. It is not so easy to establish properties of continuity and differentiability for such quantities as the percolation probability $\theta(p)$ or the number of open clusters per vertex

$$(4.18) \qquad \kappa(p) = \sum_{n=1}^{\infty} n^{-1} P_p(|C| = n),$$

and it is with this type of problem in mind that we study the behaviour of a_{nmb} when n, m, and b are large.

Suppose $0 < p = 1 - q < 1$ and differentiate (4.17) with respect to p to obtain

$$(4.19) \qquad \frac{d}{dp} P_p(|C| = n) = \sum_{m,b} a_{nmb} \left(\frac{m}{p} - \frac{b}{q} \right) p^m q^b.$$

For large n, the behaviour of this sum depends on the typical behaviour of $p^{-1}|A_e| - q^{-1}|\triangle A|$ as A ranges over the set of animals with n vertices. We shall estimate this quantity with the aid of the following 'large deviation' estimate.

(4.20) Theorem. *There exists* $\varepsilon > 0$ *such that, for* $0 < x < \varepsilon$,

(4.21)
$$\sum_{m,b:\left|\frac{m}{p}-\frac{b}{q}\right|>dxn} a_{nmb}p^m q^b \le 3d^2n^2 \exp(-\tfrac{1}{3}nx^2p^2q)$$

whenever $n \ge 1$ *and* $0 < p = 1 - q < 1$.

Rewritten in terms of probabilities, Theorem (4.20) asserts that

(4.22)
$$P_p(|h(C)| > dxn, |C| = n) \le 3d^2n^2 \exp(-\tfrac{1}{3}nx^2p^2q)$$

for all n and all sufficiently small x, where

(4.23)
$$h(C) = \frac{1}{q}|\Delta C| - \frac{1}{p}|C_e|,$$

C_e and ΔC being the edge set and edge boundary of C, respectively. This is tantamount to a large-deviation result, since it concerns the probability that $h(C)$ is comparable with the size of C.

We shall not use probabilistic arguments to demonstrate this result; instead, we emphasize the intrinsically combinatorial aspect of percolation theory by using estimates for the numbers of animals. In Section 6.2 we shall have an opportunity to show how such techniques may be rewritten in a form more reminiscent of the classical theory of large deviations.

It is well known that the number of animals with n vertices grows at most exponentially as $n \to \infty$. More explicitly, we shall see in the proof of Theorem (4.20) that the total number of animals with n vertices satisfies

(4.24)
$$\sum_{m,b} a_{nmb} \le 7^{dn} \qquad \text{for all } n.$$

As an application of Theorem (4.20), we shall prove in the next section that the number of open clusters per vertex is differentiable at all values of p, including the critical point $p = p_c$.

Proof of Theorem (4.20). We follow Kesten (1982, p. 85). From (4.17),

(4.25)
$$P_p(|C| < \infty) = \sum_{n,m,b} a_{nmb}p^m q^b \le 1,$$

whenever $0 \le p = 1 - q \le 1$. We prove (4.24) first. For each fixed n,

$$\begin{aligned}\sum_{m,b} a_{nmb}p^m q^b &\ge \sum_{m,b} a_{nmb}p^{dn}q^{2dn} \qquad \text{by (4.14)–(4.15)}\\ &= (pq^2)^{dn}\sum_{m,b} a_{nmb}\end{aligned}$$

for all values of p, giving by (4.25) that

$$\sum_{m,b} a_{nmb} \le (pq^2)^{-dn}$$

for all p. We choose p to maximize $pq^2 = p(1-p)^2$, and obtain

$$\sum_{m,b} a_{nmb} \leq (27/4)^{dn},$$

which is ample. Now we return to the proof of the theorem.

We have from (4.25) that $a_{nmb}p^m(1-p)^b \leq 1$ for all n, m, b, and all p. We may choose $p = m(m+b)^{-1}$, $q = b(m+b)^{-1}$, to obtain

$$a_{nmb} \leq p^{-m}q^{-b} = \left(\frac{m+b}{m}\right)^m \left(\frac{m+b}{b}\right)^b \quad \text{for} \quad n \geq 2. \tag{4.26}$$

Actually, the right-hand side here is an upper bound for the number $\sum_n a_{nmb}$ of animals with m edges and b boundary edges, but we shall not use this fact.

Now, for fixed $n \geq 2$,

$$\begin{aligned} \sum_{m,b:\left|\frac{m}{p}-\frac{b}{q}\right|>dxn} a_{nmb}p^m q^b &\leq \sum_{m,b:\left|\frac{m}{p}-\frac{b}{q}\right|>dxn} \left(\frac{m+b}{m}\right)^m \left(\frac{m+b}{b}\right)^b p^m q^b \qquad \text{by (4.26)} \\ &\leq 2d^2n^2 \max\{f(m,b)\} \end{aligned} \tag{4.27}$$

by (4.14) and (4.15), where

$$f(m,b) = \left(\frac{(m+b)p}{m}\right)^m \left(\frac{(m+b)q}{b}\right)^b \tag{4.28}$$

and the maximum is over all m and b satisfying (4.14), (4.15), and $|mq - bp| > dxnpq$. From (4.15), this maximum is no larger than the maximum of $f(m, b)$ over m and b satisfying $|mq - bp| \geq xmpq$. We fix m and think of $f(m, y)$ as a function of the continuous variable y. Let

$$\begin{aligned} g_m(y) &= \log f(m,y) \\ &= m \log\left(\frac{(m+y)p}{m}\right) + y \log\left(\frac{(m+y)q}{y}\right). \end{aligned} \tag{4.29}$$

We differentiate g_m with respect to y to find that

$$g_m'(y) = \log\left(\frac{(m+y)q}{y}\right) = \log\left(1 + \frac{mq - yp}{y}\right),$$

giving that g_m is an increasing function of y if $mq - yp > 0$ and a decreasing function of y if $mq - yp < 0$. Thus the maximum value of $g_m(y)$ over values of y satisfying $|mq - yp| \geq xmpq$ occurs either when $mq - yp = xmpq$ or when $mq - yp = -xmpq$. These two possibilities correspond to

$$y = mqp^{-1}(1 \pm xp). \tag{4.30}$$

We substitute these two values into (4.29) and expand the logarithms to find that

$$g_m(y) = m\log(1 \pm xpq) + mqp^{-1}(1 \pm xp)\log\left(\frac{1 \pm xpq}{1 \pm xp}\right)$$
$$= -\tfrac{1}{2}mx^2p^2q(1 + O(x)),$$

where the $O(x)$ term does not depend on m or p. We have proved that, for fixed m and n (≥ 2) satisfying $m \leq dn$, it is the case that

$$\max\{f(m, b): |mp^{-1} - bq^{-1}| > dxn\} \leq \exp(-\tfrac{1}{3}mx^2p^2q)$$

for all small positive values of x. We substitute this into (4.27), and we use (4.15) to obtain

$$\sum_{m,b:\left|\frac{m}{p}-\frac{b}{q}\right|>dxn} a_{nmb}p^mq^b \leq 2d^2n^2\max\{\exp(-\tfrac{1}{3}mx^2p^2q): m \geq n-1\}$$
$$\leq 3d^2n^2\exp(-\tfrac{1}{3}nx^2p^2q) \qquad \text{if} \quad n \geq 2$$

for all small positive values of x. The term in $n = 1$ presents no difficulty. ■

4.3 Differentiability of κ

As an application of the techniques of the last section concerning the counting of lattice animals, we prove next that κ is a continuously differentiable function of p throughout the interval $[0, 1]$.

(4.31) Theorem. Differentiability of κ. *The number $\kappa(p)$ of open clusters per vertex is a continuously differentiable function on* $[0, 1]$.

We shall prove that the derivative of κ is the term by term derivative of the series in (4.18):

$$\kappa'(p) = \sum_{n,m,b}\frac{1}{n}a_{nmb}(mp^{m-1}q^b - bp^mq^{b-1}), \tag{4.32}$$

where $q = 1 - p$.

Rather more than the conclusion of Theorem (4.31) is true. We shall see in Sections 5.4 and 6.8 that κ is analytic on $[0, p_c)$ and infinitely differentiable on $(p_c, 1]$; it is believed that κ is analytic on $(p_c, 1]$ also. Furthermore, it is thought that κ is twice differentiable at p_c but not three times differentiable; this question is related to the problem of critical exponents, to which we return in Chapter 7. Somewhat more is known in two dimensions; for example, Kesten (1982) has shown that κ is twice differentiable at $p_c(2)$ in this case.

Proof of Theorem (4.31). If we differentiate the series in (4.18) term by term we obtain

$$\sum_{n=1}^{\infty} n^{-1}\frac{d}{dp}P_p(|C| = n);$$

we show first that this series converges uniformly for values of p in every interval of the form $[p_1, p_2]$ where $0 < p_1 < p_2 < 1$. Suppose that $0 < p = 1 - q < 1$. Then

$$\frac{d}{dp} P_p(|C| = n) = \sum_{m,b} a_{nmb} \left(\frac{m}{p} - \frac{b}{q} \right) p^m q^b \tag{4.33}$$

from (4.19). We split this sum into two parts depending on whether or not $|mp^{-1} - bq^{-1}| > dxn$, where x is small and will be chosen later. When $|mp^{-1} - bq^{-1}| > dxn$ we use the inequality

$$\left| \frac{m}{p} - \frac{b}{q} \right| \leq \frac{m+b}{pq} \leq \frac{3dn}{pq}, \tag{4.34}$$

valid because of (4.14) and (4.15). By the result of Theorem (4.20), when x is small,

$$\left| n^{-1} \frac{d}{dp} P_p(|C| = n) \right| \leq dx P_p(|C| = n) + \left(\frac{3d}{pq} \right) 3d^2 n^2 \exp(-\tfrac{1}{3} n x^2 p^2 q).$$

We set $x = a(n^{-1} \log n)^{1/2}$ to obtain, for $n \geq N \geq 3$,

$$\left| n^{-1} \frac{d}{dp} P_p(|C| = n) \right| \leq da \sqrt{\frac{\log N}{N}}\, P_p(|C| = n) + \gamma n^{2 - a^2 \eta}$$

for appropriate $\gamma = \gamma(p, d) < \infty$ and $\eta = \eta(p) > 0$. Choose p_1 and p_2 satisfying $0 < p_1 < p_2 < 1$, write

$$\hat{\gamma} = \sup\{\gamma(p, d) : p_1 \leq p \leq p_2\}, \qquad \hat{\eta} = \inf\{\eta(p) : p_1 \leq p \leq p_2\},$$

and choose a such that $2 - a^2 \hat{\eta} < -2$. Then

$$\sum_{n=N}^{\infty} \left| n^{-1} \frac{d}{dp} P_p(|C| = n) \right| \leq da \sqrt{\frac{\log N}{N}} + \hat{\gamma} \sum_{n=N}^{\infty} \frac{1}{n^2} \quad \text{if } p_1 \leq p \leq p_2, \tag{4.35}$$

since

$$\sum_{n=N}^{\infty} P_p(|C| = n) \leq P_p(|C| < \infty) \leq 1.$$

We see that the left-hand side of (4.35) converges to 0 as $N \to \infty$, uniformly in $p \in [p_1, p_2]$, and it follows that κ is continuously differentiable on (0, 1) with derivative given by (4.32).

It remains to show that κ is continuously differentiable at $p = 0$ and $p = 1$. We shall not spend much time on this small point, since it is not particularly important and furthermore we shall prove much more later (see Theorems (5.108) and (6.120)). Here is a sketch of a primitive argument for the case $p = 0$; an analogous argument is valid for $p = 1$. When $p < p_c$ we have that $P_p(|C| < \infty) = 1$ so that

$$\sum_{n,m,b} a_{nmb} p^m q^b = 1 \qquad \text{for} \quad p < p_c.$$

Clearly $\kappa(0) = 1$, so that

$$\frac{\kappa(p) - \kappa(0)}{p} = \sum_{n,m,b:n \geq 2} \left(\frac{1}{n} - 1\right) a_{nmb} p^{m-1} q^b.$$

We pick out the term in p^0:

$$\frac{\kappa(p) - \kappa(0)}{p} = A(p) + \sum_{n,m,b:m=1} \left(\frac{1}{n} - 1\right) a_{nmb} p^{m-1} q^b, \tag{4.36}$$

where

$$A(p) = \sum_{n,m,b:m \geq 2} \left(\frac{1}{n} - 1\right) a_{nmb} p^{m-1} q^b$$
$$\to 0 \quad \text{as} \quad p \to 0;$$

this is not difficult to check (use (4.24)). There are exactly $2d$ animals containing the origin which have $m = 1$, and each such animal has $n = 2$ and $b = 2(2d - 1)$. We let $p \to 0$ in (4.36) to find that $\kappa'(0) = -d$. On the other hand, we have from the first part of the proof that

$$\kappa'(p) = \sum_{n,m,b} \frac{1}{n} a_{nmb}(mp^{m-1}q^b - bp^m q^{b-1}) \qquad \text{if} \quad 0 < p < 1. \tag{4.37}$$

We may use an argument similar to the above to see that $\kappa'(p) \to -d$ as $p \to 0$. ∎

4.4 Notes

Section 4.1. Sykes and Essam (1964) used the number of open clusters per vertex in their beautiful but non-rigorous derivation of exact values for the critical probabilities of certain two-dimensional lattices (see Section 9.2). The existence of the limit of Theorem (4.2) was proved by Grimmett (1976), and the form of the limit by Wierman (1978); the proof here is essentially that of Grimmett (1981b) with an improvement due to Kesten (1982, p. 240). Brånvall (1980) and Cox and Grimmett (1981, 1984) have proved central limit theorems for such quantities as the number of open clusters within a large box.

Section 4.2. The large deviation result of Theorem (4.20) is due to Kesten (1982, p. 85). Aizenman, Kesten, and Newman (1987) attain a similar goal by related methods, and we shall use their argument in Section 6.2.

Finally, we mention the beautiful theorem of Delyon (1980). In a general context, he explores the asymptotic behaviour of the number of animals with m edges and αm boundary edges, showing that this number grows at a rate which depends on neither the dimension of the graph nor the lattice structure, so long as α is no larger than some value *which depends* on the choice of lattice.

Section 4.3. Grimmett (1981b) first used lattice animals to show that κ is differentiable in two dimensions; Kesten (1982) carried such calculations further. See Aizenman, Kesten, and Newman (1987) and Nguyen (1988) also.

5

The Subcritical Phase

5.1 The Radius of an Open Cluster

In this chapter we consider the subcritical phase of bond percolation on $\mathbb{L}^d$ when $d \geq 2$; that is, we suppose that the edge-probability p satisfies $p < p_c$. In this phase, all open clusters are finite almost surely and furthermore have finite mean size. We are interested in such quantities as (i) estimates for the probability of an open path joining two vertices x and y when the distance between x and y is large, and (ii) estimates for the rate of decay of $P_p(|C| = n)$ as $n \to \infty$. Such estimates contain information about the structure of the process over long ranges, and as applications of such estimates we shall prove that $\chi(p)$ and $\kappa(p)$ are analytic functions of p when $p < p_c$.

The first estimate of this type was proved by Hammersley (1957a); he used branching process arguments to show in effect that the probability of an open path joining the origin to a vertex on the surface of the box $B(n)$ decays exponentially as $n \to \infty$ if $\chi(p) < \infty$. Hammersley's technique is illuminating and his result is the first and perhaps simplest in a class of such results. We begin with some notation.

As before, we write $\partial B(n)$ for the surface of $B(n)$:

$$\partial B(n) = \{x \in \mathbb{Z}^d : \|x\| = n\}$$

where $\|x\| = \max\{|x_i| : 1 \leq i \leq d\}$ as usual. For $x \in \mathbb{Z}^d$ we write $B(n, x)$ for the box $x + B(n)$ with side-length $2n$ and centre at x, and $\partial B(n, x) = x + \partial B(n)$ for the surface of this box. As usual, we write $P_p(0 \leftrightarrow \partial B(n))$ for the probability of an open path from the origin to a vertex in $\partial B(n)$.

(5.1) Theorem. Exponential tail decay of the radius of an open cluster. *Suppose that $\chi(p) < \infty$. Then there exists $\sigma(p) > 0$ such that*

(5.2) $$P_p(0 \leftrightarrow \partial B(n)) \leq e^{-n\sigma(p)} \qquad \textit{for all } n.$$

It is largely for historical and pedagogical reasons that we state and prove Hammersley's theorem. We have already proved a stronger version of the result (Theorem (3.4)), valid under the 'milder' (but actually equivalent) condition that $p < p_c$, but the proof of Theorem (5.1) is considerably easier and is a good example of a certain type of argument.

We think of this result as demonstrating the exponential decay of the distribution of the radius of open clusters, when $\chi(p) < \infty$. We define the *radius* of the open cluster $C(x)$ at x by

$$\text{rad}(C(x)) = \max\{\delta(x, y): y \in C(x)\}. \tag{5.3}$$

This is the usual graph-theoretic definition of radius, although some authors (Kesten (1987e), for example) prefer to use the distance function $\|x - y\|$ in place of $\delta(x, y)$. It is easy to see that

$$\|x - y\| \leq \delta(x, y) \leq d\|x - y\|, \tag{5.4}$$

so that the two definitions are equivalent within a bounded factor. We see now that

$$P_p(\text{rad}(C) \geq m) \leq P_p(0 \leftrightarrow \partial S(m))$$

where $S(m) = \{y \in \mathbb{Z}^d: \delta(0, y) = m\}$ as before, giving from (5.4) and the conclusion of the theorem that

$$\begin{aligned} P_p(\text{rad}(C) \geq m) &\leq P_p(0 \leftrightarrow \partial B(\lfloor m/d \rfloor)) \\ &\leq \exp(-\sigma(p)\lfloor m/d \rfloor) \end{aligned}$$

when $\chi(p) < \infty$. Thus the tail of $\text{rad}(C)$ decays at least exponentially when $\chi(p) < \infty$.

Proof of Theorem (5.1). Let $x \in \partial B(n)$, and let $\tau_p(0, x) = P_p(0 \leftrightarrow x)$ be the probability that there is an open path of $\mathbb{L}^d$ joining the origin to x. We write N_n for the number of vertices in $\partial B(n)$ with this property, so that the mean value of N_n is

$$E_p(N_n) = \sum_{x \in \partial B(n)} \tau_p(0, x). \tag{5.5}$$

We note that

$$\begin{aligned} \sum_{n=0}^{\infty} E_p(N_n) &= \sum_{n=0}^{\infty} \sum_{x \in \partial B(n)} \tau_p(0, x) \\ &= \sum_{x \in \mathbb{Z}^d} \tau_p(0, x) \qquad \text{since } (\partial B(n): n \geq 0) \text{ partitions } \mathbb{Z}^d \\ &= E_p|\{x \in \mathbb{Z}^d: 0 \leftrightarrow x\}| \\ &= \chi(p). \end{aligned} \tag{5.6}$$

If there is an open path from the origin to a vertex of $\partial B(m + k)$, then there exists a vertex x in $\partial B(m)$ which is connected by disjoint open

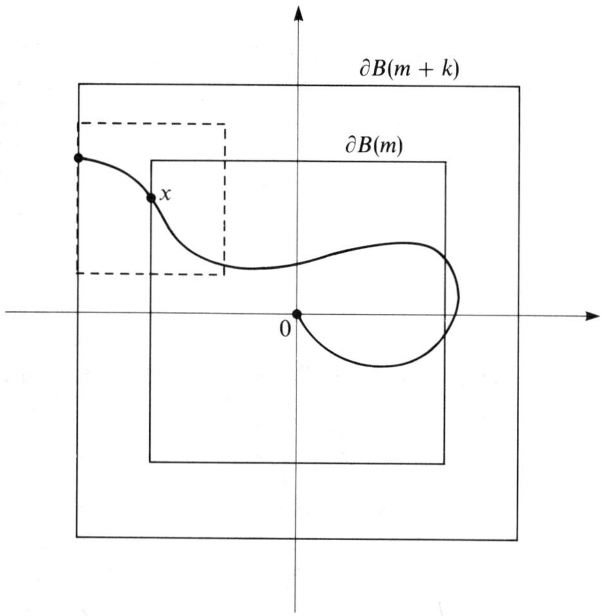

Figure 5.1. The vertex x is joined by disjoint open paths to the origin and to the surface of the box $B(k, x)$, indicated by the dashed lines.

paths both to the origin and to a vertex on the surface $\partial B(k, x)$ of the box with side-length $2k$ and centre at x; see Figure 5.1. We use the BK inequality to find that

$$(5.7) \qquad P_p(0 \leftrightarrow \partial B(m+k)) \le \sum_{x \in \partial B(m)} P_p(0 \leftrightarrow x) P_p(x \leftrightarrow \partial B(k, x))$$
$$= \sum_{x \in \partial B(m)} \tau_p(0, x) P_p(0 \leftrightarrow \partial B(k))$$

by translation invariance. Thus

$$(5.8) \qquad P_p(0 \leftrightarrow \partial B(m+k)) \le E_p(N_m) P_p(0 \leftrightarrow \partial B(k)) \qquad \text{for} \quad m, k \ge 1.$$

The BK inequality makes this calculation simple; Hammersley (1957a) used a different argument.

Suppose now that $\chi(p) < \infty$, so that $\sum_{m=0}^{\infty} E_p(N_m) < \infty$ from (5.6). In this case, $E_p(N_m) \to 0$ as $m \to \infty$, and we may choose m such that $\eta = E_p(N_m)$ satisfies $\eta < 1$. Let n be a positive integer and write $n = mr + s$ where r and s are non-negative integers and $0 \le s < m$. Then

$$\begin{aligned} P_p(0 \leftrightarrow \partial B(n)) &\le P_p(0 \leftrightarrow \partial B(mr)) && \text{since} \quad n \ge mr \\ &\le \eta^r && \text{from (5.8)} \\ &\le \eta^{-1+n/m} && \text{since} \quad n < m(r+1), \end{aligned}$$

which provides an exponentially decaying bound of the form of (5.2). ∎

The conclusion of Theorem (5.1) may be sharpened to show that, to the 'logarithmic' order, $P_p(0 \leftrightarrow \partial B(n))$ decays exactly as an exponential function of n, whenever $0 < p < p_c$. That is to say,

(5.9) $$P_p(0 \leftrightarrow \partial B(n)) \approx e^{-n\varphi(p)} \quad \text{as} \quad n \to \infty$$

for some $\varphi(p)$ satisfying $\varphi(p) > 0$ when $p < p_c$ (we recall that $a_n \approx b_n$ means $\log a_n / \log b_n \to 1$). Such asymptotic results are common, and the usual strategy is to use the limit theorem for subadditive sequences (see Appendix II) to establish the limit appropriate for (5.9), and then to use a bound such as (3.5) or (5.2) to show that the quantity $\varphi(p)$ in the exponent is strictly positive.

(5.10) Theorem. Asymptotic tail behaviour of the radius of an open cluster. *Suppose that $0 < p \leq 1$. There exist strictly positive constants ρ and σ, independent of p, and a unique function $\varphi(p)$, such that*

(5.11) $$\rho n^{1-d} e^{-n\varphi(p)} \leq P_p(0 \leftrightarrow \partial B(n)) \leq \sigma n^{d-1} e^{-n\varphi(p)}$$

for all $n \geq 1$.

It is believed that $P_p(0 \leftrightarrow B(n))$ behaves in the manner of an exponential term multiplied by a power of n.

Estimates such as (5.11) are of no value when $p > p_c$, since in this case

(5.12) $$P_p(0 \leftrightarrow \partial B(n)) \geq \theta(p) > 0 \qquad \text{for all } n;$$

thus $\varphi(p) = 0$ whenever $p > p_c$. On the other hand, $P_p(0 \leftrightarrow \partial B(n))$ decays exponentially whenever $0 < p < p_c$ (see Theorem (5.1) or Theorem (3.4)), so that $\varphi(p) > 0$ when $0 < p < p_c$. Thus

(5.13) $$\varphi(p) \begin{cases} > 0 & \text{if} \quad 0 < p < p_c, \\ = 0 & \text{if} \quad p > p_c. \end{cases}$$

It follows from (5.9) and (5.11) that $\varphi(p)$ is a non-increasing function of p. Other properties of φ are not so easy to ascertain, but we present some of them in the next theorem.

(5.14) Theorem. *The function φ in (5.11) satisfies*

(5.15) *φ is continuous and non-increasing on* $(0, 1]$,

(5.16) *φ is strictly decreasing and positive on* $(0, p_c)$,

(5.17) $\varphi(p) \to \infty$ *as* $p \downarrow 0$,

(5.18) $\varphi(p_c) = 0$.

See Figure 5.2 for a sketch of φ.

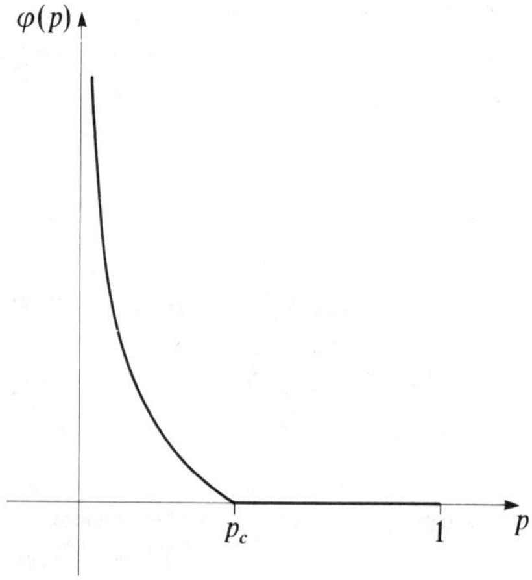

Figure 5.2. A sketch of the function φ.

We turn now to the proofs of Theorems (5.10) and (5.14). Our principal techniques are the FKG and BK inequalities, combined with the limit theorems for subadditive sequences; see Appendix II for an account of the latter technique.

Proof of Theorem (5.10). Suppose that $0 < p \leq 1$, and let $\beta(n) = P_p(0 \leftrightarrow \partial B(n))$. Our strategy is to show that the sequence $(\log \beta(n): n \geq 1)$ is both subadditive and superadditive, subject to small error terms. The limit theorem for the subadditive inequality then delivers the goods.

Fix m, $n \geq 1$. It is not difficult to find a good upper bound for $\beta(m+n)$ involving the product $\beta(m)\beta(n)$. We recall from (5.7) and (5.8) that

(5.19) $$\beta(m+n) \leq \sum_{x \in \partial B(m)} \tau_p(0, x)\beta(n),$$

where $\tau_p(0, x)$ is the probability of an open path from the origin to the vertex x. However,

(5.20) $$\tau_p(0, x) \leq \beta(m) \qquad \text{if} \quad x \in \partial B(m)$$

so that

(5.21) $$\beta(m+n) \leq |\partial B(m)|\, \beta(m)\beta(n).$$

Next we establish a similar *lower* bound for $\beta(m+n)$. This is slightly more complicated and requires an estimate of the probability that there is an open path of $B(n)$ joining the origin to a specified face on the surface of $B(n)$. More specifically, let $\gamma(n)$ be the probability that there is an open path of $B(n)$ joining the origin to some vertex x in $\partial B(n)$ satisfying $x_1 = n$. We have defined $\gamma(n)$ in terms of a particular face of

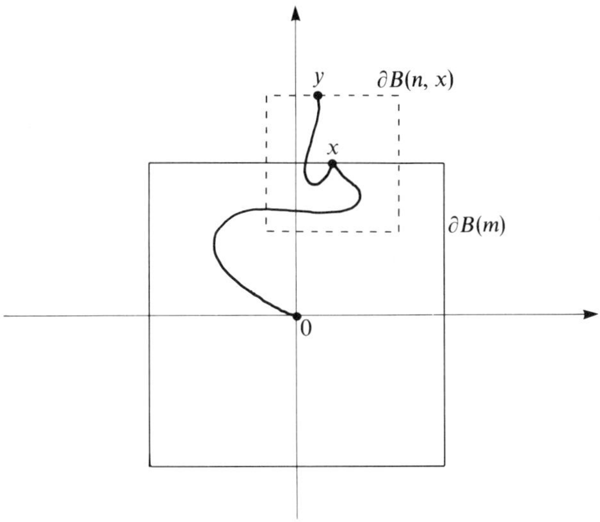

Figure 5.3. A sketch of the events (i) that x is joined by an open path of $B(m)$ to the origin, and (ii) that x is joined by an open path of $B(n, x)$ to some vertex y on the 'upper' face of $\partial B(n, x)$.

$B(n)$, but clearly this probability is the same for each of the $2d$ faces of $B(n)$. Thus

$$\gamma(n) \leq \beta(n) \leq 2d\gamma(n). \tag{5.22}$$

Let $x \in \partial B(m)$, and choose k such that $x_k = \pm m$; we shall suppose that $x_k = m$, and an analogous argument is valid if $x_k = -m$. Let U_x be the event that there is an open path in $B(m)$ joining x to the origin, and let V_x be the event that there is an open path in $B(n, x)$ joining x to some vertex y of $\partial B(n, x)$ for which $y_k = m + n$; see Figure 5.3. Now

$$\beta(m + n) \geq P_p(U_x \cap V_x) \tag{5.23}$$

since the union of two such paths contains a connection from the origin to $\partial B(m + n)$. We use the FKG inequality to find that

$$\beta(m + n) \geq P_p(U_x)P_p(V_x). \tag{5.24}$$

Now $P_p(V_x) = \gamma(n)$, and

$$\begin{aligned} \beta(m) &= P_p\left(\bigcup_{x \in \partial B(m)} U_x\right) \\ &\leq \sum_{x \in \partial B(m)} P_p(U_x), \end{aligned} \tag{5.25}$$

giving that there exists $x \in \partial B(m)$ such that

$$P_p(U_x) \geq \frac{1}{|\partial B(m)|}\beta(m). \tag{5.26}$$

With this choice of x, (5.24) becomes

$$\begin{aligned}(5.27)\qquad \beta(m+n) &\geq \frac{1}{|\partial B(m)|}\beta(m)\gamma(n)\\ &\geq \frac{1}{2d|\partial B(m)|}\beta(m)\beta(n) \qquad \text{by (5.22)},\end{aligned}$$

as required.

The remaining part of the argument is devoted to analysing the consequences of inequalities (5.21) and (5.27). First, we note that

$$\begin{aligned}(5.28)\qquad |\partial B(m)| &\leq 2d|\{x \in \mathbb{Z}^d: x_1 = m, |x_i| \leq m \text{ for } 2 \leq i \leq d\}|\\ &= 2d(2m+1)^{d-1}\\ &\leq d3^d m^{d-1} \qquad \text{for} \quad m \geq 1,\end{aligned}$$

and then we rewrite the inequalities in the form

$$(5.29)\qquad \log\beta(m+n) \leq \log\beta(m) + \log\beta(n) + g(m),$$

$$(5.30)\qquad \log\beta(m+n) \geq \log\beta(m) + \log\beta(n) - g(m),$$

where

$$\begin{aligned}(5.31)\qquad g(r) &= \log(d^2 3^{d+1} r^{d-1})\\ &= \log(d^2 3^{d+1}) + (d-1)\log r.\end{aligned}$$

There are at least two ways of applying the subadditive limit theorem to these inequalities, and one such possibility is to note from the monotonicity of g that the sequences $(\log\beta(n): n \geq 1)$ and $(-\log\beta(n): n \geq 1)$ satisfy the generalized subadditive inequality with error function g. An alternative argument obviates the need to appeal to this general version of the subadditive limit theorem. Suppose that $m \leq n$, and add $g(n)$ to both sides of (5.29) to obtain

$$g(n) + \log\beta(m+n) \leq \{g(m) + \log\beta(m)\} + \{g(n) + \log\beta(n)\}.$$

Now, by (5.31),

$$\begin{aligned}(5.32)\qquad g(m+n) - g(n) &= (d-1)\log\left(1 + \frac{m}{n}\right)\\ &\leq (d-1)\log 2\end{aligned}$$

since $m \leq n$. Adding these two inequalities, we find that the sequence $(a_k: k \geq 1)$, defined by

$$a_k = g(k) + (d-1)\log 2 + \log\beta(k),$$

satisfies the subadditive inequality

$$a_{m+n} \leq a_m + a_n \qquad \text{for} \quad m, n \geq 1.$$

We have from (5.30) in a similar fashion that the sequence $(g(k) + (d-1)\log 2 - \log \beta(k)\colon k \geq 1)$ satisfies the subadditive inequality also. Now $k^{-1}g(k) \to 0$ as $k \to \infty$, and it follows by the subadditive limit theorem (Theorem (II.2) of Appendix II) that

(5.33) $$\varphi(p) = \lim_{k\to\infty} \left\{ -\frac{1}{k} \log \beta(k) \right\}$$

exists, and furthermore that

(5.34) $$g(m) + (d-1)\log 2 + \log \beta(m) \geq -m\varphi(p)$$

and

(5.35) $$g(m) + (d-1)\log 2 - \log \beta(m) \geq m\varphi(p)$$

for all $m \geq 1$. We combine these two inequalities to obtain

(5.36) $$\begin{aligned} |m\varphi(p) + \log \beta(m)| &\leq g(m) + (d-1)\log 2 \\ &\leq c + (d-1)\log m \end{aligned}$$

by (5.31), for some constant $c = c(d)$. This proves (5.11). ∎

Proof of Theorem (5.14). Once we have taken into account the remarks preceding the statement of the theorem, it remains to prove that

(5.37) φ is continuous on $(0, 1]$, and

(5.38) φ is strictly decreasing on $(0, p_c)$,

in addition to (5.17) and (5.18).

First we prove that φ is continuous on $(0, 1]$. Let $b_m(p) = -m^{-1} \log \beta(m)$, and write (5.36) in the form

(5.39) $$|\varphi(p) - b_m(p)| \leq \frac{1}{m}\Big(c + (d-1)\log m\Big).$$

Now $b_m(p) \to \varphi(p)$ as $m \to \infty$, and this convergence is uniform on $(0, 1]$ since the right-hand side of (5.39) does not involve p. Also, $b_m(p)$ is a continuous function of p, since the event $\{0 \leftrightarrow \partial B(m)\}$ depends only on the finite collection of edges within $B(m)$. Thus $\varphi(p)$ is the uniform limit of continuous functions, and therefore φ is continuous also. We have proved also that $\varphi(p_c) = 0$, since

$$\begin{aligned} \varphi(p_c) &= \lim_{p \downarrow p_c} \varphi(p) && \text{by continuity} \\ &= 0 && \text{by (5.13).} \end{aligned}$$

The strict monotonicity of φ on $(0, p_c]$ follows from Theorem (2.38), applied to the event $\{0 \leftrightarrow \partial B(n)\}$. This is an increasing event which depends only on the edges in $B(n)$, and it is a consequence of Theorem

(2.38) that

$$\frac{\log P_a(0 \leftrightarrow \partial B(n))}{\log a} \geq \frac{\log P_b(0 \leftrightarrow \partial B(n))}{\log b} \quad \text{if} \quad a \leq b.$$

We divide by n and take the limit as $n \to \infty$ to deduce from (5.11) that

$$\frac{\varphi(a)}{\log a} \leq \frac{\varphi(b)}{\log b} \quad \text{if} \quad 0 < a \leq b \leq 1;$$

hence

(5.40) $$\varphi(a) \geq \varphi(b)\frac{\log(1/a)}{\log(1/b)} \quad \text{if} \quad 0 < a \leq b \leq 1.$$

Finally, we show that $\varphi(p) \to \infty$ as $p \downarrow 0$, and we use the 'Peierls argument' of the proof of Theorem (1.10). From (1.15) and the remarks thereafter,

(5.41) $$\begin{aligned} P_p(0 \leftrightarrow \partial B(n)) &\leq P_p(N(n) \geq 1) \\ &\leq \{p\lambda(d) + o(1)\}^n, \end{aligned}$$

where $N(n)$ is the number of open paths starting at the origin and $\lambda(d)$ is the connective constant of $\mathbb{L}^d$. We combine this with (5.11) to find that

$$\begin{aligned} \varphi(p) &\geq -\log\{p\lambda(d)\} \\ &\to \infty \quad \text{as} \quad p \downarrow 0. \end{aligned}$$

■

5.2 Connectivity Functions and Correlation Length

Let $x_1, x_2, \ldots, x_n$ be vertices of $\mathbb{L}^d$, and define $\tau_p(x_1, x_2, \ldots, x_n)$ to be the probability that $x_1, x_2, \ldots, x_n$ are in the same open cluster of the percolation process. Such functions τ_p are called *connectivity functions.* We have from translation invariance that

(5.42) $$\tau_p(x_1, x_2, \ldots, x_n) = \tau_p(0, x_2 - x_1, \ldots, x_n - x_1).$$

Of particular interest is the 'two-point' connectivity function given by

(5.43) $$\tau_p(x, y) = P_p(x \leftrightarrow y).$$

In the light of Theorems (5.1) and (5.10), it is not surprising that $\tau_p(x, y)$ behaves more or less in the manner of an exponential function of the distance $\delta(x, y)$ separating x and y, when $p < p_c$. Of especial interest is the case when x is the origin and $y = e_n$ where e_n is defined by $e_n = (n, 0, \ldots, 0)$.

(5.44) Theorem. Asymptotic behaviour of the connectivity function. *Suppose that* $0 < p \leq 1$, *and let* $\varphi(p)$ *be as given in Theorem*

(5.10). *Then*

$$\lim_{n\to\infty}\left\{-\frac{1}{n}\log \tau_p(0, e_n)\right\} = \varphi(p) \tag{5.45}$$

and there exists a strictly positive constant ζ, *independent of* p, *such that*

$$\zeta p n^{4(1-d)} e^{-n\varphi(p)} \le \tau_p(0, e_n) \le e^{-n\varphi(p)} \qquad \textit{for all } n. \tag{5.46}$$

Thus $\tau_p(0, e_n)$ behaves roughly as $e^{-n\varphi(p)}$ when n is large. Some care is needed in estimating $\tau_p(0, x)$ for vertices x of $\mathbb{L}^d$ which are distant from the origin and not lying on an axis of $\mathbb{L}^d$. It is easy to adapt the argument leading to (5.45) to conclude that

$$\varphi(p, x) = \lim_{n\to\infty}\left\{-\frac{1}{n}\log \tau_p(0, nx)\right\}$$

exists for all $x \in \mathbb{Z}^d$, but not much is known about how $\varphi(p, x)$ depends on the choice of x. On the other hand, it is not too difficult to find upper and lower bounds for $\tau_p(0, x)$ which are valid for all x.

(5.47) Proposition. *Suppose* $0 < p \le 1$, *and let* $\varphi(p)$ *be as given in Theorem* (5.10). *Then there exists a strictly positive constant* λ, *independent of* p, *such that*

$$\lambda p^d |x|^{4d(1-d)} e^{-|x|\varphi(p)} \le \tau_p(0, x) \le e^{-\|x\|\varphi(p)} \tag{5.48}$$

for all vertices x.

We recall that $|x| = \delta(0, x)$, $\|x\| = \max\{|x_i|: 1 \le i \le d\}$, and $\|x\| \le |x| \le d\|x\|$ for all x. Thus the inequalities in (5.48) specify the decay rate of $\tau_p(0, x)$ as $|x| \to \infty$, up to a numerical factor in the exponent which depends on the number of dimensions.

The above remarks make extensive reference to the function φ, but how may we relate φ to other relevant quantities? The following concrete upper bound on $\tau_p(0, x)$ is of some value here.

(5.49) Proposition. *Suppose* $0 < p < p_c$, *and let* $\chi(p)$ *be the mean cluster size and* $\varphi(p)$ *be as given in Theorem* (5.10). *Then*

$$\tau_p(0, x) \le \{1 - \chi(p)^{-1}\}^{|x|} \qquad \textit{for all } x, \tag{5.50}$$

giving that

$$\varphi(p)^{-1} \le \chi(p) \qquad \textit{if} \quad 0 < p < p_c. \tag{5.51}$$

We recall from Theorem (5.14) that $\varphi(p) \to 0$ as $p \uparrow p_c$, and it follows from (5.51) that

$$\chi(p) \to \infty \qquad \text{as} \quad p \uparrow p_c. \tag{5.52}$$

We have seen that probabilities such as $P_p(0 \leftrightarrow \partial B(n))$ and $\tau_p(0, e_n)$ decay roughly as $e^{-n\varphi(p)}$ where $\varphi(p) > 0$, if $0 < p < p_c$. It is likely that such probabilities behave roughly like a power of n multiplied by the exponential factor $e^{-n\varphi(p)}$; remember (5.11) and (5.46). On the other hand, there is no such exponential decay when $p = p_c$, since $\varphi(p_c) = 0$; actually, it is likely that such probabilities behave like negative powers of n when $p = p_c$. Thus a subcritical process with edge-probability p behaves qualitatively differently from the critical process with edge-probability p_c only when the spatial scale n on which we observe the process is such that $n\varphi(p)$ is large. This observation is particularly important when the difference between p and p_c is small, since then $\varphi(p)$ is small also, so that the two processes differ qualitatively only over a large spatial scale. As p approaches p_c from beneath, we have that $\varphi(p) \to 0$, so that the required scale becomes larger and larger. This notion of scaling seems to be central to understanding the behaviour of the percolation process at and near the critical point p_c. In such discussions it is usual to emphasize not the function $\varphi(p)$ but its reciprocal $\xi(p) = \varphi(p)^{-1}$, where we define $\xi(p_c) = \infty$ since $\varphi(p_c) = 0$; the quantity $\xi(p)$ is called the *correlation length* of the percolation process with edge-probability p, since only when n is of order $\xi(p)$ or greater do the exponential terms in expressions such as

$$P_p(0 \leftrightarrow \partial B(n)) \approx e^{-n/\xi(p)} \tag{5.53}$$

and

$$\tau_p(0, e_n) \approx e^{-n/\xi(p)} \tag{5.54}$$

become significant. Henceforth, we shall describe the asymptotics of such probabilities in terms of $\xi(p)$ rather than $\varphi(p)$. We note from Theorem (5.14) and equation (5.51) that ξ is continuous and strictly increasing on $(0, p_c)$, and satisfies

$$\xi(p) \to 0 \quad \text{as} \quad p \downarrow 0, \tag{5.55}$$

$$\xi(p) \to \infty \quad \text{as} \quad p \uparrow p_c, \tag{5.56}$$

and

$$\xi(p) \leq \chi(p) \quad \text{if} \quad 0 < p < p_c. \tag{5.57}$$

Note also that the correlation length $\xi(p)$ is defined only for $p \leq p_c$. In Chapter 6, we shall encounter the natural counterpart of ξ when $p > p_c$.

Proof of Theorem (5.44). This is similar to but simpler than the proof of Theorem (5.10). Clearly

$$\{0 \leftrightarrow e_{m+n}\} \supseteq \{0 \leftrightarrow e_m\} \cap \{e_m \leftrightarrow e_{m+n}\}; \tag{5.58}$$

see Figure 5.4 for an illustration of this. By the FKG inequality and

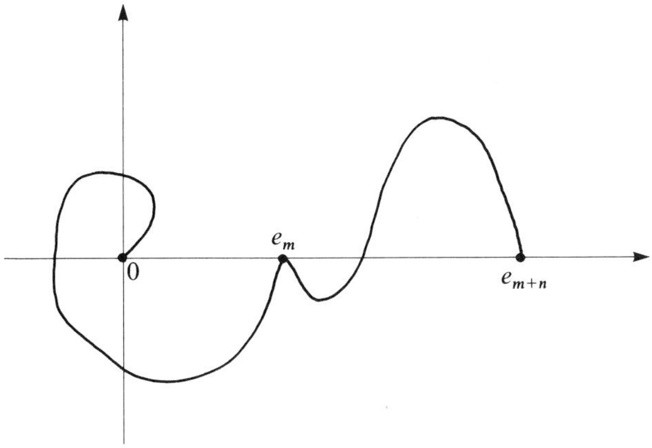

Figure 5.4. If $0 \leftrightarrow e_m$ and $e_m \leftrightarrow e_{m+n}$, then $0 \leftrightarrow e_{m+n}$.

translation invariance,

$$\tau_p(0, e_{m+n}) \geq \tau_p(0, e_m)\tau_p(e_m, e_{m+n}) = \tau_p(0, e_m)\tau_p(0, e_n). \tag{5.59}$$

Thus $t(k) = -\log \tau_p(0, e_k)$ satisfies

$$t(m+n) \leq t(m) + t(n),$$

and the subadditive limit theorem (Theorem (II.2) of Appendix II) gives us that

$$\eta(p) = \lim_{n\to\infty} \left\{-\frac{1}{n} \log \tau_p(0, e_n)\right\}$$

exists. Furthermore $\log \tau_p(0, e_n) = t(n) \leq -n\eta(p)$ for all n.

We show next that $\eta(p) = \varphi(p)$. First note that $\tau_p(0, e_n) \leq P_p(0 \leftrightarrow \partial B(n))$, as in (5.20); this implies that $\eta(p) \geq \varphi(p)$. It requires slightly more work to show the opposite inequality.

We have from (5.26) that there exists a vertex x in $\partial B(m)$ such that

$$P_p(0 \leftrightarrow x \text{ in } B(m)) \geq \frac{1}{|\partial B(m)|} P_p(0 \leftrightarrow \partial B(m)). \tag{5.60}$$

Using the rotation invariance of the lattice, we may pick such a vertex x for which $x_1 = m$. By symmetry, the events $\{0 \leftrightarrow x\}$ and $\{x \leftrightarrow e_{2m}\}$ are equally likely (just reflect the set of paths from 0 to x in the hyperplane $\{y \in \mathbb{Z}^d: y_1 = m\}$ to obtain the isomorphic set of paths from e_{2m} to x). On the other hand,

$$\{0 \leftrightarrow e_{2m}\} \supseteq \{0 \leftrightarrow x\} \cap \{x \leftrightarrow e_{2m}\}$$

as in (5.58). We use the FKG inequality here to obtain

(5.61) $$\tau_p(0, e_{2m}) \geq \tau_p(0, x)^2,$$

which we combine with (5.60) to obtain

$$\begin{aligned}\tau_p(0, e_{2m}) &\geq A_1 m^{2(1-d)} P_p(0 \leftrightarrow \partial B(m))^2 \\ &\geq A_2 m^{4(1-d)} e^{-2m\varphi(p)} \qquad \text{by (5.11)}\end{aligned}$$

for appropriate positive constants A_1 and A_2; we have used the fact that $|\partial B(m)|$ is of order m^{d-1}. Not only does it follow that $\eta(p) = \varphi(p)$, but also we have obtained the left-hand inequality in (5.46) for even values of n. It is only a minor nuisance that we need this inequality for odd n also. If $n = 2m + 1$ then, with the same choice of x,

$$\{0 \leftrightarrow e_{2m+1}\} \supseteq \{0 \leftrightarrow x\} \cap \{\langle x, x + e_1\rangle \text{ is open}\} \cap \{x + e_1 \leftrightarrow e_{2m+1}\},$$

the intersection of increasing events. Applying the FKG inequality, we obtain

$$\tau_p(0, e_{2m+1}) \geq p\tau_p(0, x)^2$$

and the result follows as before. ■

Proof of Proposition (5.47). Suppose that $\|x\| = n$, so that $x \in \partial B(n)$. We have from (5.61) that

$$\begin{aligned}\tau_p(0, x) &\leq \tau_p(0, e_{2n})^{1/2} \\ &\leq e^{-n\varphi(p)} \qquad \text{by (5.46),}\end{aligned}$$

and this is the second inequality which we are required to prove.

Turning to the first inequality in (5.48), let us suppose that $|x| = n$, so that

$$\sum_{i=1}^{d} |x_i| = n.$$

Let $y_i = (|x_1|, |x_2|, \dots, |x_i|, 0, \dots, 0)$ for $0 \leq i \leq d$. By symmetry, $\tau_p(0, x) = \tau_p(0, y_d)$. Now,

(5.62) $$\{0 \leftrightarrow y_d\} \supseteq \bigcap_{0<i\leq d} \{y_{i-1} \leftrightarrow y_i\}$$

by the usual argument. By the FKG inequality and the symmetry of the process,

(5.63) $$\begin{aligned}\tau_p(0, y_d) &\geq \prod_{i=1}^{d} \tau_p(y_{i-1}, y_i) \\ &= \prod_{i=1}^{d} \tau_p(0, e_{|x_i|}).\end{aligned}$$

For large $|x_i|$, $\tau_p(0, e_{|x_i|})$ behaves approximately like $\exp(-\varphi(p)|x_i|)$, so that this last product behaves approximately as $\exp(-\varphi(p)\sum_i |x_i|) =$

$e^{-\varphi(p)n}$ as required. Arguing more rigorously, it follows from (5.46) that

(5.64) $$\tau_p(0, y_d) \geq Ap^d\{|x_1||x_2|\dots|x_d|\}^{4(1-d)}e^{-n\varphi(p)}$$

for some positive constant A, and the required conclusion is obtained by way of the standard inequality for geometric and arithmetic means:

(5.65) $$\begin{aligned}\{|x_1||x_2|\dots|x_d|\}^{1/d} &\leq \frac{1}{d}\sum_{i=1}^{d}|x_i| \\ &= n/d.\end{aligned}$$ ∎

Proof of Proposition (5.49). We follow Aizenman and Newman (1984). The argument is closely related to the proof of Theorem (5.1), except that we seek an inequality involving the norm $\delta(0, x) = |x|$ rather than $\|x\|$. Instead of the box $B(n)$, we consider the sphere $S(n) = \{x \in \mathbb{Z}^d: |x| \leq n\}$ and its boundary $\partial S(n) = \{x \in \mathbb{Z}^d: |x| = n\}$; similarly, we write $S(n, x) = x + S(n)$ and $\partial S(n, x) = x + \partial S(n)$ for the corresponding sets centred at x. Let M_n be the number of vertices in $\partial S(n)$ which are connected to the origin by open paths, so that

(5.66) $$E_p(M_n) = \sum_{x \in \partial S(n)} \tau_p(0, x),$$

as in (5.5). Similarly to (5.6), we have that

(5.67) $$\sum_{n=0}^{\infty} E_p(M_n) = \chi(p).$$

The argument leading to (5.7) is easily adapted to give

(5.68) $$\tau_p(0, z) \leq \sum_{x \in \partial S(m)} \tau_p(0, x)\tau_p(x, z)$$

whenever $|z| \geq m$. The term $\tau_p(x, z)$ is no greater than 1, so that

(5.69) $$\tau_p(0, z) \leq E_p(M_m) \qquad \text{when} \quad |z| \geq m,$$

by comparison with (5.66). We can do much better than this. We fix m and suppose that $|z| = n$. We write $n = mr + s$ for non-negative integers r, s satisfying $0 \leq s < m$. We now iterate (5.68) to obtain

(5.70) $$\begin{aligned}\tau_p(0, z) &\leq \sum_{x \in \partial S(m)} \tau_p(0, x) \sum_{y \in \partial S(m, x)} \tau_p(x, y)\tau_p(y, z) \\ &\leq \cdots \\ &\leq E_p(M_m)^r \qquad \text{as in (5.69)} \\ &= E_p(M_m)^{\lfloor |z|/m \rfloor} \qquad \text{for all } z \text{ and } m.\end{aligned}$$

We may improve this still further as follows. Let $u \in \mathbb{Z}^d$ and k be a positive integer. By the usual argument, we have that

$$\begin{aligned}\tau_p(0, ku) &\geq P_p(ju \leftrightarrow (j+1)u \text{ for } 0 \leq j < k) \\ &\geq \tau_p(0, u)^k\end{aligned}$$

by the FKG inequality. Thus

$$\begin{aligned}
(5.71)\qquad \tau_p(0, u) &\le \tau_p(0, ku)^{1/k} \\
&\le E_p(M_m)^{\lfloor |ku|/m \rfloor /k} && \text{by (5.70)} \\
&\to E_p(M_m)^{|u|/m} && \text{as} \quad k \to \infty
\end{aligned}$$

for all vertices u and all integers m.

Suppose now that $0 < p < p_c$, so that $\chi(p) < \infty$. There exists $m \ge 1$ such that

$$(5.72)\qquad E_p(M_m) \le \{1 - \chi(p)^{-1}\}^m$$

since if $E_p(M_m) > \{1 - \chi(p)^{-1}\}^m$ for all $m \ge 1$, then (5.67) implies that

$$\chi(p) > \sum_{m=0}^{\infty} \{1 - \chi(p)^{-1}\}^m = \chi(p),$$

a contradiction. We choose $m \ge 1$ such that $E_p(M_m) \le \{1 - \chi(p)^{-1}\}^m$ and substitute this into (5.71) to obtain

$$(5.73)\qquad \tau_p(0, u) \le \{1 - \chi(p)^{-1}\}^{|u|}$$

as claimed. Inequality (5.51) is a consequence of the fact that $1 - r \le e^{-r}$ when $r \ge 0$. ■

5.3 Exponential Decay of the Cluster Size Distribution

We turn our attention now to the number $|C|$ of vertices in the open cluster C at the origin. If $p < p_c$, then this cluster is almost surely finite. We require an estimate on the tail of the distribution of $|C|$. It is easy to see (as in Section 3.2) that

$$(5.74)\qquad P_p(0 \leftrightarrow \partial B(n)) \le P_p(|C| \ge n) \le P_p(0 \leftrightarrow \partial B(\alpha n^{1/d}))$$

for some positive constant $\alpha = \alpha(d)$; thus the exponential decay of $P_p(0 \leftrightarrow \partial B(n))$ as $n \to \infty$ implies that $P_p(|C| \ge n)$ decays at least as fast as $\exp(-\gamma n^{1/d})$ for some positive $\gamma = \gamma(p, d)$. This is not the best possible result, as the next theorem indicates.

(5.75) Theorem. Exponential decay of the cluster size distribution. *Suppose that $0 < p < p_c$. There exists $\lambda(p) > 0$ such that*

$$(5.76)\qquad P_p(|C| \ge n) \le e^{-n\lambda(p)} \qquad \textit{for all} \quad n \ge 1.$$

We shall actually prove a more concrete inequality than this, in which $\lambda(p)$ is expressed in terms of the mean cluster size χ. More precisely, we shall prove that

$$(5.77)\qquad P_p(|C| \ge n) \le 2\exp(-\tfrac{1}{2}n/\chi(p)^2) \qquad \text{if} \quad n > \chi(p)^2.$$

We may use subadditivity to show that $-n^{-1} \log P_p(|C| = n)$ converges as $n \to \infty$, and the inequalities above show that the limit is finite and strictly positive whenever $0 < p < p_c$.

(5.78) Theorem. *Suppose $0 < p < 1$. The limit*

(5.79) $$\zeta(p) = \lim_{n\to\infty} \left\{ -\frac{1}{n} \log P_p(|C| = n) \right\}$$

exists and satisfies $0 \le \zeta(p) < \infty$ for all $p \in (0, 1)$. Also

(5.80) $$P_p(|C| = n) \le \frac{(1-p)^2}{p} n e^{-n\zeta(p)} \qquad \text{for all} \quad n \ge 1.$$

The limit $\zeta(p)$ is strictly positive if $0 < p < p_c$.

We note that

(5.81) $$P_p(n \le |C| < \infty) = \sum_{m=n}^{\infty} P_p(|C| = m),$$

so that the limit ζ in (5.79) satisfies

(5.82) $$\zeta(p) = \lim_{n\to\infty} \left\{ -\frac{1}{n} \log P_p(n \le |C| < \infty) \right\}$$

also (this is seen easiest by considering the cases $\zeta(p) = 0$ and $\zeta(p) > 0$ separately). We have from (5.74) that

(5.83) $$\zeta(p) \le \varphi(p) \qquad \text{for} \quad 0 < p < p_c,$$

where φ is the decay rate of $P_p(0 \leftrightarrow \partial B(n))$ given in Theorem (5.10). We shall see in Theorem (6.91) that the tail of $|C|$ decays slower than exponentially when $p > p_c$, so that $\zeta(p) = 0$ if $p > p_c$.

Proof of Theorem (5.75). We use the method of Aizenman and Newman (1984) aided by the BK inequality. We shall obtain an exponentially decaying upper bound for $P_p(|C| \ge n)$ by showing that $v(t) = E_p(|C| \exp(t|C|))$ is finite for small positive values of t, and then using Markov's inequality. For $t > 0$, $v(t)$ is the mean of an increasing random variable, and this suggests that the BK inequality may be useful in finding an upper bound for $v(t)$. We shall suppose henceforth that $0 < p < p_c$ so that $\chi(p) < \infty$, and we shall concentrate on finding good upper bounds for the moments $E_p(|C|^n)$ of $|C|$ for $n \ge 1$. Clearly

(5.84) $$|C| = \sum_{x \in \mathbb{Z}^d} I_{\{0 \leftrightarrow x\}},$$

the sum over x of the indicator function of the event that $x \in C$. Thus

(5.85) $$\begin{aligned} E_p(|C|^n) &= E_p\left(\sum_{x_1, \ldots, x_n} I_{\{0 \leftrightarrow x_1\}} I_{\{0 \leftrightarrow x_2\}} \cdots I_{\{0 \leftrightarrow x_n\}} \right) \\ &= \sum_{x_1, \ldots, x_n} \tau_p(0, x_1, x_2, \ldots, x_n), \end{aligned}$$

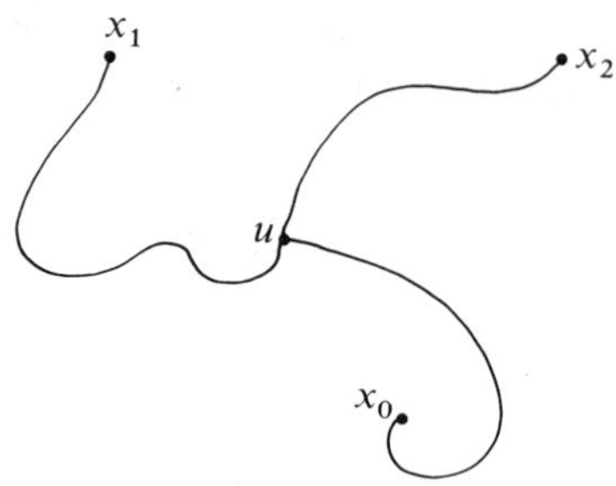

Figure 5.5. If x_0, x_1, and x_2 are in the same open cluster then there exists a vertex u which is joined by edge-disjoint open paths to x_0, x_1, and x_2.

where the summations are over all n-tuples $(x_1, x_2, \ldots, x_n)$ of (possibly non-distinct) vertices of $\mathbb{L}^d$, and the connectivity function

$$\tau_p(y_0, y_1, \ldots, y_n) = P_p(y_0, y_1, \ldots, y_n \text{ are in the same open cluster})$$

is the probability that $y_0, y_1, \ldots, y_n$ are in the same open cluster of the percolation process. We need upper bounds for these connectivity functions. Note that

$$\text{(5.86)} \qquad E_p(|C|^0) = 1, \qquad E_p(|C|) = \chi(p),$$

so that the first case of interest is when $n = 2$. Suppose first that x_0, x_1, and x_2 are distinct vertices which are in the same open cluster of the process. We claim that there exists a vertex u (possibly equal to x_0, x_1, or x_2) such that u is joined by three edge-disjoint open paths to x_0, x_1, and x_2, respectively; if u equals one of the three original vertices, say $u = x_0$, then the required path joining u to x_0 may be taken to be the trivial path with zero length. See Figure 5.5. This observation is an elementary fact of graph theory, and is a consequence of the following lemma.

(5.87) Lemma. *Let G be a connected graph and let W be a non-empty subset of the vertex set of G. There exists a vertex $w \in W$ such that, in the graph obtained from G by deleting w and its incident edges, all vertices in $W \setminus \{w\}$ are in the same connected component.*

Proof. Let T be a spanning tree of G and let u be any given vertex of G; we think of T as being a tree with root at u. We may suppose that $|W| \geq 2$, so that at least one of the branches of T contains a vertex in W. Let $u = x_0, x_1, x_2, \ldots, x_m$ be the vertices on such a branch, listed in increasing order of distance from u in the tree T. We define $J = \max\{j : x_j \in W\}$ and $w = x_J$. A glance at Figure 5.6 confirms that w has the required property. ∎

We apply this lemma with $G = C(x_0)$ and $W = \{x_0, x_1, x_2\}$, to deduce

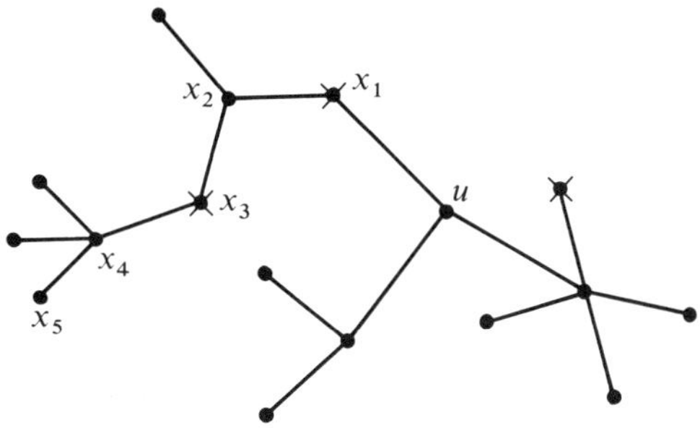

Figure 5.6. An illustration of the proof of Lemma (5.87). The crosses indicate vertices in W, so that $J = 3$ and $w = x_3$.

that there is an open path π_1 joining two of the vertices of W which does not use the third such vertex. However, $C(x_0)$ is connected, so that there exists an open path π_2 from the third vertex to some vertex in π_1, using no other vertex in π_1; we may take u to be the vertex common to π_1 and π_2. We have assumed above that x_0, x_1, x_2 are distinct, but this assumption is not essential. If $x_0 = x_1 = x_2$ say, we may take $u = x_0$, and if $x_0 = x_1 \neq x_2$ we may take $u = x_0$. We have proved that

(5.88) $$\tau_p(x_0, x_1, x_2) \leq \sum_{u \in \mathbb{Z}^d} P_p(\{u \leftrightarrow x_0\} \circ \{u \leftrightarrow x_1\} \circ \{u \leftrightarrow x_2\}),$$

and the BK inequality (particularly inequality (2.17)) yields the next lemma.

(5.89) Lemma. *For all values of p and vertices x_0, x_1, x_2 we have that*

(5.90) $$\tau_p(x_0, x_1, x_2) \leq \sum_{u \in \mathbb{Z}^d} \tau_p(u, x_0)\tau_p(u, x_1)\tau_p(u, x_2).$$

We may apply this to (5.85) with $n = 2$ to find that

(5.91) $$\begin{aligned} E_p(|C|^2) &\leq \sum_{u, x_1, x_2} \tau_p(u, 0)\tau_p(u, x_1)\tau_p(u, x_2) \\ &= \left(\sum_y \tau_p(0, y)\right)^3 \\ &= (E_p|C|)^3 \\ &= \chi(p)^3. \end{aligned}$$

We have treated the case $n = 2$ in some detail, since it indicates a route to an upper bound for $E_p(|C|^n)$ for general n.

We begin with some notation. A *tree* is a connected graph with no circuits; a tree on n vertices has $n - 1$ edges. We call a tree a *skeleton* if each vertex has degree either 1 or 3. It is an easy calculation to see that if a skeleton has k vertices of degree 1 (called *exterior* vertices) then it

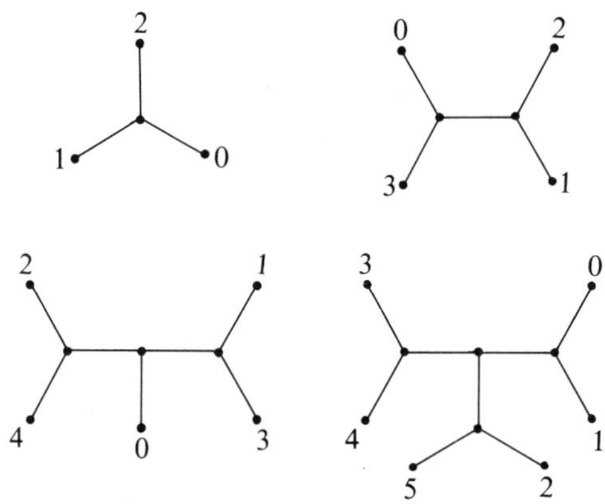

Figure 5.7. Some labelled skeletons.

has $k-2$ vertices of degree 3 (called *interior* vertices). We denote by $I(S)$ the set of interior vertices of the skeleton S. A skeleton with k exterior vertices is called *labelled* if there is an assignment of the numbers 0, 1, 2, ..., $k-1$ to its exterior vertices. Two labelled skeletons S_1 and S_2 are called *isomorphic* if there is a one–one correspondence between the vertex sets of S_1 and S_2 under which both the adjacency relation and the labellings of the exterior vertices are preserved. We shall be interested in labelled skeletons, of which some examples are presented in Figure 5.7.

We saw above that, if x_0, x_1, and x_2 are in the same open cluster, then there exists a labelled skeleton S with three exterior vertices together with a mapping $\psi_{\mathbf{x}}$ from the vertex set of S into $\mathbb{Z}^d$ such that the following hold:

(a) the exterior vertex of S with label i is mapped to x_i by $\psi_{\mathbf{x}}$, for $i = 0$, 1, 2; and

(b) the edges of S correspond to edge-disjoint open paths of the lattice, in that there exist three edge-disjoint open paths joining the three pairs $(\psi_{\mathbf{x}}(v), x_0)$, $(\psi_{\mathbf{x}}(v), x_1)$, and $(\psi_{\mathbf{x}}(v), x_2)$, where v is the unique interior vertex of S.

We now make this construction more general. Suppose that x_0, x_1, ..., x_k are (possibly non-distinct) vertices of $\mathbb{L}^d$ which are in the same open cluster. We claim that there exists a labelled skeleton S with $k+1$ exterior vertices together with a mapping $\psi_{\mathbf{x}}$ from the vertex set of S into $\mathbb{Z}^d$, such that the following hold:

(c) the exterior vertex of S with label i is mapped to x_i by $\psi_{\mathbf{x}}$, for $i = 0$, 1, ..., k; and

(d) the edges of S correspond to edge-disjoint open paths of the lattice, in that there exist $2k-1$ edge-disjoint open paths joining the $2k-1$ pairs $((\psi_{\mathbf{x}}(u), \psi_{\mathbf{x}}(v)) : \langle u, v\rangle \in S)$.

Here, $\langle u, v\rangle$ is the edge of S with endvertices u and v. We prove this claim by induction on k, noting that it is valid for $k = 2$. Suppose that it is valid for $k = n$ and suppose that $x_0, x_1, \ldots, x_{n+1}$ are in the same open cluster of $\mathbb{L}^d$. From Lemma (5.87), there exists j such that every vertex in $\{x_i: 0 \leq i \leq n+1\}\backslash\{x_j\}$ is in the same connected component of the graph obtained from this cluster by deleting x_j and its incident edges; we may assume without loss of generality that $j = n + 1$. Suppose for the moment that none of $x_0, x_1, \ldots, x_n$ equals x_{n+1}. By the induction hypothesis, there exists a labelled skeleton S with $n + 1$ exterior vertices, together with a mapping $\psi_{\mathbf{x}}$, satisfying (c) and (d) above with $k = n$, such that x_{n+1} lies in none of the $2n - 1$ edge-disjoint open paths referred to in (d). However, x_{n+1} is in the same open cluster as $x_0, x_1, \ldots, x_n$, and so there exists an open path π joining x_{n+1} to some vertex in these $2n - 1$ edge-disjoint paths, using exactly one vertex (z, say) in these paths. We amend the skeleton S accordingly (see Figure 5.8), adding an additional interior vertex v at the appropriate place, and extending the domain of the mapping $\psi_{\mathbf{x}}$ by defining $\psi_{\mathbf{x}}(v) = z$; thus the claim is valid for $k = n + 1$ also. We have assumed that x_{n+1} appears only once in the sequence $x_0, x_1, \ldots, x_{n+1}$, but this assumption is in no way essential. Suppose, for example, that x_{n+1} appears twice and $x_0 = x_{n+1}$, say. The above argument remains valid, and in this case π is the path containing the single vertex x_{n+1} $(= x_0 = z)$ and no edges.

It follows that, for $k \geq 2$,

(5.92) $$\tau_p(x_0, x_1, \ldots, x_k) \leq \sum_S \sum_{\psi_{\mathbf{x}}} P_p(\text{there exist edge-disjoint paths joining } \psi_{\mathbf{x}}(u) \text{ to } \psi_{\mathbf{x}}(v) \text{ for each edge } \langle u, v\rangle \text{ of } S),$$

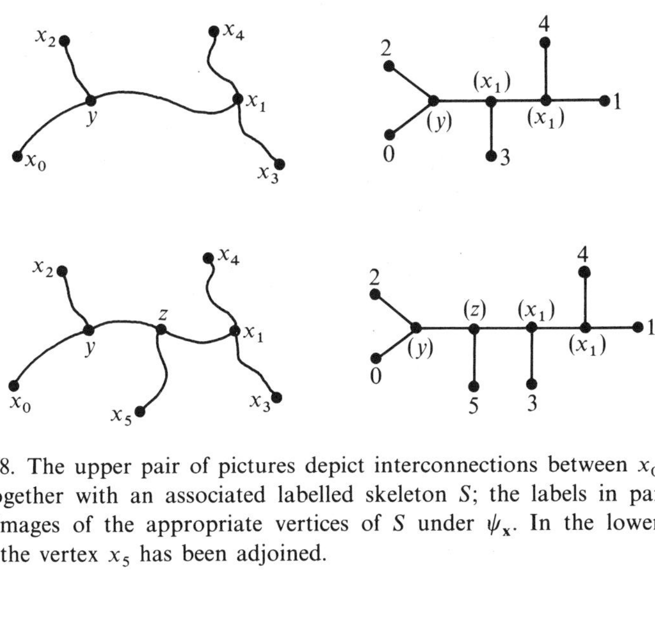

Figure 5.8. The upper pair of pictures depict interconnections between x_0, x_1, x_2, x_3, x_4 together with an associated labelled skeleton S; the labels in parentheses are the images of the appropriate vertices of S under $\psi_{\mathbf{x}}$. In the lower pair of pictures, the vertex x_5 has been adjoined.

where the first sum is over all labelled skeletons with $k+1$ exterior vertices and the second sum is over all admissible mappings $\psi_{\mathbf{x}}$ from the vertex set of S into $\mathbb{Z}^d$. The second sum amounts to the summation over all possible choices for $\psi_{\mathbf{x}}(v)$ as v ranges over $I(S)$, the set of interior vertices of S. The above inequality reduces to (5.88) when $k = 2$. We use the BK inequality to find that

$$\tau_p(x_0, x_1, \ldots, x_k) \leq \sum_S \sum_{\psi_{\mathbf{x}}} \prod_{\langle u,v\rangle \in S} \tau_p(\psi_{\mathbf{x}}(u), \psi_{\mathbf{x}}(v)), \tag{5.93}$$

in agreement with (5.90). Hence, for $n \geq 2$,

$$\begin{aligned} E_p(|C|^n) &= \sum_{x_1, x_2, \ldots, x_n} \tau_p(x_0, x_1, x_2, \ldots, x_n) \\ &\leq \sum_{x_1, x_2, \ldots, x_n} \sum_S \sum_{\psi_{\mathbf{x}}} \prod_{\langle u,v\rangle \in S} \tau_p(\psi_{\mathbf{x}}(u), \psi_{\mathbf{x}}(v)), \end{aligned}$$

by (5.85) and (5.93); we have written $x_0 = 0$. When we perform the summation over $x_1, x_2, \ldots, x_n$, the right-hand side becomes

$$\sum_S \sum_\psi \left\{ \prod_{\langle u,v\rangle}' \tau_p(\psi(u), \psi(v)) \right\} \left\{ \sum_x \tau_p(z, x) \right\}^n = \chi(p)^n \sum_S \sum_\psi \prod_{\langle u,v\rangle}' \tau_p(\psi(u), \psi(v)),$$

where z is a typical (but fixed) vertex, the second summation is over all mappings $\psi: I(S) \cup \{0\} \to \mathbb{Z}^d$ satisfying $\psi(0) = x_0$, and the primed product is over all edges $\langle u, v\rangle$ of S as u and v range over the set $I(S) \cup \{0\}$; we have written 0 for the exterior vertex of S with label 0. We now perform the summation over admissible mappings ψ to obtain

$$\begin{aligned} E_p(|C|^n) &\leq \chi(p)^n \sum_S \left\{ \sum_y \tau_p(z, y) \right\}^{n-1} \\ &= N_{n+1} \chi(p)^{2n-1}, \end{aligned} \tag{5.94}$$

where N_{n+1} is the number of labelled skeletons with $n+1$ exterior vertices. Now, any labelled skeleton with $n+1$ exterior vertices has $2n-1$ edges. If we remove the exterior vertex labelled n, together with the incident edge and corresponding interior vertex, we obtain a labelled skeleton with n exterior vertices and $2n-3$ edges (see Figure 5.9). Exactly $2n-3$ labelled skeletons with $n+1$ exterior vertices correspond to the resulting labelled skeleton with n exterior vertices, since the $(n+1)$th exterior vertex may be connected into any of its $2n-3$ edges. It is not difficult to convince oneself that these $2n-3$ labelled skeletons are non-isomorphic, and it follows that

$$N_{n+1} = (2n-3)N_n. \tag{5.95}$$

However $N_3 = 1$, so that

$$N_{n+1} = (2n-3)(2n-5)\ldots 5 \cdot 3 \cdot 1 = \frac{(2n-2)!}{2^{n-1}(n-1)!}. \tag{5.96}$$

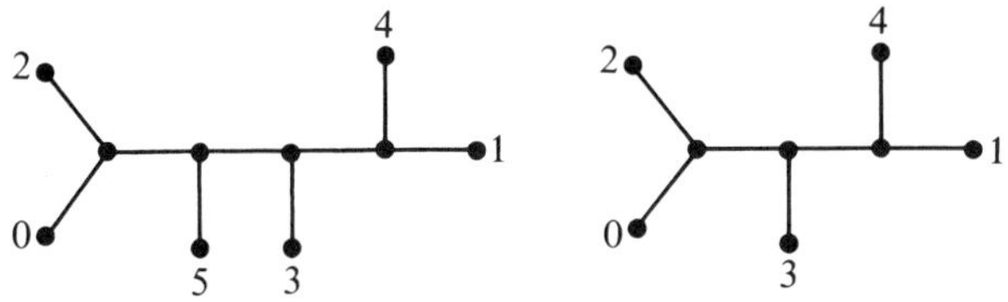

Figure 5.9. A labelled skeleton with six exterior vertices, and the skeleton obtained by deleting the vertex labelled 5 together with the incident edge. The smaller skeleton has five exterior vertices and seven edges, and it is not difficult to check that the seven corresponding labelled skeletons having seven exterior vertices are non-isomorphic.

We combine this with (5.86) and (5.94) to find that

$$E_p(|C|e^{t|C|}) = \sum_{n=0}^{\infty} \frac{1}{n!} t^n E_p(|C|^{n+1}) \tag{5.97}$$

$$\leq \chi(p)\left\{1 + \sum_{n=1}^{\infty} \frac{1}{n!} t^n N_{n+2}\chi(p)^{2n}\right\} \qquad \text{by (5.94)}$$

$$= \chi(p)(1 - 2t\chi(p)^2)^{-1/2} \qquad \text{by (5.96)}$$

for t satisfying $0 \leq t < \frac{1}{2}\chi(p)^{-2}$. We now use Markov's inequality (see Grimmett and Stirzaker (1982, p. 181)) to obtain

$$P_p(|C| \geq n) = P_p(|C|e^{t|C|} \geq ne^{tn}) \tag{5.98}$$

$$\leq \frac{1}{ne^{tn}} E_p(|C|e^{t|C|})$$

for $t \geq 0$. Applying (5.97), we find that

$$P_p(|C| \geq n) \leq \frac{\chi(p)}{ne^{tn}}(1 - 2t\chi(p)^2)^{-1/2} \tag{5.99}$$

for values of t satisfying $0 \leq t < \frac{1}{2}\chi(p)^{-2}$. We choose

$$t = \frac{1}{2\chi(p)^2} - \frac{1}{2n},$$

noting that $t > 0$ if $n > \chi(p)^2$, to obtain

$$P_p(|C| \geq n) \leq (e/n)^{1/2} \exp(-\tfrac{1}{2}n/\chi(p)^2) \tag{5.100}$$

in agreement with (5.77). We have proved Theorem (5.75). ∎

Proof of Theorem (5.78). We follow Kunz and Souillard (1978, p. 91). For ease of description we restrict ourselves to the two-dimensional case, but the same argument is valid whenever $d \geq 2$. Suppose that $0 <$

$p = 1 - q < 1$, and let

$$\pi_n = P_p(|C| = n). \tag{5.101}$$

We shall show the following lemma.

(5.102) Lemma. *If* $0 < p = 1 - q < 1$, *then* $\pi_k = P_p(|C| = k)$ *satisfies*

$$\frac{1}{m+n}\pi_{m+n} \geq pq^{-2}\frac{1}{m}\pi_m\frac{1}{n}\pi_n \qquad \textit{for all} \quad m, n \geq 1. \tag{5.103}$$

The limit theorem for subadditive sequences will then yield the result.

Proof of Lemma (5.102). For any connected subgraph S of $\mathbb{L}^2$, we denote by tr(S) the top vertex on the right-hand side of S. More specifically, $\text{tr}(S) = x$ if $x_1 = \max\{y_1 : y \in S\}$ and $x_2 = \max\{y_2 : (x_1, y_2) \in S\}$. We define bl($S$) similarly as the bottom vertex on the left-hand side of S. See Figure 5.10.

Let π'_k be the probability that the open cluster C at the origin has exactly k vertices and $\text{bl}(C) = 0$. The origin may generally be any of the k vertices of C, and each such possibility has the same probability, so that

$$\pi'_k = \frac{1}{k}\pi_k. \tag{5.104}$$

Let m and n be positive integers. Suppose that σ and τ are connected subgraphs of $\mathbb{L}^2$ with $\text{bl}(\sigma) = \text{bl}(\tau) = 0$, such that σ has m vertices and τ has n vertices. From σ and τ we may construct another subgraph, written $\sigma * \tau$, by translating τ to a new position in $\mathbb{L}^2$ so that $\text{bl}(\tau)$ is translated to the vertex immediately to the right of $\text{tr}(\sigma)$; we add the edge joining $\text{tr}(\sigma)$ to the new position of $\text{bl}(\tau)$ so that the new subgraph $\sigma * \tau$ is connected and has $m + n$ vertices. See Figure 5.11 for an illustration of this construction. It is easy to see that

$$P_p(C = \sigma * \tau) = pq^{-2}P_p(C = \sigma)P_p(C = \tau), \tag{5.105}$$

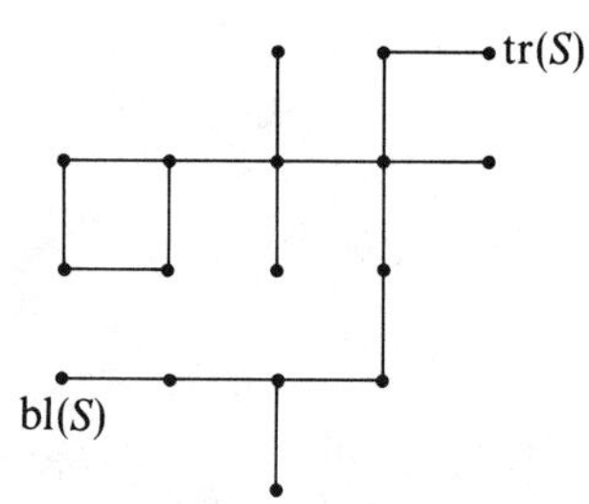

Figure 5.10. A connected subgraph S of $\mathbb{L}^2$, with the 'top right' and 'bottom left' vertices indicated.

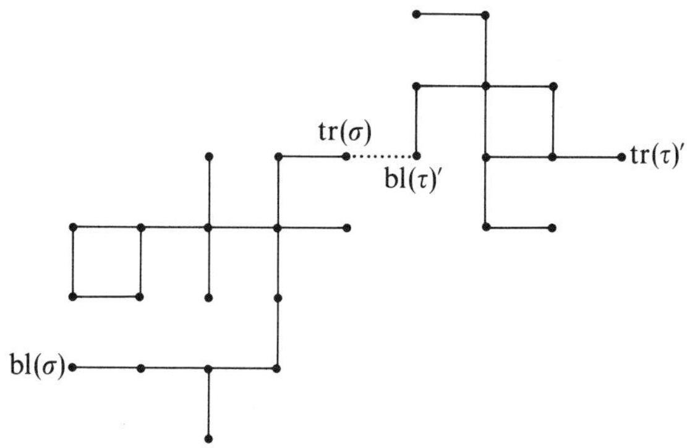

Figure 5.11. The graph σ together with a translation of the graph τ, the bottom left vertex of τ being translated to the immediate right of the top right vertex of σ. The new positions of $\mathrm{bl}(\tau)$ and $\mathrm{tr}(\tau)$ are as indicated by the primed quantities.

where the term pq^{-2} arises from the fact that we have declared open an edge which was counted as closed for both σ and τ. Thus two clusters with sizes m and n may be combined to form a cluster with size $m + n$. Furthermore, this composite cluster is unique in the sense that $\sigma * \tau$ may be decomposed in a unique appropriate way into two clusters with sizes m and n. Now

$$(5.106) \qquad \sum_{\sigma} P_p(C = \sigma) = P_p(|C| = m, \mathrm{bl}(C) = 0) = \pi'_m,$$

and similarly

$$\sum_{\tau} P_p(C = \tau) = \pi'_n,$$

where the sums are over all σ and τ with $\mathrm{bl}(\sigma) = \mathrm{bl}(\tau) = 0$ and having m and n vertices, respectively. Thus

$$\begin{aligned} pq^{-2}\pi'_m\pi'_n &= \sum_{\sigma,\tau} pq^{-2} P_p(C = \sigma) P_p(C = \tau) \\ &= \sum_{\sigma,\tau} P_p(C = \sigma * \tau) \qquad \text{by (5.105)} \\ &\le P_p(|C| = m + n, \mathrm{bl}(C) = 0) \\ &= \pi'_{m+n}, \end{aligned}$$

and (5.103) follows from (5.104). ■

We have shown that the sequence $(a - \log(\pi_n/n)\colon n \ge 1)$ is subadditive, where $a = -\log(pq^{-2})$. We apply the subadditive limit theorem (see Appendix II) to find that

$$(5.107) \qquad \zeta(p) = \lim_{n\to\infty} \left\{ -\frac{1}{n} \log(\pi_n/n) \right\}$$

exists, and

$$a - \log(\pi_n/n) \geq n\zeta(p) \qquad \text{for all } n$$

as required. It is obvious that $p^{n-1}q^{2n+2} \leq \pi_n \leq 1$, giving that $0 \leq \zeta(p) \leq -\log(pq^2)$. The strict positivity of $\zeta(p)$ when $0 < p < p_c$ follows from Theorem (5.75). ■

5.4 Analyticity of κ and χ

All is peaceful when p is less than p_c. In particular, estimates on the tail of the sizes of open clusters imply the analyticity of macroscopic quantities such as the number of open clusters per vertex and the mean cluster size.

(5.108) Theorem. Analyticity of κ and χ below p_c. *The number $\kappa(p)$ of open clusters per vertex and the mean size $\chi(p)$ of the open cluster at the origin are analytic functions of p on the interval $[0, p_c)$.*

Proof. We shall prove that χ is analytic on $[0, p_c)$; the same proof works for κ. Suppose that $p = 1 - q < p_c$. We write $\chi(p)$ in the form (1.20),

(5.109) $$\chi(p) = \sum_{n=1}^{\infty} nP_p(|C| = n),$$

valid since $p < p_c$, where

(5.110) $$P_p(|C| = n) = \sum_{m,b} a_{nmb}p^m q^b$$

as in (4.17). Here, a_{nmb} is the number of animals with n vertices, m edges, and b 'boundary edges'; furthermore, the series in (5.110) is finite.

We follow the calculation of Kesten (1982, p. 250). We recall first that $a_{nmb} = 0$ unless n, m, and b satisfy

(5.111) $$1 \leq b \leq 2dn, \qquad n - 1 \leq m \leq dn$$

(see (4.14) and (4.15)), and we assume therefore that these inequalities are satisfied. We may write (5.109) in the form

(5.112) $$\chi(p) = \sum_n \sum_{m=n-1}^{dn} \sum_{b=1}^{2dn} na_{nmb}p^m q^b.$$

Let z be a complex number, and write

(5.113) $$K(z) = \sum_n \sum_{m=n-1}^{dn} \sum_{b=1}^{2dn} na_{nmb}z^m(1 - z)^b$$

whenever this sum converges. K is the sum of polynomials in z, and thus K is analytic on any disc in which the series in (5.113) is uniformly convergent. We show first that K is analytic on a neighbourhood of the origin of the complex plane. We have from (4.24) that, for fixed n,

$$\sum_{m,b} a_{nmb} \le 7^{dn},$$

and therefore,

$$\left|\sum_{m,b} na_{nmb}z^m(1-z)^b\right| \le \sum_{m=n-1}^{dn}\sum_{b=1}^{2dn} na_{nmb}|z|^m(1+|z|)^{2dn} \tag{5.114}$$

$$\le n7^{dn}|z|^{n-1}(1+|z|)^{2dn}$$

$$\le Anc(z)^{n-1}$$

if $|z| < \frac{1}{2}$, where A depends on d only and

$$c(z) = |z|\{7(1+|z|)^2\}^d.$$

However, $c(z) < 1$ if $|z|$ is sufficiently small, giving from (5.113) and (5.114) that $K(z)$ is uniformly convergent when $|z|$ is sufficiently small; thus K is analytic on a neighbourhood of the origin.

Suppose now that $0 < \alpha < \beta < p_c$. We shall show that K is uniformly convergent on a region of the complex plane containing $[\alpha, \beta]$ in its interior. Suppose $p \in [\alpha, \beta]$. If $|z - p| < \delta$ and δ is small then

$$|a_{nmb}z^m(1-z)^b| \le a_{nmb}p^mq^b\left(\frac{p+\delta}{p}\right)^m\left(\frac{q+\delta}{q}\right)^b \tag{5.115}$$

$$\le a_{nmb}p^mq^bc(p, \delta)^{dn}$$

by (5.111), where

$$c(p, \delta) = \frac{p+\delta}{p}\left(\frac{q+\delta}{q}\right)^2. \tag{5.116}$$

Hence, for fixed n,

$$\sum_{m,b} n|a_{nmb}z^m(1-z)^b| \le nc(p, \delta)^{dn}\sum_{m,b} a_{nmb}p^mq^b$$

$$= nc(p, \delta)^{dn}P_p(|C| = n)$$

$$\le nc(p, \delta)^{dn}P_p(|C| \ge n).$$

The right-hand side here is no larger than

$$nc(p, \delta)^{dn}P_\beta(|C| \ge n) \le nc(p, \delta)^{dn}e^{-n\lambda(\beta)}$$

for some $\lambda(\beta) > 0$, by Theorem (5.75), giving that

$$\sum_{m,b} n|a_{nmb}z^m(1-z)^b| \le n\{c(p, \delta)^d e^{-\lambda(\beta)}\}^n.$$

There exists a positive value of δ which is sufficiently small to ensure that $c(p, \delta)^d e^{-\lambda(\beta)} < 1 - \delta$ for all $p \in [\alpha, \beta]$, giving from (5.113) that K converges uniformly on a region of the complex plane containing $[\alpha, \beta]$, so that K is analytic on $[0, p_c)$ as required. ■

5.5 Notes

Section 5.1. Theorem (5.1) was proved by Hammersley (1957a), who did not have the advantage of the BK inequality. Inequality (5.8) appears in van den Berg and Kesten (1985). Theorem (5.10) was proved by Grimmett (1981, unpublished); see also Kesten (1987e). Related inequalities for two dimensions appear in Grimmett (1981a). The proof of Theorem (5.14) comes mostly from Grimmett (1983), the principal exception being the proof of the strict monotonicity of φ on $(0, p_c)$. Van den Berg (1983, unpublished) pointed out that the ad hoc two-dimensional arguments of Grimmett (1983) may be replaced by the more general argument from reliability theory. Chayes and Chayes (1986a, b) have dealt with similar material.

Section 5.2. Excepting the lower bound for $\tau_p(0, e_n)$, Theorem (5.44) is in the folklore of percolation theory. Proposition (5.49) comes directly from Aizenman and Newman (1984); see also Durrett (1985a). For the origins of correlation length, see Chayes, Chayes, and Fröhlich (1985) and Chayes and Chayes (1986a, b). Inequalities such as (5.7) and (5.68) are sometimes named after Simon (1980) and Lieb (1980); similar inequalities have been in use since Kesten (1980b) at least.

Section 5.3. Theorem (5.75) was proved first by Kesten (1981) by a 'block' approach; previously the exponential decay of the size of C was known only in a 'Peierls regime' when p is sufficiently close to 0. Our proof is a re-working of the argument of Aizenman and Newman (1984), using the BK inequality. The (near-) supermultiplicativity of $P_p(|C| = n)$ was observed by Kunz and Souillard (1978). Theorem (5.78) and its ramifications may be used (see Grimmett (1985b, 1987a)) to estimate the size of the largest open cluster, longest open path, and longest open circuit of $B(n)$ for large n.

Section 5.4. The analyticity of κ was explored by Kunz and Souillard (1978) when p is sufficiently small. Grimmett (1981b) showed that κ is infinitely differentiable for $0 < p < p_c$ when $d = 2$, and Kesten (1981) proved analyticity for all d. All such results apply to χ also.

6

The Supercritical Phase

6.1 Introduction

In this chapter we shall consider bond percolation on $\mathbb{L}^d$ where $d \geq 2$, and we shall suppose that $p > p_c$. In this case, there exists at least one infinite open cluster, and the first main result of the chapter is that there exists a unique such cluster.

In the subcritical phase we were interested in such quantities as $P_p(|C(0)| \geq n)$ and $P_p(0 \leftrightarrow x)$, and particularly in their asymptotic behaviour for large n and $|x|$. Such quantities are of less interest in the supercritical phase, since their asymptotic behaviour is dominated to the first order by the probabilities that the origin is in the infinite open cluster, or that both the origin and x are in this cluster, respectively. Of greater interest are the corresponding probabilities defined in terms of *finite* open clusters. That is to say, we shall study the asymptotic behaviour of $P_p(n \leq |C(0)| < \infty)$ and $P_p(0 \leftrightarrow x, |C(0)| < \infty)$.

Here is a guide to the chapter. We begin by proving that, if there exists an infinite open cluster, then there is exactly one such cluster. As a consequence of this, the percolation probability $\theta(p)$ is a continuous function of p on $(p_c, 1]$. There follow later two sections devoted respectively to the exponential decay of (a) the tail of the radius of a finite open cluster, and (b) the truncated connectivity function $\tau_p^f(0, x) = P_p(0 \leftrightarrow x, |C| < \infty)$ when x is distant from the origin; we prove that these quantities decay *strictly* exponentially (that is, with *strictly* positive rate) at least when p exceeds p_c^+, the critical probability of the half-space $\{x \in \mathbb{Z}^d: x_1 \geq 0\}$. This is followed by a proof that the cluster size distribution decays slower than exponentially in the supercritical phase. In Section 6.8 we prove that the percolation probability $\theta(p)$, the mean size $\chi^f(p)$ of the finite open cluster at the origin, and the number $\kappa(p)$ of open clusters per vertex are infinitely differentiable functions of p when $p > p_c^+$. It is conjectured that $p_c^+ = p_c$, and it is a major open problem to prove or disprove this. Finally, we reach some simple conclusions about the geometry of the infinite open cluster.

6.2 Uniqueness of the Infinite Open Cluster

The principal result of this section is the following: for any value of p for which there is positive probability of an infinite open cluster, there exists almost surely exactly one infinite open cluster.

(6.1) Theorem. Uniqueness of the infinite open cluster. *If p is such that $\theta(p) > 0$ then*

$$P_p(\text{there is exactly one infinite open cluster}) = 1.$$

In proving the uniqueness of the infinite open cluster, we shall make use of the following lemma. Let I_{uv} be the event that the vertices u and v are in different infinite open clusters of $\mathbb{L}^d$.

(6.2) Lemma. *Suppose that $\theta(p) > 0$, and let x and y be neighbours in $\mathbb{L}^d$. If*

$$P_p(I_{xy}) = 0, \tag{6.3}$$

then

$$P_p(\text{there is a unique infinite open cluster}) = 1. \tag{6.4}$$

Clearly

$$P_p(\text{there are two or more infinite open clusters}) \le \sum_{u,v \in \mathbb{Z}^d} P_p(I_{uv}), \tag{6.5}$$

so that the infinite open cluster is almost surely unique whenever $P_p(I_{uv}) = 0$ for all pairs u, v of distinct vertices; by the lemma, it is sufficient to check this condition for a pair u, v of neighbours.

Proof of Lemma (6.2). Of the several ways of proving this, the following is possibly the simplest. It is trivial that the infinite open cluster is unique almost surely if $p = 1$, and so we assume that $0 < p < 1$. Let x and y be two neighbours of $\mathbb{L}^d$. For any box B containing x and y in its interior, and any pair u, v of distinct vertices in the surface ∂B of B, we denote by $I_{uv}(B)$ and $J_{uvxy}(B)$ the two events illustrated in Figure 6.1. More precisely, $I_{uv}(B)$ is the event that u and v are in different infinite open clusters of the graph obtained from $\mathbb{L}^d$ by declaring every edge in B to be closed, and $J_{uvxy}(B)$ is the event that every edge of B is closed

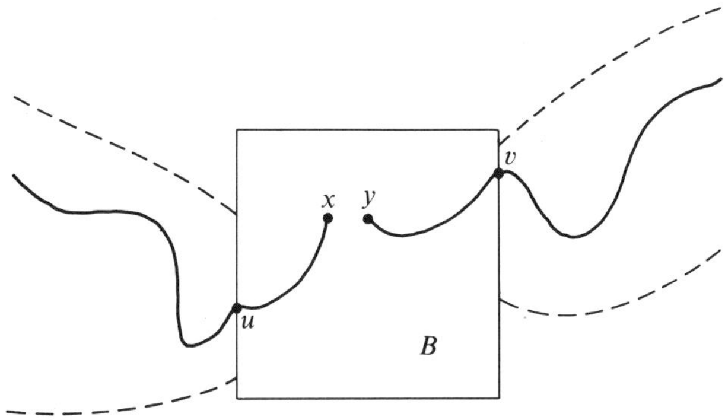

Figure 6.1. Representations of the events (i) $I_{uv}(B)$, that u and v are in disjoint infinite open clusters of $\mathbb{L}^d$ when every edge in B is closed, and (ii) $J_{uvxy}(B)$, that all edges in B are closed except for those edges in two disjoint paths, the first joining u to x and the second joining v to y.

except for two disjoint open paths joining u to x and v to y, respectively. We note that $I_{uv}(B)$ and $J_{uvxy}(B)$ are independent events since they are defined in terms of disjoint sets of edges. Furthermore x and y are in different infinite open clusters if both $I_{uv}(B)$ and $J_{uvxy}(B)$ occur, so that

$$P_p(I_{xy}) \geq P_p(I_{uv}(B))P_p(J_{uvxy}(B)). \tag{6.6}$$

However, $P_p(J_{uvxy}(B)) > 0$ since the event in question depends on only finitely many edges, and thus $P_p(I_{uv}(B)) = 0$ for all suitable u, v, and B whenever $P_p(I_{xy}) = 0$. If there exist two or more infinite open clusters then $I_{uv}(B)$ occurs for some u, v, B, giving that

$$P_p(\text{there are two or more infinite open clusters}) \leq \sum_{\substack{u,v \in \partial B \\ B}} P_p(I_{uv}(B)) = 0 \tag{6.7}$$

if $P_p(I_{xy}) = 0$, where the sum is over all appropriate u, v, and B. We have proved the lemma. ∎

Proof of Theorem (6.1). We follow Gandolfi, Grimmett, and Russo (1988), and we suppose henceforth that $0 < p < 1$. We fix a positive integer n; later we shall take the limit as $n \to \infty$. As usual, $B(n)$ denotes the box with side-length $2n$ and centre at the origin; we think of $B(n)$ as being a subgraph of $\mathbb{L}^d$ endowed with all edges of the lattice which join vertices of $B(n)$. Let $f = \langle x, y \rangle$ be an edge of $B(n)$. It is our target to show that there is probability 0 that x and y are in distinct infinite open clusters; we shall approximate to this event by the event that x and y are in different open clusters of $B(n)$ both of which intersect the surface $\partial B(n)$

of $B(n)$, and then we shall take the limit as $n \to \infty$ to obtain the desired conclusion. The conclusion of the theorem will then follow by an appeal to Lemma (6.2).

Let $\mathscr{C}_n$ be the collection of open clusters of $B(n)$ which contain vertices in its surface $\partial B(n)$. We shall consider the following random sets of edges of $B(n)$:

(6.8) $$F_n = \bigcup_{D \in \mathscr{C}_n} D_e,$$

(6.9) $$G_n = \bigcup_{D \in \mathscr{C}_n} \Delta D,$$

(6.10) $$H_n = \bigcup_{\substack{D_1, D_2 \in \mathscr{C}_n \\ D_1 \neq D_2}} \{\Delta D_1 \cap \Delta D_2\},$$

where J_e is the set of edges and ΔJ is the edge boundary of a subgraph J of $B(n)$. Thus F_n is the set of edges of $B(n)$ which are contained in clusters in $\mathscr{C}_n$, G_n is the set of edges in the edge boundaries of such clusters, and H_n is the set of edges which are in the edge boundaries of two different such clusters; see Figure 6.2. The set H_n contains exactly those closed edges of $B(n)$ whose endvertices are in distinct open clusters of $B(n)$ both of which intersect the surface of $B(n)$. We shall be interested in the cardinality of H_n relative to the number of edges in $B(n)$, and the following lemma is not difficult to prove.

(6.11) Lemma. *Let x and y be neighbours in $\mathbb{L}^d$. Then*

(6.12) $$\lim_{n \to \infty} \left\{ \frac{1}{|B(n)_e|} E_p |H_n| \right\} = P_p(I_{xy}).$$

We shall not use the full conclusion of this lemma, but only (6.12) with '$=$' replaced by '$\geq$'.

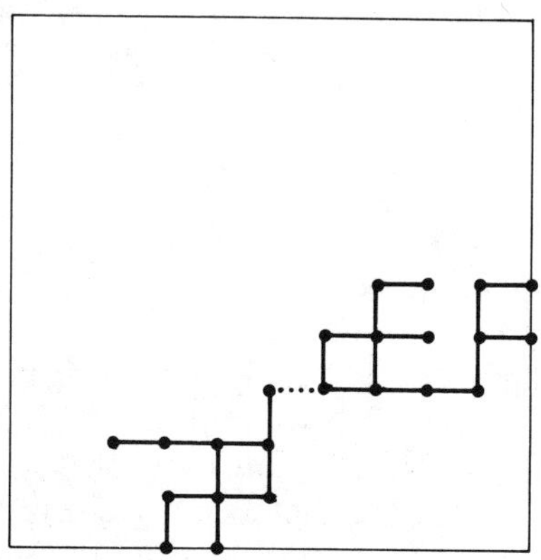

Figure 6.2. The dotted edge is in the edge boundaries of two disjoint open clusters of $B(n)$ both of which intersect the surface of $B(n)$.

Proof. We bound $E_p|H_n|$ beneath as follows:

$$E_p|H_n| = \sum_{f \in B(n)_e} P_p(f \in H_n) \tag{6.13}$$
$$\geq \sum_{f \in B(n)_e} P_p(\text{the endvertices of } f \text{ are in different infinite open clusters}),$$

where the summations are over all edges f of $B(n)$, and we have used the fact that $f \in H_n$ whenever the two endvertices of f are in different infinite open clusters of $\mathbb{L}^d$. The final summand is constant for all f, so that

$$E_p|H_n| \geq |B(n)_e| P_p(I_{xy}) \qquad \text{as} \quad n \to \infty$$

for any pair x, y of neighbours, and we have proved one half of (6.12).

Slightly more care is required for the other half. Let m be a positive integer satisfying $m < n$ and, for each edge f of $B(n)$, we write $B(m, f)$ for a box with side-length $2m$ and centre at one of the endvertices of f. Let $I_m(f)$ be the event that the endvertices of f are in different open clusters of $B(m, f)$ both of which intersect the surface $\partial B(m, f)$. If $f \in B(n-m)_e$ then $B(m, f) \subseteq B(n)$; for such an edge f, $I_m(f)$ occurs whenever $f \in H_n$, and therefore

$$E_p|H_n| = \sum_{f \in B(n)_e} P_p(f \in H_n) \tag{6.14}$$
$$\leq |B(n)_e \backslash B(n-m)_e| + \sum_{f \in B(n-m)_e} P_p(I_m(f)).$$

However, $|B(n)_e \backslash B(n-m)_e| = o(|B(n)_e|)$ as $n \to \infty$, and $P_p(I_m(f))$ does not depend on the choice of f; thus

$$\limsup_{n\to\infty} \left\{ \frac{1}{|B(n)_e|} E_p|H_n| \right\} \leq P_p(I_m(f)). \tag{6.15}$$

We note that $I_{m+1}(f) \subseteq I_m(f)$ for all m, and

$$\lim_{m\to\infty} I_m(f) = \{\text{the endvertices of } f \text{ are in different infinite open clusters}\}.$$

It follows that, as $m \to \infty$,

$$P_p(I_m(f)) \to P_p(\text{the endvertices of } f \text{ are in different open clusters})$$

and the result follows by taking the limit as $m \to \infty$ in (6.15). ■

Returning to the proof of the theorem, in the light of Lemmas (6.2) and (6.11) it suffices to prove that

$$\limsup_{n\to\infty} \left\{ \frac{1}{|B(n)_e|} E_p|H_n| \right\} \leq 0, \tag{6.16}$$

and the remainder of the proof is devoted to demonstrating this. From the definition of G_n and H_n, we have that

$$|H_n| = \left\{ \sum_{D \in \mathscr{C}_n} |\Delta D| \right\} - |G_n|. \tag{6.17}$$

On the other hand, it is the case that

$$P_p(f \in G_n \mid f \in F_n \cup G_n) = 1 - p; \tag{6.18}$$

to see this, note that the event

$$\{f \in F_n \cup G_n\} = \{\text{at least one endvertex of } f \text{ is joined to } \partial B(n) \text{ by an open path}\}$$

does not depend on the state of f, so that, for edges f in $F_n \cup G_n$, it is the case that f belongs to G_n if and only if f is closed. It follows that

$$\begin{aligned} E_p|G_n| &= \sum_{f \in B(n)_e} P_p(f \in G_n) \\ &= \sum_{f \in B(n)_e} (1-p)P_p(f \in F_n \cup G_n) \\ &= (1-p)\{E_p|F_n| + E_p|G_n|\}. \end{aligned}$$

Solving this equation, we obtain

$$E_p|G_n| = \frac{1}{p}(1-p)E_p|F_n|. \tag{6.19}$$

We substitute this into (6.17) and use the definition (6.8) of F_n to obtain

$$E_p|H_n| = E_p\left(\sum_{D \in \mathscr{C}_n} \left\{|\Delta D| - \frac{1-p}{p}|D_e|\right\}\right). \tag{6.20}$$

In order to estimate the right-hand side of this inequality we require a large-deviation estimate somewhat akin to the conclusion of Theorem (4.20) and equation (4.22). Let $C_n(f)$ be the open cluster of $B(n)$ containing the edge f; if f is closed, we set $C_n(f) = \varnothing$. We write

$$h(C_n(f)) = \frac{1}{1-p}|\Delta C_n(f)| - \frac{1}{p}|C_n(f)_e| \tag{6.21}$$

and

$$\bar{C}_n(f) = \Delta C_n(f) \cup C_n(f)_e, \tag{6.22}$$

the set of edges with at least one endvertex in $C_n(f)$. Suppose that we are told that $|\bar{C}_n(f)| = k$ for some k. We may think of $C_n(f)$ as being built by the process of examining k edges in turn and observing whether they are open or closed. If we score $(1-p)^{-1}$ for each such edge which is closed, and $-p^{-1}$ for each such edge which is open, then our total score is $h(C_n(f))$. Assuming the scores are independent (which they are not), then $h(C_n(f))$ is the sum of k independent and identically distributed random variables with zero means, so that the chance that $h(C_n(f))$ is larger than $k\varepsilon$ should decay exponentially as $k \to \infty$. A more formal statement and proof follow.

(6.23) Lemma. *There exists a function $H(p)$ which is strictly positive on $(0, 1)$ such that*

$$P_p(h(C_n(f)) \geq k\varepsilon, |\bar{C}_n(f)| = k) \leq e^{-k\varepsilon^2 H(p)} \tag{6.24}$$

for all $\varepsilon > 0$, n, $k \geq 1$, and for all edges f of $B(n)$.

Proof. Our proof of this lemma proceeds by a direct calculation, but we precede this with a justification of the intuitive approach described above. We may think of the process of building the open cluster $C_n(f)$, containing a specified edge f, as the progressive observation of the states of a sequence of edges. First we observe the state of f. If f is open, we next observe those edges which share an endvertex with f, and so on. At each stage we observe the state of a *new* edge, so that we proceed by a sequence of independent trials. This process terminates if and only if $|C_n(f)| < \infty$. The event $\{|\triangle C_n(f)| = b, |C_n(f)_e| = m\}$ occurs if and only if:

(i) in the first $b + m$ independent trials, there are exactly m successes (corresponding to open edges); and
(ii) the process stops after $b + m$ trials, owing to the geometry of the construction so far.

Neglecting the latter condition, we obtain that

$$\begin{aligned} P_p(|\triangle C_n(f)| = b, |C_n(f)_e| = m) &\leq P(m \text{ successes in } b + m \text{ trials}) \\ &= \binom{b+m}{m} p^m (1-p)^b. \end{aligned} \tag{6.25}$$

From this inequality, (6.24) may be derived via Markov's inequality as in the classical theory of large deviations. An alternative route to (6.25) is to count lattice animals as in (4.26).

We follow Aizenman, Kesten, and Newman (1987). Suppose that n, $k \geq 1$, $\varepsilon > 0$, and $0 < p = 1 - q < 1$. Let $a_{mb}(f)$ be the number of connected subgraphs of $B(n)$ containing m edges and with b edges in their edge boundary, and containing the edge f. For any $r \geq 0$, we have by Markov's inequality that

$$P_p(h(C_n(f)) \geq k\varepsilon, |\bar{C}_n(f)| = k) \leq e^{-\varepsilon k r} E_p(e^{rh(C_n(f))}; |\bar{C}_n(f)| = k). \tag{6.26}$$

However,

$$\begin{aligned} E_p(e^{rh(C_n(f))}; |\bar{C}_n(f)| = k) &= \sum_{m,b: m+b=k} e^{brq^{-1} - mrp^{-1}} a_{mb}(f) p^m q^b \\ &= (qe^{rq^{-1}} + pe^{-rp^{-1}})^k \sum_{m,b: m+b=k} a_{mb}(f) \pi^m (1-\pi)^b, \end{aligned}$$

where

$$\pi = \frac{pe^{-rp^{-1}}}{qe^{rq^{-1}} + pe^{-rp^{-1}}}.$$

Thus

$$P_p(h(C_n(f)) \geq k\varepsilon, |\bar{C}_n(f)| = k) \leq \{e^{-\varepsilon r} f(p, r)\}^k P_\pi(|\bar{C}_n(f)| = k) \leq \{e^{-\varepsilon r} f(p, r)\}^k \tag{6.27}$$

by (6.26), where

$$f(p, r) = qe^{rq^{-1}} + pe^{-rp^{-1}}.$$

Now

$$f(p, r) = 1 + O(r^2) \qquad \text{as} \quad r \downarrow 0,$$

giving that there exists $a(p)$ satisfying $0 < a(p) < \infty$ such that

$$f(p, r) \leq e^{a(p)r^2} \qquad \text{for all} \quad r > 0.$$

We insert this into (6.27) to obtain

$$P_p(h(C_n(f)) \geq k\varepsilon, |\bar{C}_n(f)| = k) \leq e^{k\{-\varepsilon r + a(p)r^2\}}$$

for $r > 0$. Setting $r = \frac{1}{2}\varepsilon/a(p)$ we deduce (6.24) with $H(p) = \frac{1}{4}a(p)^{-1}$. ∎

Finally, we show that (6.20) and Lemma (6.23) imply (6.16). We fix ε and δ such that $\varepsilon > 0$ and $0 < \delta < 1$, and we write

$$\mathscr{C}_n' = \{D \in \mathscr{C}_n : |\bar{D}| \geq n^\delta\}, \tag{6.28}$$

the set of open clusters D of $B(n)$ which intersect $\partial B(n)$ and satisfy $|\bar{D}| \geq n^\delta$. Let A_n be the event that $h(D) < \varepsilon|\bar{D}|$ for all $D \in \mathscr{C}_n'$. From Lemma (6.23),

$$1 - P_p(A_n) \leq |B(n)_e| \sum_{k=n^\delta}^{\infty} e^{-k\varepsilon^2 H(p)} \to 0 \qquad \text{as} \quad n \to \infty, \tag{6.29}$$

since, if A_n does not occur, then there exists an edge f of $B(n)$ for which $h(C_n(f)) \geq \varepsilon|\bar{C}_n(f)|$ and $|\bar{C}_n(f)| \geq n^\delta$. Hence

$$E_p\left(\sum_{D \in \mathscr{C}_n'} \left\{\frac{1}{1-p}|\Delta D| - \frac{1}{p}|D_e|\right\}\right) \leq E_p\left(\sum_{D \in \mathscr{C}_n'} \varepsilon|\bar{D}|\right) + \frac{2}{1-p}|B(n)_e|(1 - P_p(A_n)), \tag{6.30}$$

since the left summation is bounded above by $2|B(n)_e|/(1-p)$. However,

$$\sum_{D \in \mathscr{C}_n'} |\bar{D}| \leq 2|B(n)_e|; \tag{6.31}$$

we substitute this into (6.30), divide by $|B(n)_e|$, and take the limits as $n \to \infty$ and $\varepsilon \downarrow 0$, obtaining

$$\limsup_{n \to \infty} \left[\frac{1}{|B(n)_e|} E_p\left(\sum_{D \in \mathscr{C}_n'} \left\{|\Delta D| - \frac{1-p}{p}|D_e|\right\}\right)\right] \leq 0. \tag{6.32}$$

Finally, any cluster in $\mathscr{C}_n \setminus \mathscr{C}'_n$ is contained in $B(n) \setminus B(n - n^\delta)$, giving that

$$\begin{aligned}(6.33) \quad \frac{1}{|B(n)_e|} E_p\left(\sum_{D \in \mathscr{C}_n \setminus \mathscr{C}'_n} \left\{ |\Delta D| - \frac{1-p}{p} |D_e| \right\} \right) &\leq 2 \frac{|B(n)_e \setminus B(n - n^\delta)_e|}{|B(n)_e|} \\ &= 2\{1 - (1 + o(1))(1 - n^{\delta - 1})^d\} \\ &\to 0 \quad \text{as} \quad n \to \infty\end{aligned}$$

since $\delta < 1$. We combine this with (6.32) to deduce by (6.20) that

$$(6.34) \qquad \limsup_{n \to \infty} \left\{ \frac{1}{|B(n)_e|} E_p |H_n| \right\} \leq 0$$

as required. ∎

6.3 Continuity of the Percolation Probability

Certainly the percolation probability is continuous when $p < p_c$, since $\theta(p) = 0$ in this case. It is a consequence of the uniqueness of the infinite open cluster that θ is continuous when $p > p_c$ also.

(6.35) Theorem. Continuity of θ in the supercritical phase. *The percolation probability $\theta(p)$ is a continuous function of p on $(p_c, 1]$.*

It is an open question whether or not θ is continuous *at* p_c. It is not difficult to show that θ is continuous from the right on the whole interval $[0, 1]$, and thus $\theta(p_c) = \lim_{p \downarrow p_c} \theta(p)$. On the other hand, $\theta(p) = 0$ for $p < p_c$, giving that θ is continuous at p_c if and only if $\theta(p_c) = 0$, which is to say that all open clusters are finite almost surely when $p = p_c$. It is believed that $\theta(p_c) = 0$ for all $d \geq 2$, but this is known to hold in the case of two dimensions only (see the remarks after Lemma (9.12)). In this direction, it is known that the percolation probability is continuous *at* the critical point for percolation on a half-space—see Section 6.4. In Section 6.8 we shall see that θ is infinitely differentiable for large enough values of p; it is believed that this holds whenever $p > p_c$.

We prove the above theorem in two parts.

(6.36) Lemma. *The percolation probability function $\theta(p)$ is continuous from the right on* $[0, 1]$.

(6.37) Lemma. *The percolation probability function $\theta(p)$ is continuous from the left on* $(p_c, 1]$.

Proof of Lemma (6.36). We follow Russo (1978), using an argument previewed in the proof of Theorem (3.8). Clearly

$$\theta(p) = \lim_{n\to\infty} P_p(0 \leftrightarrow \partial B(n)).$$

However, $P_p(0 \leftrightarrow \partial B(n))$ is a continuous function of p, since the event $\{0 \leftrightarrow \partial B(n)\}$ depends only on the edges in $B(n)$; moreover,

$$P_p(0 \leftrightarrow \partial B(n)) \geq P_p(0 \leftrightarrow \partial B(n+1)) \qquad \text{for all } n.$$

Thus θ is the decreasing limit of continuous functions, and therefore θ is upper semi-continuous. On the other hand, θ is monotonic non-decreasing, and thus θ is continuous from the right. ∎

Before moving to the proof of Lemma (6.37), we note that the above argument may be used to establish the continuity in p of the connectivity function $\tau_p(x, y)$. Let

$$\tau_p(x, y; B(n)) = P_p(x \leftrightarrow y \text{ in } B(n))$$

be the probability that x and y are joined by an open path of $B(n)$. It is clear that

$$\tau_p(x, y; B(n)) \leq \tau_p(x, y; B(n+1)),$$

so that

$$\tau_p(x, y; B(n)) \uparrow \tau_p(x, y) \qquad \text{as} \quad n \to \infty.$$

Now $\tau_p(x, y; B(n))$ is a continuous function of p, and therefore $\tau_p(x, y)$ is lower semi-continuous. For the upper bound on $\tau_p(x, y)$ we use

$$\tau_p^+(x, y; B(n)) = P_p(\{x \leftrightarrow y \text{ in } B(n)\} \cup \{x \leftrightarrow \partial B(n) \text{ and } y \leftrightarrow \partial B(n)\}), \tag{6.38}$$

the probability that either x and y are in the same open cluster of $B(n)$, or both are joined by open paths to the surface $\partial B(n)$ of $B(n)$. It is easy to see that

$$\tau_p^+(x, y; B(n)) \geq \tau_p^+(x, y; B(n+1))$$

and

$$\tau_p^+(x, y; B(n)) \to \tau_p^+(x, y) \qquad \text{as} \quad n \to \infty, \tag{6.39}$$

where

$$\tau_p^+(x, y) = P_p(\text{either } x \leftrightarrow y, \text{ or } x \text{ and } y \text{ are in different infinite open clusters}). \tag{6.40}$$

There is probability 0 of two or more infinite open clusters, so that $\tau_p^+(x, y) = \tau_p(x, y)$, giving that $\tau_p(x, y)$ is the decreasing limit of $\tau_p^+(x, y; B(n))$ as $n \to \infty$. The latter quantities are continuous functions of p, and thus $\tau_p(x, y)$ is upper semi-continuous. Hence $\tau_p(x, y)$ is a continuous function of p. The same argument is valid for the m-point connectivity function $\tau_p(x_1, x_2, \ldots, x_m)$.

Proof of Lemma (6.37). It is here that we use the fact that when $p > p_c$ there exists a unique infinite open cluster. We use the argument of van den Berg and Keane (1984). Let $(X(e): e \in \mathbb{E}^d)$ be a collection of independent random variables indexed by the set of edges of $\mathbb{L}^d$, each having the uniform distribution on $[0, 1]$. For $0 \le p \le 1$, we define $\eta_p(e)$ to be 1 if $X(e) < p$ and 0 otherwise; we call an edge e *p-open* if $\eta_p(e) = 1$ and *p-closed* otherwise. We write C_p for the p-open cluster of $\mathbb{L}^d$ containing the origin, and note that $C_\pi \subseteq C_p$ if $\pi \le p$. Now,

$$\theta(p) = P(|C_p| = \infty) \tag{6.41}$$

and

$$\lim_{\pi \uparrow p} \theta(\pi) = \lim_{\pi \uparrow p} P(|C_\pi| = \infty) = P(|C_\pi| = \infty \text{ for some } \pi < p), \tag{6.42}$$

since the events $\{|C_\pi| = \infty\}$ are increasing in π. Bearing in mind the monotonicity of θ, we are required to prove that

$$P(|C_p| = \infty, |C_\pi| < \infty \text{ for all } \pi < p) = 0 \qquad \text{for} \quad p > p_c. \tag{6.43}$$

Let $p > p_c$, and suppose that $|C_p| = \infty$. If $p_c < \alpha < p$, there exists (almost surely) an infinite α-open cluster I_α, and furthermore I_α is almost surely a subgraph of C_p, since otherwise there would exist at least two infinite p-open clusters. It follows that there exists a p-open path l joining the origin to some vertex of I_α. Such a path l has finite length and each edge e in l satisfies $X(e) < p$; therefore $\mu = \max\{X(e): e \in l\}$ satisfies $\mu < p$. If π satisfies $\pi \ge \alpha$ and $\mu < \pi < p$ then there exists a π-open path joining the origin to some vertex of I_α, so that $|C_\pi| = \infty$. However $\pi < p$, implying that (6.43) holds as required. ∎

6.4 Percolation on Half-Spaces

When $p > p_c$ there exists a unique infinite open cluster, and each vertex has strictly positive probability $\theta(p)$ of belonging to this cluster. What is the shape of the open cluster at the origin if (i) it is the infinite cluster, or (ii) it is a finite cluster? There are gaps in the present state of knowledge in this regard. For example, it is believed that the probability that the origin is in a finite cluster of size n decays (logarithmically) in the manner of $\exp(-\gamma(p)n^{(d-1)/d})$ as $n \to \infty$; no proof of this is known. Certain results of value are known for values of p exceeding p_c^+, the critical probability of bond percolation on the half-space $H = \{x \in \mathbb{Z}^d: x_1 \ge 0\}$. Certainly $p_c^+ \ge p_c$ since $H \subset \mathbb{L}^d$, and it is believed that $p_c^+ = p_c$. This section is devoted to a discussion of this conjecture together with certain related results.

We begin with some notation. Let G be an infinite connected subgraph of $\mathbb{L}^d$ containing the origin, and let us consider bond percolation on G. For a vertex x of G, we define $\theta(p, x; G)$ to be the probability that x is in an infinite open cluster of G when each edge is open with probability p. We shall be interested in the critical probability of G, and to this end we note that, for given p, either $\theta(p, x; G) = 0$ for all x, or $\theta(p, x; G) > 0$ for all x. To see this, we use the FKG inequality to show that, for vertices x and y,

$$\theta(p, x; G) \geq P_p(\text{in } G,\ x \leftrightarrow y \text{ and } y \text{ is in an infinite open cluster})$$
$$\geq P_p(x \leftrightarrow y \text{ in } G)\theta(p, y; G).$$

We may therefore concentrate on the event that the origin is in an infinite open cluster of G. The critical probability $p_c(G)$ is defined by

$$p_c(G) = \sup\{p: \theta(p; G) = 0\},$$

where $\theta(p; G) = \theta(p, 0; G)$. It is not difficult to adapt earlier proofs to see that there exists almost surely no infinite open cluster in G when $p < p_c(G)$, and there exists almost surely an infinite open cluster when $p > p_c(G)$.

Let us suppose for the moment that $d \geq 3$. Of particular interest are those subgraphs of $\mathbb{L}^d$ induced by the following sets of vertices:

$$\text{the 'half-space'} \quad H = \mathbb{Z}_+ \times \mathbb{Z}^{d-1},$$
$$\text{the 'slab'} \quad S_k = \{0, 1, 2, \dots, k-1\} \times \mathbb{Z}^{d-1},$$
$$\text{the 'slice'} \quad T_k = \{0, 1, 2, \dots, k-1\}^{d-2} \times \mathbb{Z}_+^2,$$

for $k \geq 1$, where $\mathbb{Z}_+ = \{0, 1, 2, \dots\}$. Clearly $T_k \subset S_k \subset H \subset \mathbb{L}^d$, so that

$$p_c(T_k) \geq p_c(S_k) \geq p_c^+ \geq p_c, \tag{6.44}$$

where p_c^+ is defined to be $p_c(H)$. Furthermore $T_k \subset T_{k+1}$, and thus the limit

$$p_c(T) = \lim_{k \to \infty} p_c(T_k)$$

exists. Clearly $p_c(T) \geq p_c$, and it is believed that equality is valid here. It is currently known only that $p_c(T) = p_c^+$.

(6.45) Theorem. *The critical probability p_c^+ of the half-space equals the limit $p_c(T)$ of the critical probabilities of slices.*

It remains to show that $p_c = p_c^+$. As a consequence of (6.44) and this theorem, we have that the limit

$$p_c(S) = \lim_{k \to \infty} p_c(S_k)$$

of the critical probabilities of slabs satisfies $p_c(S) = p_c^+$.

No proof of Theorem (6.45) is presented here, although the method is sketched at the end of this section. One obtains similarly the continuity of the percolation probability for the half-space.

(6.46) Theorem. Continuity of the percolation probability on half-spaces. *For bond percolation on the half-space H, the percolation probability $\theta(p; H)$ is a continuous function of p at the critical point p_c^+, which is to say that $\theta(p_c^+; H) = 0$.*

When combined with an argument of Aizenman, Chayes, Chayes, Fröhlich, and Russo (1983), the two results above imply immediately the (almost sure) uniqueness of the infinite open cluster when $p > p_c^+$, on any subspace of $\mathbb{L}^d$ of the form $\mathbb{Z}_+^e \times \mathbb{Z}^{d-e}$ for $1 \le e \le d$. Thus, for the half-space and substantial subgraphs thereof, one obtains three observations of substance, only one of which is currently known for the whole lattice: the uniqueness of the infinite open cluster, the equality of critical probabilities, and the absence of an infinite open cluster *at* the critical point.

It is extremely plausible that $p_c^+ = p_c$, and the following observation is convincing. If p were to satisfy $p_c < p < p_c^+$ then there would exist a unique infinite open cluster of the whole lattice whose intersection with *any* half-space is (almost surely) the union of only finite clusters.

Here is a sketch of the proofs of Theorems (6.45) and (6.46). Suppose that p is such that $\theta(p; H) > 0$. For $\varepsilon > 0$ and any large integer N, it is possible to find a large box of the form $B_{RS} = \{x \in \mathbb{Z}^d : 0 \le x_1 \le R, |x_i| \le S \text{ for } 2 \le i \le d\}$ such that (with probability exceeding $1 - \varepsilon$) some vertex of the hyperplane $x_1 = 0$ which is near to the origin is joined by open paths of B_{RS} to at least N vertices on each of the other faces of B_{RS}. We may now use these new vertices as starting points for similar open paths contained in copies of B_{RS} positioned so as not to overlap B_{RS} itself. Continuing in like manner, it is possible to build (with positive probability) an infinite open cluster in a copy of the slice T_k, for sufficiently large values of k. Thus $\theta(p; T_k) > 0$ for all large k, implying that $p \ge p_c(T_k)$ and therefore $p_c^+ \ge \lim_{k\to\infty} p_c(T_k)$ as required for Theorem (6.45). Furthermore, if $\theta(p_c^+; H) > 0$ then the above construction of an infinite open cluster succeeds with positive probability for values of p slightly smaller than p_c^+ (the box B_{RS} is finite, so that if $P_{p_c^+}(A) > 1 - \varepsilon$ then $P_p(A) > 1 - \varepsilon$ for p slightly smaller than p_c^+, for any event A defined in terms of the states of edges inside B_{RS}). This contradicts the definition of p_c^+, and the assertion of Theorem (6.46) follows. See Barsky, Grimmett, and Newman (1988) for more details.

The picture is much clearer if $d = 2$, since the technique of planar duality is available in this case. We shall see in Section 9.5 results which reach considerably further than the mere statement that $p_c^+ = p_c$ for $\mathbb{L}^2$. Furthermore, the percolation probability $\theta(p)$ is continuous at the critical point. Nothing is gained in two dimensions by studying slabs or slices; after all, such slabs are essentially one-dimensional sets on which no interesting critical phenomena can occur.

6.5 The Radius of a Finite Open Cluster

When $p < p_c$, the probability of an open path from the origin to a vertex on the surface $\partial B(n)$ of the box $B(n)$ decays exponentially as $n \to \infty$. This is of course false when $p > p_c$, since this probability is then at least as large as the percolation probability $\theta(p)$. We therefore turn our attention to the probability that there is an open path from the origin to $\partial B(n)$ *not lying in the infinite open cluster.* We may expect this probability to decay exponentially as $n \to \infty$, in that there exists $\sigma(p)$ such that

(6.47) $$P_p(0 \leftrightarrow \partial B(n), |C| < \infty) \approx e^{-n\sigma(p)} \quad \text{as} \quad n \to \infty$$

(remember that $a_n \approx b_n$ means $\log a_n / \log b_n \to 1$). We may expect further that $\sigma(p) > 0$ when $p > p_c$. The next theorem establishes the existence of the limit required for (6.47).

(6.48) Theorem. Asymptotic tail behaviour of the radius of a finite open cluster. *Suppose that* $0 < p < 1$. *The limit*

(6.49) $$\sigma(p) = \lim_{n \to \infty} \left\{ -\frac{1}{n} \log P_p(0 \leftrightarrow \partial B(n), |C| < \infty) \right\}$$

exists and satisfies $\sigma(p) < \infty$. *Furthermore, there exists a constant* $A(p, d)$ *which is finite for* $d \geq 2$, $0 < p < 1$, *such that*

(6.50) $$P_p(0 \leftrightarrow \partial B(n), |C| < \infty) \leq A(p, d) n^d e^{-n\sigma(p)} \qquad \textit{for all } n.$$

This theorem is proved using subadditivity in a way not dissimilar to the proof of Lemma (5.102), although there are certain extra complications owing to the condition that the open cluster at the origin be finite. The main question is, of course, to determine those values of p for which it is the case that $\sigma(p)$ is strictly positive, since only then does the tail of the radius distribution decay exponentially. It is conjectured that $\sigma(p) > 0$ whenever $p_c < p < 1$, and it is a major open problem to prove this. At present we know only that $\sigma(p) > 0$ when p exceeds p_c^+, the critical probability of the half-space.

(6.51) Theorem. *Suppose that* $0 < p < 1$. *Then*

(6.52) $$\sigma(p) = \lim_{n \to \infty} \left\{ -\frac{1}{n} \log P_p(0 \leftrightarrow \partial B(n), |C| < \infty) \right\}$$

satisfies $\sigma(p) > 0$ *if* $p > p_c^+$.

It is conjectured that $p_c^+ = p_c$ if $d \geq 3$; the corresponding result is known for two dimensions. We shall prove Theorem (6.51) under the assumption that $d \geq 3$, returning in Theorem (9.23) to the case $d = 2$.

One important technique in the proof of Theorem (6.51) is a method for constructing the open cluster C at the origin in a step by step manner, and we do this as follows. First we order the edges of $\mathbb{L}^d$ in some arbitrary but deterministic way, and we write e_i for the ith edge in this ordering. We shall construct an increasing sequence $C_1, C_2, \ldots$ of (random) subgraphs of $\mathbb{L}^d$. We define $C_1 = \{0\}$, the graph containing the origin only. Having found C_m, we let C_{m+1} be the graph obtained by adding to C_m the earliest open edge which lies in the edge boundary of C_m, if such an edge exists. That is to say, we define

$$C_{m+1} = C_m \cup \{e_j\}, \tag{6.53}$$

where $j = \min\{i: e_i \text{ is open, } e_i \notin C_m, \text{ but } e_i \text{ is incident to a vertex of } C_m\}$. Thus, we add to C_m the edge e_j together with any endvertex of e_j which does not already belong to C_m. A single edge is added at each stage, until the stage when no more suitable open edges can be found. This occurs when we have exhausted the entire open cluster C at the origin. If $C_m = C$ for some C then we define $C_l = C$ for $l \geq m$, so that

$$C = \lim_{m\to\infty} C_m. \tag{6.54}$$

The main property of this construction is that the C_m's have a certain property similar to that of stopping times for a Markov process: for any m and any connected subgraph Σ of $\mathbb{L}^d$ containing the origin, the event $\{C_m = \Sigma\}$ depends only on the states of edges in Σ and its edge boundary, and not on the state of any edge having both endvertices outside Σ. We note that, if x is any vertex of the open cluster C at the origin, then there exists a random integer $m = m(x)$ such that x lies in C_m, regardless of whether C is finite or infinite.

Proof of Theorem (6.48). We follow Chayes, Chayes, Grimmett, Kesten, and Schonmann (1988). First we shall prove the existence of the limit in (6.49), and we begin with some notation. For $i = 1, 2, \ldots, d$, we define

$$R_i = \max\{x_i: x \in C\}, \qquad L_i = \min\{x_i: x \in C\}, \tag{6.55}$$

the maximum and minimum values of the ith coordinates of vertices in the open cluster C at the origin. We write $D_i = R_i - L_i$ for the *width* of C in the ith coordinate direction, and we define the *diameter* of C by

$$\operatorname{diam}(C) = \max\{D_i: 1 \leq i \leq d\}. \tag{6.56}$$

(6.57) Lemma. *For $m, n \geq 0$,*

$$P_p(\operatorname{diam}(C) = m + n + 2) \geq \frac{p^2(1-p)^{2d-2}}{d^2(2n+1)^d} P_p(\operatorname{diam}(C) = m) P_p(\operatorname{diam}(C) = n). \tag{6.58}$$

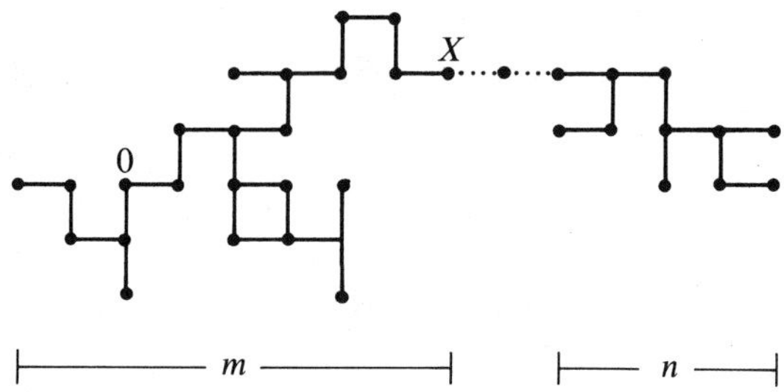

Figure 6.3. Two open clusters having widths m and n in the first coordinate direction may be used to build an open cluster with width $m + n + 2$.

Proof. The idea of the proof is as follows. If we are provided with two finite open clusters whose widths in the first coordinate direction are m and n, then we may position one of them to the right of the other in order to obtain a finite open cluster with width $m + n + 2$, the extra 2 arising out of the geometry of the construction, illustrated in Figure 6.3.

First we note that, if $\operatorname{diam}(C) = k$, then we do not lose a great deal by assuming that the width of C is maximal in the first coordinate direction. Certainly

$$P_p(\operatorname{diam}(C) = k) \geq P_p(D_1 = \operatorname{diam}(C) = k), \tag{6.59}$$

and we claim also that

$$P_p(D_1 = \operatorname{diam}(C) = k) \geq \frac{1}{d} P_p(\operatorname{diam}(C) = k). \tag{6.60}$$

It is the case that $\operatorname{diam}(C) = D_i$ for some $i = 1, 2, \ldots, d$, and therefore, if (6.60) were false, it would follow that

$$\begin{aligned} P_p(\operatorname{diam}(C) = k) &\leq \sum_{i=1}^{d} P_p(D_i = \operatorname{diam}(C) = k) \\ &< P_p(\operatorname{diam}(C) = k), \end{aligned} \tag{6.61}$$

a contradiction.

Suppose that $D_1 = \operatorname{diam}(C) = k$. Then $C \subseteq B(k)$, the box with side-length $2k$ and centre at the origin. The cluster C has at least one *leftmost* vertex (that is, a vertex v with $v_1 \leq w_1$ for all vertices w in C), and it follows that there exists a (non-random) vertex x of $B(k)$ with the property that

$$\begin{aligned} &P_p(x \text{ is a leftmost vertex of } C,\ D_1 = \operatorname{diam}(C) = k) \\ &\qquad \geq \frac{1}{|B(k)|} P_p(D_1 = \operatorname{diam}(C) = k). \end{aligned} \tag{6.62}$$

To see this we argue as before. If (6.62) were false for all $x \in B(k)$ then

$$\begin{aligned} P_p(D_1 = \text{diam}(C) = k) &\leq \sum_{x \in B(k)} P_p(x \text{ is a leftmost vertex of } C,\ D_1 = \text{diam}(C) = k) \\ &< P_p(D_1 = \text{diam}(C) = k), \end{aligned}$$

a contradiction. If $D_1 = \text{diam}(C) = k$ and x is a leftmost vertex of C, then x is a leftmost vertex of an open cluster with diameter k whose width is maximal in the first coordinate direction. The probability of this event is constant for all vertices x, and thus we may take x to be the origin, obtaining that

$$\text{(6.63)} \qquad \begin{aligned} &P_p(D_1 = \text{diam}(C) = k,\ L_1 = 0,\ R_1 = k) \\ &\qquad \geq \frac{1}{|B(k)|} P_p(D_1 = \text{diam}(C) = k). \end{aligned}$$

When combined with (6.60), this implies that

$$\text{(6.64)} \qquad P_p(\text{diam}(C) = k,\ L_1 = 0,\ R_1 = k) \geq \frac{1}{d\,|B(k)|} P_p(\text{diam}(C) = k).$$

We perform next the required geometrical construction, illustrated in Figure 6.3. Let m and n be non-negative integers, and suppose that the open cluster C at the origin is such that $D_1 = \text{diam}(C) = m$; that is, we assume that C has diameter m and the width of C is maximal in the first coordinate direction. We call a vertex x of C *rightmost* if $y_1 \leq x_1$ for all $y \in C$, and we denote by X a rightmost vertex of C; if there is more than one such vertex, we pick that with greatest second coordinate, and if there is more than one of these then we pick the one with greatest third coordinate, and so on. Having found X, we consider the event that the vertex $X + (2, 0, 0, \ldots, 0)$ is a leftmost vertex of the open cluster containing it, and that this open cluster has diameter n and its width is maximal in the first coordinate direction. By (6.64), the probability of this event, conditional on X, is at least $(d\,|B(n)|)^{-1} P_p(\text{diam}(C) = n)$; we have used here the observation that, if we are given that $X = x$, then we have no extra information about the states of edges for which both endvertices z satisfy $z_1 > x_1$. We now change the states of the edges incident to $X + (1, 0, 0, \ldots, 0)$ so that the edges leading to X and $X + (2, 0, 0, \ldots, 0)$ are open but all other incident edges are closed. The outcome of this construction is a finite open cluster containing the origin and having diameter $m + n + 2$; furthermore, any such outcome arises in a unique way from suitable clusters with diameters m and n. Thus, the probability that C has diameter $m + n + 2$ is at least

$$\text{(6.65)} \qquad P_p(D_1 = \text{diam}(C) = m) \frac{1}{d\,|B(n)|} P_p(\text{diam}(C) = n) p^2 (1 - p)^{2d-2},$$

where the term $p^2(1-p)^{2d-2}$ is the worst penalty payable for the changes in the states of edges incident to $X + (1, 0, 0, \ldots, 0)$. Thus

$$\begin{aligned} P_p(\operatorname{diam}(C) = m + n + 2) \\ &\geq \frac{p^2(1-p)^{2d-2}}{d|B(n)|} P_p(D_1 = \operatorname{diam}(C) = m) P_p(\operatorname{diam}(C) = n) \\ &\geq \frac{p^2(1-p)^{2d-2}}{d^2|B(n)|} P_p(\operatorname{diam}(C) = m) P_p(\operatorname{diam}(C) = n) \end{aligned}$$

by (6.60). Of course $|B(n)| = (2n+1)^d$. ∎

Let us now concentrate on inequality (6.58). Taking logarithms, we find that the sequence $\delta(k) = -\log P_p(\operatorname{diam}(C) = k)$ satisfies

(6.66) $$\delta(m + n + 2) \leq \delta(m) + \delta(n) + \log\left\{\frac{d^2(2n+1)^d}{p^2(1-p)^{2d-2}}\right\}.$$

This differs slightly from the usual form of the generalized subadditive inequality in that the argument in the left-hand side is $m + n + 2$ instead of $m + n$, and the error term is a function n rather than of $m + n$. These minor aberrations are easily taken into account in the proof of the subadditive limit theorem. As noted at the end of Appendix II, inequality (6.66) implies the existence of the limit

(6.67) $$\begin{aligned} \sigma(p) &= \lim_{n\to\infty} \left\{\frac{1}{n}\delta(n)\right\} \\ &= \lim_{n\to\infty} \left\{-\frac{1}{n}\log P_p(\operatorname{diam}(C) = n)\right\} \end{aligned}$$

together with the inequality

$$\delta(k) \geq (k+2)\sigma(p) - \log\left\{\frac{d^2(2k+1)^d}{p^2(1-p)^{2d-2}}\right\} \qquad \text{for all } k,$$

which we rewrite as

(6.68) $$P_p(\operatorname{diam}(C) = k) \leq \frac{d^2}{p^2(1-p)^{2d-2}}(2k+1)^d e^{-(k+2)\sigma(p)}.$$

In particular, $\sigma(p) < \infty$ if $0 < p < 1$.

Having proved the exponential behaviour of the tail of the diameter distribution, it is easy to show the same for the radius. It is clear that

(6.69) $$P_p(0 \leftrightarrow \partial B(n), |C| < \infty) \leq P_p(n \leq \operatorname{diam}(C) < \infty),$$

since the diameter of C is at least as large as its radius. On the other hand,

$$P_p(0 \leftrightarrow \partial B(n), |C| < \infty) \geq P_p(\operatorname{diam}(C) = n, L_1 = 0, R_1 = n),$$

since, if the origin is a leftmost vertex of a finite cluster with width n in the first coordinate direction, then there is an open path joining the origin to $\partial B(n)$. Hence

$$(6.70)\qquad P_p(0 \leftrightarrow \partial B(n), |C| < \infty) \geq \frac{1}{d(2n+1)^d} P_p(\operatorname{diam}(C) = n)$$

by (6.64). It follows from (6.69) and (6.70), in conjunction with (6.67), that

$$(6.71)\qquad -\frac{1}{n} \log P_p(0 \leftrightarrow \partial B(n), |C| < \infty) \to \sigma(p) \qquad \text{as} \quad n \to \infty.$$

The only case requiring special attention is when $\sigma(p) = 0$, in which instance (6.70) alone implies (6.71).

Finally, we show that there exists a constant $A(p, d)$ such that (6.50) holds. Suppose that $\sigma(p) > 0$. From (6.68) and (6.69),

$$(6.72)\quad P_p(0 \leftrightarrow \partial B(n), |C| < \infty) \leq \frac{d^2}{p^2(1-p)^{2d-2}} \sum_{m=n}^{\infty} (2m+1)^d e^{-(m+2)\sigma(p)},$$

of which (6.50) is an immediate consequence with a suitable choice for $A(p, d)$. If $\sigma(p) = 0$ then (6.50) is satisfied with $A(p, d) = 1$. ∎

Proof of Theorem (6.51). We turn to the proof that $\sigma(p) > 0$ when $p > p_c^+$ and $d \geq 3$, and we shall follow Chayes, Chayes, and Newman (1987). Let $H(n)$ be the hyperplane $\{x \in \mathbb{Z}^d : x_1 = n\}$ of $\mathbb{Z}^d$, and let G_n be the event that the origin is in a finite cluster and there is an open path in $B(n)$ joining the origin to a vertex of $H(n)$. The box $B(n)$ has $2d$ faces, and so

$$(6.73)\qquad P_p(G_n) \leq P_p(0 \leftrightarrow \partial B(n), |C| < \infty) \leq 2d P_p(G_n),$$

as in (5.22). We shall prove that, if $p > p_c^+$, there exists $\gamma(p) > 0$ such that

$$(6.74)\qquad P_p(G_n) \leq e^{-n\gamma(p)} \qquad \text{for all } n;$$

it follows from the second inequality in (6.73) that $\sigma(p) \geq \gamma(p)$, and it is an immediate consequence that $\sigma(p) > 0$ when $p > p_c^+$.

The idea of the proof is as follows. Fix a positive integer k and suppose that G_{nk} occurs for some $n \geq 1$. In this case, each of the slabs

$$(6.75)\qquad S_k(i) = \{x \in \mathbb{Z}^d : (i-1)k \leq x_1 < ik\},$$

for $1 \leq i \leq n$, is traversed by an open path from the origin, and this open path does not intersect any infinite open cluster contained within such slabs; see Figure 6.4. If $p > p_c(S_k)$, the critical probability of a slab of thickness k, then each slab contains an infinite open cluster, and the probability of avoiding it is strictly less than 1. There are n such slabs, giving that the probability of G_{nk} is no larger than α^n for some $\alpha < 1$.

We have to do a certain amount of work to make this argument

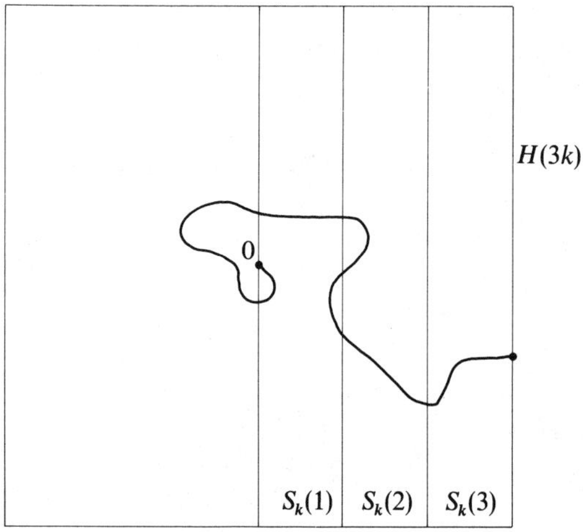

Figure 6.4. All paths from the origin to $H(3k)$ traverse the slabs $S_k(1)$, $S_k(2)$, and $S_k(3)$.

rigorous. Let k be a positive integer. First, we construct the open cluster C at the origin as the limit of the sequence $(C_m: m \geq 1)$ in the manner described just after the statement of the theorem. Next we construct a sequence $v_1, v_2, \dots$ of vertices in the following manner. We set $v_1 = 0$, the origin. We let m_i be the smallest value of m for which C_m contains a vertex of the slab $S_k(i)$, and we denote this vertex by v_i. Such a v_i exists if and only if some vertex in $S_k(i)$ lies in the open cluster at the origin. In this way, we obtain a sequence $v_1, v_2, \dots, v_T$ of vertices where

$$T = \sup\{i: C \cap S_k(i) \neq \varnothing\}.$$

Suppose that $p > p_c(S_k)$, the critical probability of percolation on $S_k(i)$, and recall that $\theta(p, 0; S_k)$ is the probability that the origin is in an infinite open cluster of $S_k(1)$. Let n be a positive integer, and write $n = kr + s$ where $0 \leq s < k$. If the event G_n occurs then $T \geq r$ and v_i does not lie in an infinite open cluster of $S_k(i)$ for $1 \leq i \leq r$. Thus

$$
\begin{aligned}
(6.76)\quad P_p(G_n) &\leq P_p(T \geq r,\ v_i \text{ not in an infinite open cluster of } S_k(i), \\
&\qquad \text{for } 1 \leq i \leq r) \\
&= P_p(A_r) \\
&= P_p(A_r \mid A_{r-1})P_p(A_{r-1}),
\end{aligned}
$$

where A_j is the event that $T \geq j$ and v_i does not lie in an infinite open cluster of $S_k(i)$ for $1 \leq i \leq j$. By the method of construction of the C_m's, the events A_{j-1} and $\{T \geq j\}$ are independent of the states of all edges joining vertices in $S_k(j)$, whereas, conditional on v_j, the event that v_j lies in the infinite open cluster of $S_k(j)$ depends *only* on edges joining vertices in $S_k(j)$. Therefore

$$\begin{aligned} P_p(A_j | A_{j-1}) &\le \sum_v P_p(v \text{ not in an infinite open cluster of } \\ &\qquad S_k(j) | v_j = v, T \ge j, A_{j-1}) P_p(v_j = v | T \ge j, A_{j-1}) \\ &\qquad \times P_p(T \ge j | A_{j-1}) \\ &\le \sum_v (1 - \theta(p, v; S_k)) P_p(v_j = v | T \ge j, A_{j-1}) \\ &= 1 - \theta(p, 0; S_k), \end{aligned}$$

where the sum is over all vertices v in the hyperplane $H((j-1)k)$. This is valid for all $j \ge 2$, giving from (6.76) that

$$\begin{aligned} P_p(G_n) &\le P_p(A_r | A_{r-1}) P_p(A_{r-1} | A_{r-2}) \dots P_p(A_2 | A_1) P_p(A_1) \\ &\le (1 - \theta(p, 0; S_k))^r. \end{aligned}$$

However, $\theta(p, 0; S_k) > 0$ by the assumption that $p > p_c(S_k)$, and thus

$$P_p(G_n) \le \exp(-r\gamma_k(p)) \tag{6.77}$$

where

$$\gamma_k(p) = -\log(1 - \theta(p, 0; S_k)) > 0.$$

Now $n \le (r+1)k$, so that $r \ge (n/k) - 1$, giving that

$$P_p(G_n) \le \exp(-n\{\gamma_k(p)/k\} + \gamma_k(p)) \tag{6.78}$$

for all $n \ge 0$. This proves the result whenever $p > p_c(S_k)$. If $p > p_c^+$ then $p > p_c(S_k)$ for some k by Theorem (6.45), and (6.78) delivers the result. ■

6.6 Truncated Connectivity Functions and Correlation Length

The two-point connectivity function τ_p is defined by $\tau_p(x, y) = P_p(x \leftrightarrow y)$. When $p < p_c$, $\tau_p(x, y)$ decays roughly as an exponential function of the distance $\delta(x, y)$ from x to y. When $p > p_c$, $\tau_p(x, y)$ does not decay at all, since

$$\begin{aligned} \tau_p(x, y) &\ge P_p(x \text{ and } y \text{ in the infinite open cluster}) \\ &= P_p(|C(x)| = \infty, |C(y)| = \infty) \\ &\ge P_p(|C(x)| = \infty) P_p(|C(y)| = \infty) \\ &= \theta(p)^2, \end{aligned}$$

where we have used the FKG inequality. It is an open question to determine the behaviour of $\tau_p(x, y)$ when $p = p_c$.

Of more interest than τ_p when $p > p_c$ is the *truncated connectivity function* τ_p^f defined by

(6.79) $\quad \tau_p^f(x, y) = P_p(x \text{ and } y \text{ are in the same finite open cluster}).$

Note that

$$\tau_p^f(x, y) = \tau_p(x, y) - P_p(x \text{ and } y \text{ in the infinite open cluster}),$$

so that

$$\tau_p(x, y) = \tau_p^f(x, y) \qquad \text{if} \quad \theta(p) = 0. \tag{6.80}$$

The superscript 'f' refers to the requirement that x and y be in *finite* open clusters, just as $\chi^f(p)$ is the mean size of the finite open cluster at the origin.

It is believed that $\tau_p^f(x, y)$ decays in the manner of an exponential function of the distance between x and y. Clearly, the exponential decay of $\tau_p^f(x, y)$ is closely related to that of $P_p(0 \leftrightarrow \partial B(n), |C(0)| < \infty)$, and so it is not surprising that such decay is known for τ_p^f only when $p > p_c^+$. It is an immediate consequence of Theorem (6.48) that there exists a constant $A(p, d)$, finite for $d \geq 2$ and $0 < p < 1$, such that

$$\tau_p^f(0, x) \leq A(p, d)\|x\|^d e^{-\|x\|\sigma(p)} \qquad \text{for all } x, \tag{6.81}$$

where $\sigma(p)$ is given in (6.49) and $\|x\| = \max\{|x_i|: 1 \leq i \leq d\}$; we recall that $\sigma(p) > 0$ if $p > p_c^+$. There is a formula for $\sigma(p)$ in two dimensions: we shall see in Theorem (9.23) that, if $d = 2$,

$$-\frac{1}{n} \log \tau_p^f(0, e_n) \to 2\xi(1-p)^{-1} \qquad \text{as} \quad n \to \infty, \tag{6.82}$$

where $e_n = (n, 0)$ and $\xi(1 - p)$ is the correlation length of the subcritical percolation process with edge-probability $1 - p$. No such simple formula is likely to be valid when $d \geq 3$.

In studying the truncated connectivity function, we shall concentrate on the event that the origin and the vertex $e_n = (n, 0, 0, \ldots, 0)$ are in the same finite open cluster. Our principal result states that the probability of this event has the same asymptotic behaviour as the probability that the radius of the open cluster at the origin is finite but at least n.

(6.83) Theorem. Asymptotic behaviour of the truncated connectivity function. *Suppose that* $0 < p < 1$, *and let* $\sigma(p)$ *and* $A(p, d)$ *be as given in Theorem* (6.48). *Then*

$$\lim_{n\to\infty} \left\{ -\frac{1}{n} \log \tau_p^f(0, e_n) \right\} = \sigma(p) \tag{6.84}$$

and

$$\tau_p^f(0, e_n) \leq A(p, d) n^d e^{-n\sigma(p)} \qquad \textit{for all } n. \tag{6.85}$$

In the subcritical phase ($p < p_c$) the correlation length $\xi(p)$ was given by

$$\xi(p)^{-1} = \lim_{n\to\infty}\left\{-\frac{1}{n}\log P_p(0 \leftrightarrow \partial B(n))\right\}$$
$$= \lim_{n\to\infty}\left\{-\frac{1}{n}\log \tau_p(0, e_n)\right\}.$$

We define the *correlation length* $\xi(p)$ analogously in the supercritical phase ($p > p_c$),

(6.86) $$\xi(p)^{-1} = \lim_{n\to\infty}\left\{-\frac{1}{n}\log P_p(0 \leftrightarrow \partial B(n), |C| < \infty)\right\}$$
$$= \lim_{n\to\infty}\left\{-\frac{1}{n}\log \tau_p^f(0, e_n)\right\},$$

by reference to the corresponding events involving *finite* open clusters. This seems to be a sensible definition, and we shall return to the notion of correlation length in Chapter 7. At this stage we note only that we are lacking a complete proof that $\sigma(p) > 0$ when $p > p_c$, which is to say that we cannot yet prove that the correlation length $\xi(p)$ is finite throughout the supercritical phase.

Proof of Theorem (6.83). We follow Chayes, Chayes, Grimmett, Kesten, and Schonmann (1988). The upper bound on $\tau_p^f(0, e_n)$ is easy. Just note that

(6.87) $$\tau_p^f(0, e_n) \le P_p(0 \leftrightarrow \partial B(n), |C| < \infty),$$

and use Theorem (6.48). Inequality (6.81) is derived similarly.

Slightly more work is required for the lower bound, and we follow the line of argument in the proof of Theorem (5.44). Let m be a positive integer and suppose that the open cluster C at the origin has the following properties:

(a) $\operatorname{diam}(C) = D_1$, so that the width of C is maximal in the first coordinate direction;
(b) $\operatorname{diam}(C) = m$;
(c) $L_1 = 0$ and $R_1 = m$, so that the origin is a leftmost vertex of C in the sense that $x_1 \ge 0$ for all $x \in C$;

where we use the notation of the last section. Under these conditions, there exists a vertex in the hyperplane $H(m) = \{x \in \mathbb{Z}^d : x_1 = m\}$ which is in the same finite open cluster as the origin, and furthermore this vertex is a rightmost vertex of C. In a similar way to (6.60) and (6.62), there exists a (non-random) vertex x in $H(m) \cap \partial B(m)$ for which

$$(6.88) \qquad P_p(x \in C, \operatorname{diam}(C) = m, L_1 = 0, R_1 = m) \geq \frac{1}{|\partial B(m)|} P_p(\operatorname{diam}(C) = m, L_1 = 0, R_1 = m),$$

and we choose x accordingly. We now put two such events together in such a way as to force the origin and the vertex e_{2m+2} to be in the same finite open cluster. Let us consider the event that e_{2m+2} and the vertex $x + (2, 0, 0, \dots, 0)$ are (respectively) rightmost and leftmost vertices of the same finite open cluster having diameter m; by symmetry, this event has the same probability as the left-hand side of equation (6.88), and furthermore it is independent of the event that that the origin and x are leftmost and rightmost vertices of the same finite open cluster having diameter m. If we change the states of the edges incident to $x + (1, 0, 0, \dots, 0)$ so that the two edges leading to x and $x + (2, 0, 0, \dots, 0)$ are open but all others are closed, then we force the origin and e_{2m+2} to be in the same finite open cluster. We have proved that

(6.89)

$$\begin{aligned} \tau_p^f(0, e_{2m+2}) &\geq p^2(1-p)^{2d-2} P_p(x \in C, \operatorname{diam}(C) = m, L_1 = 0, R_1 = m)^2 \\ &\geq \frac{p^2(1-p)^{2d-2}}{|\partial B(m)|^2} P_p(\operatorname{diam}(C) = m, L_1 = 0, R_1 = m)^2 \\ &\geq \frac{p^2(1-p)^{2d-2}}{d^2 |\partial B(m)|^2 |B(m)|^2} P_p(\operatorname{diam}(C) = m)^2 \end{aligned}$$

by (6.88) and (6.64). In a similar way, we have that

$$(6.90) \qquad \tau_p^f(0, e_{2m+3}) \geq \frac{p^3(1-p)^{4d-4}}{d^2 |\partial B(m)|^2 |B(m)|^2} P_p(\operatorname{diam}(C) = m)^2.$$

With the aid of (6.49) and (6.67), inequalities (6.87), (6.89), and (6.90) imply jointly that (6.84) holds. ∎

6.7 Sub-Exponential Decay of the Cluster Size Distribution

We turn our attention next to the number of vertices in the open cluster at the origin, when this cluster is finite. Whereas $P_p(|C| = n)$ decays exponentially in n when $p < p_c$, it has a quite different behaviour when $p > p_c$. It is fairly easy to see why the following result should be valid, although some technical difficulties arise in its proof.

(6.91) Theorem. Sub-exponential decay of cluster size distribution. *Suppose that $p_c < p < 1$. There exists $\gamma(p) < \infty$ such that*

$$P_p(|C| = n) \geq \exp(-\gamma(p)n^{(d-1)/d}) \qquad \textit{for all } n. \tag{6.92}$$

We recall that $n^{(d-1)/d}$ is the order of the surface area of a ball of $\mathbb{R}^d$ with volume n. We say that the cluster size distribution decays 'sub-exponentially' since it decays to 0 at a rate which is slower than exponential decay.

It is believed that this theorem describes the correct order of magnitude of $P_p(|C| = n)$, and we therefore make the conjecture that

$$\delta(p) = \lim_{n\to\infty} \{-n^{-(d-1)/d} \log P_p(|C| = n)\} \tag{6.93}$$

exists and satisfies $0 < \delta(p) < \infty$ when $p_c < p < 1$. Assuming the existence of $\delta(p)$, one would like to understand how it behaves as $p \downarrow p_c$; at what rate does $\delta(p) \to 0$ as $p \downarrow p_c$?

What about upper bounds for $P_p(|C| = n)$? As in Chapters 3 and 5, the upper bound of the last section for the radius of the finite open cluster at the origin provides a weak upper bound for $P_p(|C| = n)$. It is easy to deduce from Theorems (6.48) and (6.51) that

$$P_p(n \leq |C| < \infty) \leq \exp(-\eta(p)n^{1/d}) \qquad \text{for all } n, \tag{6.94}$$

for some $\eta(p)$ which is strictly positive when $p > p_c^+$. We note that, for $d = 2$, the orders of the exponents in the lower bound (6.92) and the upper bound (6.94) are both $n^{1/2}$.

Here is a better upper bound for the tail of the size of the finite cluster at the origin.

(6.95) Theorem. *Suppose that $d \geq 3$. If $p > p_c^+$, there exists $\eta(p) > 0$ such that*

$$P_p(|C| = n) \leq \exp(-\eta(p)n^{(d-1)/d}) \qquad \textit{for all } n. \tag{6.96}$$

This provides an upper bound for the tail probabilities having the same order (in the exponent) as that of the lower bound (6.92), but which is valid only for values of p exceeding p_c^+. We conjectured in Section 6.4 that $p_c^+ = p_c$ when $d \geq 3$. Once this conjecture has been proved, the above theorem will provide a decent upper bound for the cluster size distribution throughout the supercritical phase in three and more dimensions. We shall not give a complete proof of this theorem. Instead, we shall sketch a 'Peierls argument' which leads to an upper bound of this form when p is close enough to 1, and then we shall indicate how to use this to obtain the full result as quoted above.

Proof of Theorem (6.91). We follow Aizenman, Delyon, and Souillard (1980). Here is the idea of the proof. When $p > p_c$, each vertex has a positive probability $\theta(p)$ of being in the infinite open cluster. Therefore, in any specified box $B(m)$, say, the mean proportion of vertices which are in the infinite open cluster equals $\theta(p)$. In particular, the probability $\nu(m)$ that more than $\frac{1}{2}\theta(p)$ of the vertices in $B(m)$ are joined to the surface of $B(m)$ remains bounded away from 0 as $m \to \infty$. Suppose that every edge joining two vertices in $\partial B(m)$ is open but every edge joining a vertex in $\partial B(m)$ to a vertex in $\partial B(m+1)$ is closed; the probability of this is approximately $\pi^{|\partial B(m)|}$ for some $\pi = \pi(p)$ satisfying $0 < \pi < 1$. If both of the above events occur then $B(m+1)$ contains a finite open cluster of $\mathbb{L}^d$ of size greater than $\frac{1}{2}|B(m)|\theta(p)$, and the probability of this is roughly at least $\nu(m)\pi^{|\partial B(m)|}$. See Figure 6.5 for an illustration of these events. The origin is in this cluster with probability approximately $\theta(p)$, so that

$$(6.97) \qquad P_p(\tfrac{1}{2}|B(m)|\theta(p) \leq |C| < \infty) \geq A\pi^{|\partial B(m)|}$$

for some positive constant $A = A(p, d)$. We write $n = \frac{1}{2}|B(m)|\theta(p)$, so that $|\partial B(m)|$ is of order $n^{(d-1)/d}$, and we obtain an inequality of the correct form. We shall now make this argument rigorous.

Suppose that $p_c < p < 1$ and let m be a positive integer. Let R_m be the number of vertices of $B(m)$ which are in the infinite open cluster.

(6.98) Lemma.

$$P_p(R_m \geq \tfrac{1}{2}\theta(p)|B(m)|) \geq \tfrac{1}{2}\theta(p).$$

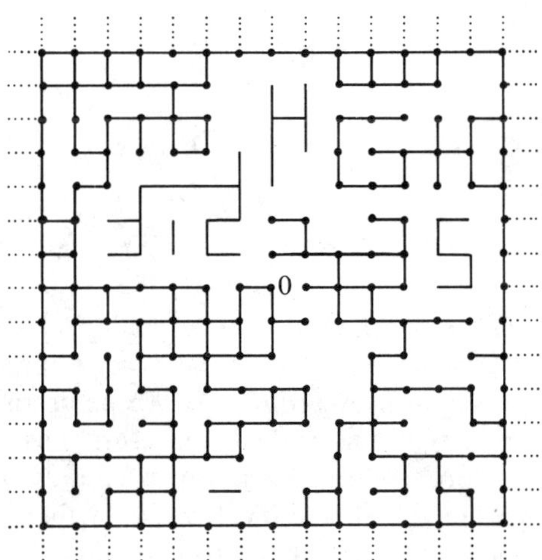

Figure 6.5. We may form a large finite open cluster by declaring all edges in $\partial B(m)$ to be open and all edges joining $\partial B(m)$ to $\partial B(m+1)$ to be closed. The cluster in question contains all vertices in $B(m)$ which are joined to $\partial B(m)$ by open paths.

Proof. Clearly $R_m \le |B(m)|$, and so

$$E_p(R_m) \le |B(m)|\,P_p(R_m \ge \tfrac{1}{2}\theta(p)|B(m)|) + \tfrac{1}{2}\theta(p)|B(m)|\,P_p(R_m < \tfrac{1}{2}\theta(p)|B(m)|)$$
$$\le |B(m)|\,P_p(R_m \ge \tfrac{1}{2}\theta(p)|B(m)|) + \tfrac{1}{2}\theta(p)|B(m)|.$$

We use the fact that $E_p(R_m) = \theta(p)|B(m)|$ and divide by $|B(m)|$ to obtain the conclusion of the lemma. ∎

Let U_m be the number of vertices inside $B(m)$ which are joined to vertices in $\partial B(m)$ by open paths. Certainly $U_m \ge R_m$, and so

$$P_p(\tfrac{1}{2}\theta(p)|B(m)| \le U_m \le |B(m)|) \ge \tfrac{1}{2}\theta(p).$$

The random variable U_m is independent of the states of all edges between pairs of vertices in $\{x \in \mathbb{Z}^d\colon \|x\| \ge m\}$; remember that $\|x\| = \max\{|x_i|\colon 1 \le i \le d\}$. Let A_m be the event that all edges joining pairs of vertices in $\partial B(m)$ are open and all edges joining vertices of $\partial B(m)$ to vertices of $\partial B(m+1)$ are closed. It is easy to check that there exists $\pi = \pi(p)$ satisfying $0 < \pi < 1$ such that

$$P_p(A_m) \ge \pi^{|\partial B(m)|} \quad \text{for all} \quad m \ge 1. \tag{6.99}$$

Furthermore, U_m is independent of A_m, so that

$$P_p(A_m, \tfrac{1}{2}\theta(p)|B(m)| \le U_m \le |B(m)|) \ge \tfrac{1}{2}\theta(p)\pi^{|\partial B(m)|}. \tag{6.100}$$

If the event on the left-hand side here occurs, then the vertex $(m, m, \dots, m)$ is contained in a finite cluster of $\mathbb{L}^d$ having between $\frac{1}{2}\theta(p)|B(m)|$ and $|B(m)|$ vertices; we translate the vertex $(m, m, \dots, m)$ to the origin to deduce that

$$P_p(\tfrac{1}{2}\theta(p)|B(m)| \le |C| \le |B(m)|) \ge \tfrac{1}{2}\theta(p)\pi^{|\partial B(m)|}. \tag{6.101}$$

It is easy to deduce from this that there exists $\eta = \eta(p) < \infty$ such that

$$P_p(n \le |C| < \infty) \ge \exp(-\eta n^{(d-1)/d}) \qquad \text{for all } n,$$

but the theorem requires more than this.

(6.102) Lemma. *There exists $\eta = \eta(p) > 0$ such that the following holds. For every $i \ge 1$, there exists a positive integer $r(i)$ such that*

$$r(1) = 1, \tag{6.103}$$

$$2 \le \frac{r(i+1)}{r(i)} \le \delta, \tag{6.104}$$

and

$$P_p(|C| = r(i)) \ge r(i)\exp(-\eta r(i)^{(d-1)/d}), \tag{6.105}$$

where $\delta = 2^{d+3}\theta(p)^{-2}$.

Proof. Remember that

$$|B(m)| = (2m+1)^d, \qquad |\partial B(m)| \le 2d(2m+1)^{d-1}. \tag{6.106}$$

From (6.101), for each $m \geq 1$, there exists $k(m)$ such that

$$\tfrac{1}{2}\theta(p)|B(m)| \leq k(m) \leq |B(m)| \tag{6.107}$$

and

$$P_p(|C| = k(m)) \geq \frac{1}{2|B(m)|}\theta(p)\pi^{|\partial B(m)|} = \frac{1}{2|B(m)|}\theta(p)\exp(-|\partial B(m)|\log(1/\pi)). \tag{6.108}$$

From (6.106) and (6.107), there exists $v = v(p) > 0$ such that

$$|\partial B(m)|\log(1/\pi) \leq vk(m)^{(d-1)/d} \qquad \text{for all } m.$$

Thus, from (6.107) and (6.108),

$$P_p(|C| = k(m)) \geq \frac{\theta(p)^2}{2k(m)}\exp(-vk(m)^{(d-1)/d}).$$

This inequality is interesting only when $k(m)$ is large, in which case the exponential term is dominating; therefore there exists $\eta = \eta(p) > 0$ for which

$$P_p(|C| = k(m)) \geq k(m)\exp(-\eta k(m)^{(d-1)/d}), \tag{6.109}$$

and

$$P_p(|C| = 1) \geq e^{-\eta}. \tag{6.110}$$

We now remove a subsequence $(r(i): i \geq 1)$ from the sequence $(1, k(1), k(2), \ldots)$ in the following manner. We set $r(1) = 1$. Having chosen $r(1), r(2), \ldots, r(i)$ with $r(i) = k(j)$ say, we let l be the smallest integer such that

$$|B(j)| \leq \tfrac{1}{4}\theta(p)|B(l)| \tag{6.111}$$

and then we set $r(i+1) = k(l)$. We have from (6.107) that

$$k(l) \geq \tfrac{1}{2}\theta(p)|B(l)| \geq 2|B(j)| \geq 2k(j)$$

so that $2 \leq r(i+1)/r(i)$. On the other hand, l is the smallest integer such that

$$(2j+1)^d \leq \tfrac{1}{4}\theta(p)(2l+1)^d$$

so that

$$(2j+1)^d > \tfrac{1}{4}\theta(p)(2l-1)^d,$$

which is to say that

$$(2l+1)^d \leq \left\{\left(\frac{4}{\theta(p)}\right)^{1/d}(2j+1) + 2\right\}^d \leq \frac{2^{d+2}}{\theta(p)}(2j+1)^d.$$

Hence, by (6.107),

$$k(l) \leq |B(l)| \leq \frac{2^{d+2}}{\theta(p)}|B(j)| \leq \frac{2^{d+3}}{\theta(p)^2}k(j),$$

giving that $r(i+1)/r(i) \leq \delta$ as required. ■

We show next that any positive integer n may be built up using the $r(i)$'s as building blocks.

(6.112) Lemma. *Let δ, η, and $(r(i): i \geq 1)$ be as given in Lemma (6.102). Any positive integer n may be expressed in the form*

$$n = \sum_{i=1}^{I} w_i r(i), \tag{6.113}$$

where $I = \max\{i: r(i) \leq n\}$ and the w_i's are integers which satisfy

$$0 \leq w_i \leq \delta \qquad \text{for all } i. \tag{6.114}$$

Furthermore, with such a representation,

$$\sum_{i=1}^{I} w_i r(i)^{(d-1)/d} \leq 4\delta n^{(d-1)/d}. \tag{6.115}$$

Proof. Certainly $n = 1$ can be expressed in the form (6.113) subject to (6.114); just take $w_I = I = 1$. Suppose then that any integer not exceeding k can be expressed in the form (6.113) subject to (6.114), and consider the integer $n = k + 1$. Let $I = \max\{i: r(i) \leq k+1\}$. Then

$$\begin{aligned} k + 1 &= r(I) + (k + 1 - r(I)) \\ &= r(I) + \sum_{i=1}^{I} w_i r(i) \end{aligned}$$

for some $(w_i: i \geq 1)$ satisfying (6.114), from the induction hypothesis applied to $k + 1 - r(I)$. It remains to check that $1 + w_I \leq \delta$. From the definition of I, $r(I) \leq k + 1 < r(I+1)$. On the other hand, $r(I+1) \leq \delta r(I)$ by (6.104), giving that

$$(1 + w_I)r(I) \leq k + 1 < \delta r(I)$$

as required.

Finally, suppose that n has been expressed in the form (6.113) subject to (6.114). We will show that (6.115) holds. We have from (6.114) that $w_i \leq \delta$, and from (6.104) that

$$r(i) \leq \tfrac{1}{2}r(i+1) \leq \frac{1}{2^{I-i}}r(I) \leq \frac{n}{2^{I-i}}$$

whenever $1 \le i \le I$. Therefore

$$\begin{aligned}\sum_{i=1}^{I} w_i r(i)^{(d-1)/d} &\le \delta n^{(d-1)/d} \sum_{i=1}^{I} (\tfrac{1}{2})^{(I-i)(d-1)/d}\\ &\le \delta n^{(d-1)/d} \sum_{i=0}^{\infty} (\tfrac{1}{2})^{i/2} \qquad \text{since} \quad d \ge 2\\ &\le 4\delta n^{(d-1)/d}\end{aligned}$$

as required. ■

We may now complete the proof of the theorem, using the sequence $(r(i): i \ge 1)$ as building blocks in the following way. Let n be a positive integer, and write n in the form

(6.116) $$n = \sum_{i=1}^{I} w_i r(i) \qquad \text{where} \quad 0 \le w_i \le \delta,$$

as in Lemma (6.112). We recall Lemma (5.102), which asserts that

$$\frac{1}{r+s} P_p(|C| = r + s) \ge pq^{-2}\frac{1}{r}P_p(|C| = r)\frac{1}{s}P_p(|C| = s) \qquad \text{if} \quad r, s \ge 1.$$

The proof of this is valid whenever $0 < p = 1 - q < 1$. It follows from (6.116) that

(6.117) $$\frac{1}{n}P_p(|C| = n) \ge (pq^{-2})^w \prod_{i=1}^{I}\left\{\frac{1}{r(i)}P_p(|C| = r(i))\right\}^{w_i},$$

where $w = \sum_{i=1}^{I} w_i$. However, $w_i \le \delta$ for all i, and $r(i) \ge 2^{i-1}$ by (6.104), giving that

$$I = \max\{i: r(i) \le n\} \le 1 + \log_2 n$$

and

$$w \le \delta(1 + \log_2 n).$$

Writing $\zeta = \min\{1, pq^{-2}\}$, we have from (6.105) and (6.115) that

$$\begin{aligned}\frac{1}{n}P_p(|C| = n) &\ge \zeta^{\delta(1+\log_2 n)} \exp\left(-\eta \sum_{i=1}^{I} w_i r(i)^{(d-1)/d}\right)\\ &\ge \zeta^{\delta(1+\log_2 n)} \exp(-4\delta\eta n^{(d-1)/d})\end{aligned}$$

and we are home. ■

Remarks on Theorem (6.95). We describe only the skeleton of the proof, due to Kesten and Zhang (1988). The first step is to show such an inequality for values of p which are sufficiently close to 1 (this was proved by Kunz and Souillard (1978); see also Kesten (1982, p. 99)). The idea is the following. There exists $\nu > 0$ such that, if the origin is in a finite cluster C of size n, then there are at least $\nu n^{(d-1)/d}$ closed edges in the external edge boundary of C; see Figure 1.6 for a drawing of the external edge boundary of C in two dimensions. Furthermore, at least one such edge has an endvertex within distance n of the origin. Proceeding roughly as in the proof of Theorem (1.10), we define a new 'neighbour relation' on the set of edges of $\mathbb{L}^d$, so that the edges in the external edge boundary form the edge set of a connected graph. We now use a 'Peierls argument', showing that the number of such connected graphs with m edges and containing a specified edge is at most $\mu(d)^m$ for some constant $\mu(d)$ depending only on the number of dimensions. The usual estimate now gives

$$P_p(|C| = n) \leq N_n\{q\mu(d)\}^{\nu n^{(d-1)/d}}, \tag{6.118}$$

where $p + q = 1$ and $N_n = O(n^d)$ is the number of edges with an endvertex within distance n of the origin. If p is sufficiently close to 1 then $q\mu(d) < 1$ and the result is proved. It is rather easy to make this argument rigorous in two dimensions, but there are certain topological complications when $d \geq 3$ (see Section 10.4 for related material). Before moving on, we note that the above argument may in principle be adapted to bond and site percolation on any lattice, and not merely bond percolation on $\mathbb{L}^d$.

The proof of the theorem continues as follows. Let N be a positive integer. From the lattice $\mathbb{L}^d$ we construct a 'renormalized' lattice $\mathscr{R}_N(\mathbb{L}^d)$ as follows. $\mathscr{R}_N(\mathbb{L}^d)$ has $\mathbb{Z}^d$ as vertex set, the vector $x \in \mathbb{Z}^d$ corresponding to the box

$$B_x = \prod_{i=1}^{d} [x_i N, (x_i + 1)N)$$

of $\mathbb{L}^d$; we declare two vertices x and y of $\mathscr{R}_N(\mathbb{L}^d)$ to be *adjacent* if B_x and B_y have a common 'face' in $\mathbb{L}^d$, and as a consequence $\mathscr{R}_N(\mathbb{L}^d)$ is isomorphic to $\mathbb{L}^d$. We declare each vertex x of $\mathscr{R}_N(\mathbb{L}^d)$ to be *closed* if either:

(a) there is no open path of $\mathbb{L}^d$ from B_x to the surface $\partial B'_x$ of $B'_x = \prod_{i=1}^d [(x_i - 2)N, (x_i + 3)N)$; or
(b) there exists a neighbour y of x in $\mathscr{R}_N(\mathbb{L}^d)$ such that there exist open paths of $\mathbb{L}^d$ from B_x to $\partial B'_x$ and from B_y to $\partial B'_x$ with the property that these paths are in disjoint open clusters of B'_x.

It may be shown that, if $p > p_c^+$, then the probability $\pi_N(p)$, that a given vertex of $\mathscr{R}_N(\mathbb{L}^d)$ is closed, may be made arbitrarily close to 0 by a choice of N sufficiently large (use a technique of Aizenman, Chayes, Chayes, Fröhlich, and Russo (1983) to prove that the probability of (b) may be made small).

Let C be the open cluster of $\mathbb{L}^d$ at the origin, as usual. We define the *external* $\mathscr{R}_N(\mathbb{L}^d)$*-boundary* of C to be the set of vertices x of $\mathscr{R}_N(\mathbb{L}^d)$ such that:

(a) $B_x \cap C = \varnothing$;
(b) x has a neighbour y such that $B_y \cap C \neq \varnothing$; and

(c) there exists an infinite path (not necessarily open) of $\mathbb{L}^d$ which has an endvertex in B_x and which is disjoint from C.

Suppose now that $|C| = n$. By the isoperimetric inequality, the external $\mathscr{R}_N(\mathbb{L}^d)$-boundary of C has order at least $n^{(d-1)/d}$. Furthermore, if n is sufficiently large compared with N, then it is not difficult to see that every vertex in the external $\mathscr{R}_N(\mathbb{L}^d)$-boundary of C is closed. We may adapt the proof leading to (6.118) to obtain that the probability of there existing such a closed 'surface' surrounding the origin of $\mathscr{R}_N(\mathbb{L}^d)$ is at most some exponentially decaying function of its order $n^{(d-1)/d}$ so long as the probability $\pi_N(p)$ is sufficiently small, and we choose N to be large enough for this to hold. A minor difficulty is that the states of the vertices of $\mathscr{R}_N(\mathbb{L}^d)$ are not independent random variables, but this difficulty is easily overcome by the observation that the state of any given vertex depends only on the states of a fixed number (that is, not depending on n and N) of vertices of the renormalized lattice. ∎

6.8 Differentiability of θ, χ^f, and κ

Once we have a decent upper bound for $P_p(n \le |C| < \infty)$ then it is easy to show the infinite differentiability of macroscopic quantities such as the percolation probability $\theta(p)$, the mean size $\chi^f(p)$ of finite open clusters, and the number $\kappa(p)$ of open clusters per vertex. It is not necessary to have a really fancy bound for $P_p(n \le |C| < \infty)$; anything which decreases faster than $n^{-\alpha}$ for all $\alpha > 0$ will suffice. We will use here the bound (6.94):

$$(6.119) \qquad P_p(n \le |C| < \infty) \le \exp(-\eta(p)n^{1/d}) \qquad \text{for all } n,$$

where $\eta(p) > 0$ if either $d = 2$ and $p_c(2) < p \le 1$, or $d \ge 3$ and $p_c^+ < p \le 1$.

(6.120) Theorem. Differentiability of θ, χ^f, and κ above p_c. *The percolation probability θ, the mean size χ^f of the finite open cluster at the origin, and the number κ of open clusters per vertex are infinitely differentiable functions of p on $(\pi(d), 1]$ where $\pi(2) = p_c(2)$ and $\pi(d) = p_c^+$ for $d \ge 3$.*

The proof shows that $E_p(f(|C|); |C| < \infty)$ is an infinitely differentiable function of p on $(\pi(d), 1]$, for any polynomial f in $|C|$ and $|C|^{-1}$. It is an open question of debatable interest to decide whether or not such functions are in general *analytic* on $(\pi(d), 1]$.

Proof. We follow the argument of Russo (1978). We shall show that θ is infinitely differentiable on $(\pi(d), 1]$, and the same argument is valid for χ^f and κ. Suppose that $\pi(d) < p = 1 - q \le 1$.

We write θ in the form

$$\theta(p) = 1 - \sum_{n=1}^{\infty} P_p(|C| = n), \tag{6.121}$$

where

$$P_p(|C| = n) = \sum_{m,b} a_{nmb} p^m q^b, \tag{6.122}$$

and a_{nmb} is the number of animals with n vertices, m edges, and b boundary edges. In order to show that θ is differentiable k times, we need to show the uniform convergence of

$$\sum_{n=N}^{\infty} \frac{d^k}{dp^k} P_p(|C| = n) = \sum_{n=N}^{\infty} \sum_{m,b} a_{nmb} \frac{d^k}{dp^k}(p^m q^b) \tag{6.123}$$

as $N \to \infty$, on suitable sub-intervals of $(\pi(d), 1]$.

Suppose first that $\pi(d) < p < 1$ and choose α and β such that $\pi(d) < \alpha < p < \beta < 1$. We shall show that the series in (6.123) is uniformly convergent on $[\alpha, \beta]$. For any $k \geq 0$, it is the case that

$$\begin{aligned} \left| \frac{d^k}{dp^k}(p^m q^b) \right| &= \left| \sum_{r=0}^{k} \binom{k}{r} m_r b_{k-r} p^{m-r} (-1)^{k-r} q^{b-(k-r)} \right| \\ &\leq p^m q^b \sum_{r=0}^{k} \binom{k}{r} (m/p)^r (b/q)^{k-r} \\ &= p^m q^b \left(\frac{m}{p} + \frac{b}{q} \right)^k \end{aligned}$$

where $x_r = x(x-1)\dots(x-r+1)$. We shall use the fact that $a_{nmb} = 0$ unless $b \leq 2dn$ and $m \leq dn$, and may therefore assume that $mp^{-1} + bq^{-1} \leq 2dn(pq)^{-1}$. By (6.123),

$$\left| \sum_{n=N}^{\infty} \frac{d^k}{dp^k} P_p(|C| = n) \right| \leq \left(\frac{2d}{pq} \right)^k \sum_{n=N}^{\infty} n^k P_p(|C| = n). \tag{6.124}$$

However $P_p(|C| = n) \leq \exp(-\eta(p) n^{1/d})$ by (6.119), where $\eta(p)$ is strictly positive for $p \in [\alpha, \beta]$; hence

$$P_p(|C| = n) \leq \exp(-\hat{\eta} n^{1/d}) \qquad \text{for all } n,$$

uniformly in $p \in [\alpha, \beta]$, where $\hat{\eta} = \inf\{\eta(p): \alpha \leq p \leq \beta\} > 0$. Therefore the sum in (6.124) is uniformly convergent on $[\alpha, \beta]$, giving that θ is k times differentiable at p.

To show (one-sided) differentiability of θ at $p = 1$ seems to require a certain amount of topology which the author is eager to avoid. Instead of using (6.119) to bound a typical term in (6.124), we use instead (6.118), an inequality which was not derived rigorously. This gives the following upper bound for the series in (6.124) when $p < 1$:

$$\left(\frac{2d\mu(d)}{p}\right)^k \sum_{n=N}^{\infty} n^k N_n \{q\mu(d)\}^{-k+\nu n^{(d-1)/d}},$$

where $N_n \le A_1 n^d$ for some constant A_1, $1 < \mu(d) < \infty$, and $\nu > 0$. If $0 < q = 1 - p \le \frac{1}{2}\mu(d)^{-1}$ say, then (6.124) is bounded by

$$A_2 \sum_{n=N}^{\infty} n^{k+d} 2^{-\nu' n^{(d-1)/d}}$$

for some constant A_2 and some $\nu' > 0$. This bound does not depend on p, and it is easy to deduce that θ is k times differentiable (from the left) at $p = 1$. ∎

6.9 Geometry of the Infinite Open Cluster

When $p > p_c$ there exists almost surely a unique infinite open cluster. What does this cluster look like? We are particularly interested in answering this question when p is only slightly greater than p_c, since the infinite open cluster has only low density in this case. Indeed, if $\theta(p_c) = 0$ as expected, then the density of the infinite open cluster decreases to 0 as p tends to p_c from above.

We study only two aspects of the geometry of the infinite open cluster. First, we shall show that, with probability near 1, there is an open path traversing the box $B(n)$ when n is large. Secondly, we shall see that the 'surface' of the infinite cluster has the same order as its volume.

Let $B(n)$ be the box with side-length $2n$ and centre at the origin. A *left-right crossing* of $B(n)$ is an open path of $B(n)$ joining some vertex x with $x_1 = -n$ to some vertex y with $y_1 = n$; we denote by $LR(n)$ the event that there exists a left–right crossing of $B(n)$. It is clear that $P_p(LR(n))$ decays exponentially to 0 as $n \to \infty$ when $0 < p < p_c$, since $LR(n)$ occurs only if one of the $(2n + 1)^{d-1}$ vertices on the left face of $B(n)$ is in an open cluster of size at least $2n + 1$. The picture is quite different when $p > p_c$, since there exists an infinite open cluster in this case.

(6.125) Theorem. *If* $\theta(p) > 0$ *then* $P_p(LR(n)) \to 1$ *as* $n \to \infty$.

The proof of this result uses only the uniqueness of the infinite open cluster when $\theta(p) > 0$, and it is for this reason that the theorem contains no estimate for the rate of convergence to 1 of $P_p(LR(n))$. It is conjectured that this takes place extremely quickly in that

$$P_p(LR(n)) \ge 1 - \exp(-\beta(p)n^{d-1}) \tag{6.126}$$

for all n and some $\beta(p) > 0$; this form of the exponent is motivated by the fact that n^{d-1} is the order of the number of possible endvertices of left–right crossings of $B(n)$. Inequality (6.126) is known to be valid if $p > p_c^+$ (see Chayes and Chayes (1986d, Prop. 3.4) and the remarks in Section 6.4); in two dimensions it is a simple consequence of the self-duality of $\mathbb{L}^2$. Related results may be found in Section 10.4.

We turn now to the ratio of boundary to volume of the infinite cluster. Let I be the set of vertices of $\mathbb{L}^d$ which are in infinite open clusters; we write I_e for the set of open edges which join pairs of vertices in I, and ΔI for the set of closed edges of $\mathbb{L}^d$ which have at least one endvertex in I. We express the boundary/volume ratio of I as

$$\lim_{n\to\infty} \frac{|\Delta I \cap B(n)|}{|I_e \cap B(n)|},$$

where $I_e \cap B(n)$ and $\Delta I \cap B(n)$ are interpreted as the subsets of I_e and ΔI containing edges which join pairs of vertices in $B(n)$.

(6.127) Theorem. *If $p_c < p < 1$ then, as $n \to \infty$,*

$$\frac{|\Delta I \cap B(n)|}{|I_e \cap B(n)|} \to \frac{1-p}{p} \qquad \textit{almost surely.}$$

Proof of Theorem (6.125). Suppose that $\theta(p) > 0$, so that there exists almost surely a unique infinite open cluster. Let ε be small and positive, and choose m large enough so that

$$P_p(I(m)) > 1 - \varepsilon, \tag{6.128}$$

where $I(m)$ is the event that some vertex of $B(m)$ is in the infinite open cluster. Let n be an integer satisfying $n > m$, and let $F_1, F_2, \dots, F_{2d}$ be a list of the $2d$ faces of $B(n)$. If $I(m)$ occurs then some vertex of $B(m)$ is joined by an open path to some face F_i. Thus

$$\begin{aligned} 1 - P_p(I(m)) &\geq 1 - P_p\left(\bigcup_{i=1}^{2d} \{B(m) \leftrightarrow F_i \text{ in } B(n)\}\right) \\ &= P_p\left(\bigcap_{i=1}^{2d} \{B(m) \leftrightarrow F_i \text{ in } B(n)\}^c\right). \end{aligned}$$

The last $2d$ events are decreasing events having equal probability. We apply the FKG inequality to deduce that

$$1 - P_p(I(m)) \geq \{1 - P_p(B(m) \leftrightarrow F \text{ in } B(n))\}^{2d}$$

for any given face F of $B(n)$. When combined with (6.128), this implies that

$$P_p(B(m) \leftrightarrow F \text{ in } B(n)) \geq 1 - \varepsilon^{1/(2d)}. \tag{6.129}$$

Writing $F_L(n)$ and $F_R(n)$ for the left and right faces of $B(n)$ (that is, the

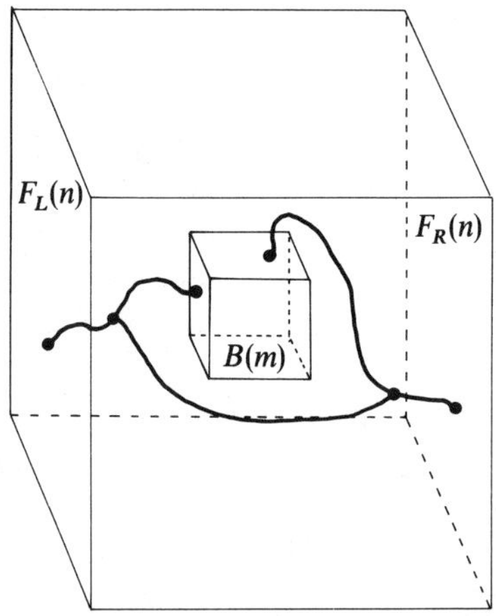

Figure 6.6. The smaller box is $B(m)$ and the larger is $B(n)$. $F_L(n)$ and $F_R(n)$ are the 'left' and 'right' faces of $B(n)$. This is a sketch of open paths joining $B(m)$ to $F_L(n)$ and to $F_R(n)$, together with an open path of $B(n)$ linking the first two paths.

sets $\{x \in \partial B(n): x_1 = -n\}$ and $\{x \in \partial B(n): x_1 = n\}$) we have by the FKG inequality that

$$(6.130) \qquad P_p(\{B(m) \leftrightarrow F_L(n) \text{ in } B(n)\} \cap \{B(m) \leftrightarrow F_R(n) \text{ in } B(n)\}) \geq \{1 - \varepsilon^{1/(2d)}\}^2.$$

See Figure 6.6 for an illustration of this event. Two open paths from $B(m)$ to $F_L(n)$ and to $F_R(n)$ may be combined to form a left–right crossing of $B(n)$ whenever they are connected in $B(n)$ by an open path (see the figure again). The chance that there exist two such paths which are not connected by an open path of $B(n)$ is no greater than the probability of the event

$$A_{m,n} = \{\text{there exist two vertices of } \partial B(m) \text{ which are in disjoint open clusters of } B(n) \text{ both of which intersect } \partial B(n)\};$$

however $A_{m,n} \supseteq A_{m,n+1}$, and it follows that

$$P_p(A_{m,n}) \to P_p(\text{there exist two vertices of } \partial B(m) \text{ in disjoint infinite open clusters}) = 0$$

as $n \to \infty$, by the uniqueness theorem for the infinite open cluster. Thus, by (6.130),

$$P_p(LR(n)) \geq \{1 - \varepsilon^{1/(2d)}\}^2 - P_p(A_{m,n}) \to \{1 - \varepsilon^{1/(2d)}\}^2 \quad \text{as} \quad n \to \infty.$$

We take the limit as $\varepsilon \downarrow 0$ to obtain the required conclusion. ∎

Proof of Theorem (6.127). We follow Grimmett (1981b). Suppose $p_c < p < 1$. For each edge e of $\mathbb{L}^d$, we let χ_e and β_e be the indicator functions of the events that e lies in I_e and ΔI, respectively. The families $(\chi_e: e \in \mathbb{E}^d)$ and $(\beta_e: e \in \mathbb{E}^d)$ are stationary under translations of the lattice; furthermore, all events which are invariant under such translations have probability either 0 or 1. By the ergodic theorem (see Dunford and Schwartz (1958, Theorem VIII.6.9) or Tempel'man (1972, Theorem 6.1 and Corollary 6.2)) we have that, as $n \to \infty$,

$$\frac{1}{E_n} \sum_{e \in B(n)} \chi_e \to E_p(\chi_e), \qquad \frac{1}{E_n} \sum_{e \in B(n)} \beta_e \to E_p(\beta_e)$$

almost surely, where E_n is the number of edges in $B(n)$ and the sums are over all such edges. However

$$E_p(\chi_e) = P_p(e \text{ is in an infinite open cluster})$$

and

$$\begin{aligned} E_p(\beta_e) &= P_p(e \text{ closed, but an endvertex of } e \text{ is in an infinite open cluster}) \\ &= qP_p^e(\text{some endvertex of } e \text{ is in an infinite open cluster}) \\ &= qp^{-1}P_p(e \text{ open, and some endvertex of } e \text{ is in an infinite open} \\ &\qquad \text{cluster}) \\ &= qp^{-1}P_p(e \text{ is in an infinite open cluster}), \end{aligned}$$

where P_p^e is the probability measure of percolation on $\mathbb{L}^d$ with e removed, and $p + q = 1$. Thus, as $n \to \infty$,

$$\begin{aligned} \frac{|\Delta I \cap B(n)|}{|I_e \cap B(n)|} &= \sum_{e \in B(n)} \beta_e \Big/ \sum_{e \in B(n)} \chi_e \\ &\to qp^{-1} \qquad \text{almost surely.} \end{aligned}$$ ∎

6.10 Notes

Section 6.2. The uniqueness of the infinite open cluster was proved by Aizenman, Kesten, and Newman (1987); Newman and Schulman (1981a, b) deal with related material. Our simpler proof of uniqueness is due to Gandolfi, Grimmett, and Russo (1988). It has been generalized by Gandolfi (1988) to yield a proof that there exists at most one infinite cluster for any stationary finite-range Gibbs state in any dimension. A similar conclusion was reached by Gandolfi, Keane, and Russo (1988) for any ergodic translation-invariant measure on the state space $\{0, 1\}^{\mathbb{E}^2}$ in two dimensions, whenever the FKG inequality remains in force. Our proof of Lemma (6.2) is due to Russo (unpublished), probably amongst others.

As for the number of infinite open clusters of bond percolation on *subgraphs* of $\mathbb{L}^d$, Barsky, Grimmett, and Newman (1988) have proved the uniqueness theorem for certain graphs including $\mathbb{Z}_+^e \times \mathbb{Z}^{d-e}$ for $1 \le e \le d$. Earlier results in this direction are due to Aizenman, Chayes, Chayes, Fröhlich, and Russo (1983), Kesten (1988a), and Newman and Wu (1988, unpublished).

The situation concerning the number N of infinite open clusters is, of course, quite different for percolation on a tree rather than on a lattice (see Section 8.1), and it is elementary that (almost surely) $N = \infty$ for such a graph whenever $N \ge 1$. For general periodic graphs G, the proof of uniqueness may be adapted successfully so long as the surface/volume ratio of the ball with radius n tends to 0 as $n \to \infty$. For graphs G which do not satisfy this condition, there exist generally three 'phases', corresponding to $N = 0$, $N = 1$, and $N = \infty$. See Grimmett and Newman (1988) for examples of such results.

Note added in proof. Burton and Keane (1988) have found a proof of the uniqueness of the infinite cluster of $\mathbb{L}^d$ which is startlingly concise and easily generalized, and the idea is as follows. By an argument of Newman and Schulman (1981a, b), it suffices to rule out the possibility that the number N of infinite open clusters is ∞ almost surely. Suppose then that $N = \infty$ almost surely. We call a vertex x an *encounter point* if $C(x)$ is infinite but the removal of x breaks $C(x)$ into the disjoint union of three infinite clusters. It may be seen that there is a strictly positive probability that x is an encounter point (use a construction similar to that of the proof of Lemma (6.2)) and therefore there is a strictly positive density of encounter points in the lattice. Consider now the intersection $I_n = I \cap B(n)$ of an infinite cluster I with the box $B(n)$. The components of I_n are tree-like structures with the encounter points of I_n as articulation points; each component has a skeletal structure resembling a regular tree with degree 3. Such trees have 'surfaces' comparable in size to the number of encounter points within them, and thus the total number of encounter points inside $B(n)$ has order no greater than $|\partial B(n)|$. On the other hand, there is a positive density of encounter points, which is to say that the number of such points has the order of $|B(n)|$, a contradiction.

Section 6.3. The right-continuity of θ was observed by Russo (1978), whilst the relationship between its left-continuity and the uniqueness of the infinite open cluster was discovered by van den Berg and Keane (1984). Results of Harris (1960) and Kesten (1980a) imply that $\theta(p_c(2)) = 0$ in two dimensions. Aizenman, Kesten, and Newman (1987) noted the continuity of the connectivity functions. Barsky, Grimmett, and Newman (1988) have proved the continuity of the percolation probability for bond percolation on subgraphs such as half-spaces. (*Added in proof.*) Hara and Slade (1988) have shown that θ is continuous at $p_c(d)$ if d is sufficiently large; see the notes for Section 8.2.

Section 6.4. Probably the first reference to slabs is to be found in Aizenman, Chayes, Chayes, Fröhlich, and Russo (1983). See also Chayes, Chayes, and Newman (1987). Theorems (6.45) and (6.46) are due to Barsky, Grimmett, and Newman (1988).

Section 6.5. Theorem (6.48) comes from Chayes, Chayes, Grimmett, Kesten, and Schonmann (1988). Our proof of Theorem (6.51) is a reworking of the corresponding result of Chayes, Chayes, and Newman (1987) with the word 'invasion' removed. Kesten (1988a) has proved the 'converse' of Theorem (6.51): his results imply that if $\sigma(p) > 0$ and $p > p_c$ then $p > p_c^+$. It is conjectured that $p_c^+ = p_c$. Hammersley (1963) grew clusters recursively; see also Klein and Shamir (1982) if you can find a copy.

Section 6.6. See Chayes, Chayes, and Newman (1987) and Chayes, Chayes, Grimmett, Kesten, and Schonmann (1988) for discussions of the truncated connectivity functions and the supercritical correlation length.

Section 6.7. Aizenman, Delyon, and Souillard (1980) proved Theorem (6.91) in a more general context than percolation. Similar estimates in two dimensions have been performed by Kesten (1982, p. 98). Theorem (6.95) is due to Kesten and Zhang (1988). Steps were taken earlier in this direction by Kunz and Souillard (1978) (with loose topological ends, later secured by Kesten (1986a, Lemma 2.24)) and Chayes, Chayes, and Newman (1987). (*Added in proof.*) K. Alexander, J. T. Chayes, and L. Chayes have answered the question posed after (6.93), when $d = 2$.

Section 6.8. This technique of proving differentiability appears first in Russo (1978), and later in Brånvall (1980), Grimmett (1981b), Kesten (1982), and Chayes, Chayes, and Newman (1987), and probably elsewhere also.

Section 6.9. Russo (1987, unpublished) has pointed out that the uniqueness of the infinite open cluster implies the existence of open crossings of large cubes; this argument was probably known to others also. A weaker form of Theorem (6.127) was anticipated in the physics literature (see Hankey (1978), for example), and appeared with complete proof in Newman and Schulman (1981a, b), as well as in its current form in Grimmett (1981b); the last paper contains also a central limit theorem for the pair $(|I_e \cap B(n)|, |\Delta I \cap B(n)|)$ in two dimensions.

7

Near the Critical Point: Scaling Theory

7.1 Power Laws and Critical Exponents

The behaviour of the percolation process on $\mathbb{L}^d$ depends dramatically on whether $p < p_c$ or $p > p_c$. In the former subcritical case, all open clusters are almost surely finite and their size-distribution has a tail which decays exponentially. In the latter supercritical case, there exists almost surely an infinite open cluster and the size-distribution of the remaining finite open clusters has a tail which decays slower than exponentially. Some of the major differences between these two phases are highlighted in the following table.

	Subcritical phase $p < p_c$	Supercritical phase $p > p_c$
Number of infinite open clusters	0 a.s.	1 a.s.
Percolation probability	$\theta(p) = 0$	$\theta(p) > 0$
Mean cluster size	$\chi(p) < \infty$	$\chi(p) = \infty$
Tail of finite clusters $P_p(n \le \|C\| < \infty)$	$\approx e^{-\zeta(p)n}$	$\ge e^{-\gamma(p)n^{(d-1)/d}}$

It is clear that interesting phenomena occur when p is near to its critical value p_c. Up to now, mathematicians have made great efforts but only limited progress towards understanding such phenomena, although they have been spurred on by the rich predictions and discoveries of more applied scientists. The physical theory of phase transitions and critical phenomena is well developed, and such techniques as scaling theory and renormalization are central to this theory. The mathematical foundations of such techniques and their applications remain uncertain, and it is a mathematical challenge of major importance to rectify this situation.

Renormalization and scaling theory make certain predictions about the behaviour of the percolation process when p is near or equal to its critical value p_c. With the present state of knowledge, many believe that these predictions cannot be appreciated fully without reference to the intuitive notions which motivate them. That is to say, whilst it may be possible to present simple and concrete statements about the conjectured behaviour of θ, χ, and other macroscopic quantities when p is near to p_c, such conjectures remain only superficial aspects of the scaling theory which gave birth to them. The challenge to mathematicians is therefore to *make sense of scaling theory*, rather than to verify a bunch of conjectures.

We shall nevertheless begin this chapter with a bunch of conjectures, and we hope that these conjectures will be sufficiently remarkable to capture the reader's attention for the remainder of this chapter at least. The material relating to the critical phenomenon is organized as follows. This chapter is devoted to a brief general introduction to scaling theory, including an innocent's guide to the bones of renormalization. This incorporates the idea of correlation length and the famous scaling relations. Very little of this material has yet been made rigorous; in the next chapter, we discuss the achievements and limitations of the current rigorous theory. We suppose throughout that $d \geq 2$.

We move on towards our bunch of conjectures. These concern largely the behaviour of θ, χ, χ^f, and κ when p is near to p_c. Consider first the percolation probability $\theta(p)$ when $p - p_c$ is small. It is believed that $\theta(p)$ decays to 0 in the manner of a power of $p - p_c$ as $p \downarrow p_c$. That is to say, we conjecture that, as $p \downarrow p_c$,

$$\theta(p) \approx (p - p_c)^\beta \qquad \text{for some} \quad \beta > 0; \tag{7.1}$$

the exponent β depends of course on the value of d. It is not clear how strong we may expect such an asymptotic relation to be, and it is for this reason that we use the logarithmic relation $\approx$. Thus we conjecture that

$$\lim_{p \downarrow p_c} \frac{\log \theta(p)}{\log(p - p_c)} = \beta. \tag{7.2}$$

We note that this conjecture requires that $\theta(p_c) = 0$, which is itself an open question when $d \geq 3$ (see Lemma (9.12) for the case $d = 2$).

We turn next to the mean cluster size $\chi(p)$ and the mean size $\chi^f(p)$ of a finite open cluster. It is believed that there exist $\gamma, \gamma' > 0$ such that

$$\chi(p) \approx (p_c - p)^{-\gamma} \qquad \text{as} \quad p \uparrow p_c \tag{7.3}$$

and

$$\chi^f(p) \approx (p - p_c)^{-\gamma'} \qquad \text{as} \quad p \downarrow p_c, \tag{7.4}$$

where the asymptotic relations are to be interpreted in the logarithmic manner of (7.2). It is believed further that $\gamma = \gamma'$. Of course $\chi(p) = \chi^f(p)$ for $p < p_c$, so that these conjectures amount to asserting that

$$\chi^f(p) \approx |p - p_c|^{-\gamma} \qquad \text{as} \quad p \to p_c \tag{7.5}$$

for some $\gamma > 0$.

It is believed that the number $\kappa(p)$ of open clusters per vertex is twice but not thrice differentiable at $p = p_c$. We thus conjecture that the third derivative of κ satisfies

$$\kappa'''(p) \approx |p - p_c|^{-1-\alpha} \qquad \text{as} \quad p \to p_c, \tag{7.6}$$

where $-1 \leq \alpha < 0$. This is not a very natural way to express the conjecture, but there are certain historical reasons for expressing the exponent in this form.

The functions θ, χ^f, and κ are all expressible as moments of $|C|$ on the event that $|C| < \infty$:

$$\theta(p) = 1 - E_p(1; |C| < \infty),$$
$$\chi^f(p) = E_p(|C|; |C| < \infty),$$
$$\kappa(p) = E_p(|C|^{-1}; |C| < \infty),$$

and thus it is natural to introduce a critical exponent for each moment $E_p(|C|^k; |C| < \infty)$. It is believed that there exists $\Delta > 0$ such that

$$\frac{E_p(|C|^{k+1}; |C| < \infty)}{E_p(|C|^k; |C| < \infty)} \approx |p - p_c|^{-\Delta} \qquad \text{as} \quad p \to p_c, \tag{7.7}$$

whenever $k \geq 1$. The hypothetical quantity Δ is called the *gap exponent.*

We move on to consider the case $p = p_c$. If $\theta(p_c) = 0$ then $\chi(p_c) = \sum_n n P_{p_c}(|C| = n)$. We saw in (5.52) that $\chi(p_c) = \infty$, and this suggests that $P_{p_c}(|C| = n)$ decays in the manner of a negative power of n. We therefore conjecture that there exists $\delta \geq 1$ such that

$$P_{p_c}(|C| = n) \approx n^{-1-1/\delta} \qquad \text{as} \quad n \to \infty; \tag{7.8}$$

there are historical reasons for writing the exponent in this form. The asymptotic relation (7.8) implies that

$$P_{p_c}(n \leq |C| < \infty) \approx n^{-1/\delta} \qquad \text{as} \quad n \to \infty. \tag{7.9}$$

We expect similarly that the probability $P_{p_c}(0 \leftrightarrow \partial B(n))$, that there is an open path from the origin to the surface of the box $B(n)$, decays in the manner of

$$P_{p_c}(0 \leftrightarrow \partial B(n)) \approx n^{-1/\rho} \qquad \text{as} \quad n \to \infty \tag{7.10}$$

for some $\rho > 0$; some authors use the notation δ_r in place of ρ. We make the slightly stronger conjecture that

(7.11) $$P_{p_c}(\mathrm{rad}(C) = n) \approx n^{-1-1/\rho} \quad \text{as} \quad n \to \infty;$$

as usual, the *radius* of C is defined as in (5.3) by $\mathrm{rad}(C) = \max\{|x|: x \in C\}$ where $|x| = \delta(0, x)$. We expect that the probability $\tau_{p_c}(0, x)$, that there exists an open path joining the origin to x, should decay in the manner of a negative power of $|x|$, and we conjecture therefore that

(7.12) $$\tau_{p_c}(0, x) \approx |x|^{2-d-\eta} \quad \text{as} \quad |x| \to \infty,$$

for some η.

We summarize these conjectures in a table.

Percolation probability	$\theta(p) \approx (p - p_c)^\beta$	as $p \downarrow p_c$
Mean cluster size	$\chi^f(p) \approx \|p - p_c\|^{-\gamma}$	as $p \to p_c$
Number of clusters per vertex	$\kappa'''(p) \approx \|p - p_c\|^{-1-\alpha}$	as $p \to p_c$
Cluster size moments	$\dfrac{E_p(\|C\|^{k+1}; \|C\| < \infty)}{E_p(\|C\|^k; \|C\| < \infty)} \approx \|p - p_c\|^{-\Delta}$	as $p \to p_c$, for $k \geq 1$
Cluster size at p_c	$P_{p_c}(\|C\| = n) \approx n^{-1-1/\delta}$	as $n \to \infty$
Connectivity at p_c	$\tau_{p_c}(0, x) \approx \|x\|^{2-d-\eta}$	as $\|x\| \to \infty$

These relations involve seven 'critical exponents' α, β, γ, δ, Δ, η, ρ, each of which depends on the value of d. No proofs are known of these relations.

In the next section, we sketch the physical theory of 'scaling', which predicts that the seven critical exponents are not independent of each other but satisfy two sets of relations, called 'scaling' and 'hyperscaling' relations. The *scaling relations* assert that α, β, γ, δ, Δ satisfy

(7.13) $$2 - \alpha = \gamma + 2\beta = \beta(\delta + 1),$$

and

(7.14) $$\Delta = \delta\beta,$$

and the validity of these equations is widely accepted. The hyperscaling relations are more questionable, and we require some preliminary discussion before we may state these relations.

A principal concept of scaling theory is the idea of 'correlation length', and we motivate this concept by studying the probability $\tau_p^f(0, e_n)$, that the origin and the vertex $e_n = (n, 0, \ldots, 0)$ are in the same finite open cluster. We believe that

(7.15) $$\tau_p^f(0, e_n) \approx \begin{cases} n^{2-d-\eta} & \text{if} \quad p = p_c, \\ e^{-n/\xi(p)} & \text{if} \quad p \neq 0, p_c, 1, \end{cases}$$

for some $\xi(p)$ satisfying $\xi(p) \to \infty$ as $p \to p_c$. How does $\tau_p^f(0, e_n)$ behave in the double limit as $p \to p_c$ and $n \to \infty$? It is not unreasonable to guess that $\tau_p^f(0, e_n)$ behaves roughly like $n^{2-d-\eta}f(n/\xi(p))$ for some function f when p is near to p_c and n is large; in the light of Theorems (5.44) and (6.83) we may guess that f is (more or less) the negative exponential function, which is to say that

(7.16) $$\frac{\tau_p^f(0, e_n)}{\tau_{p_c}^f(0, e_n)} \approx e^{-n/\xi(p)} \quad \text{as} \quad p \to p_c \quad \text{and} \quad n \to \infty,$$

where $\xi(p_c) = \infty$. That is to say, $\tau_p^f(0, e_n)$ and $\tau_{p_c}^f(0, e_n)$ differ significantly only when $n/\xi(p)$ is large. In this sense, $\xi(p)$ is the natural 'length scale' of bond percolation with edge-probability p. For example, suppose that we are told that, for some N, the origin and e_N are in the same finite open cluster of a supercritical percolation process, but we are not told the value of p. If N is small then we have little information about p, whereas if N is large then we have reason to deduce that p is near to p_c. How large need N be in order that we may deduce that $p < p_c + \varepsilon$, say? From the above argument, we require N to have at least the order of magnitude of $\xi(p_c + \varepsilon)$. That is to say, as $\varepsilon \downarrow 0$, we require that the order of N grows at least as fast as $\xi(p_c + \varepsilon)$. There is an equally valid argument for subcritical percolation.

Mathematical physicists use the term 'correlation length' rather loosely to mean the 'natural length scale' of the process: the 'correlation length' is the minimal scale on which percolation with edge-probability p differs qualitatively from percolation with edge-probability p_c. In the light of the above discussion, we define the *correlation length* $\xi(p)$ by

(7.17) $$\xi(p)^{-1} = \lim_{n \to \infty} \left\{ -\frac{1}{n} \log \tau_p^f(0, e_n) \right\},$$

in agreement with the definitions of Sections 5.2 and 6.6. It is a fundamental assumption of scaling theory that there is a unique length scale in the above sense, and it is neither clear that this holds nor is it self-evident that our definition (7.17) is the most useful definition of correlation length. We note that we are concerned only with the order of magnitude of $\xi(p)$ as p approaches p_c. It would be equally acceptable to define the correlation length to be $2\xi(p)$, and more generally to be any function $\varphi(p)$ with the property that $\varphi(p)/\xi(p)$ is bounded away from 0 and ∞ in a neighbourhood of p_c.

There is another definition of correlation length which has received considerable attention in the literature, and this is the mean radius of a finite open cluster, defined formally by

(7.18) $$\varphi(p) = \sqrt{\frac{1}{\chi^f(p)} \sum_{x \in \mathbb{Z}^d} |x|^2 \tau_p^f(0, x)} \qquad \text{for} \quad p \neq p_c.$$

We may think of $\varphi(p)$ as being $\sqrt{E(|X|^2)}$ where X is a random vertex of $\mathbb{Z}^d$ having distribution

$$P(X = x) = \frac{1}{\chi^f(p)} \tau_p^f(0, x) \qquad \text{for} \quad x \in \mathbb{Z}^d. \tag{7.19}$$

We leave it as an exercise to show that, under reasonable assumptions, $\varphi(p)/\xi(p)$ is bounded away from 0 and ∞ in a neighbourhood of p_c.

What is the scaling theory of correlation length? That is to say, what is the asymptotic behaviour of $\xi(p)$ as p approaches p_c from above or beneath? A fundamental assumption of scaling theory is that there exists $\nu > 0$ such that

$$\xi(p) \approx |p - p_c|^{-\nu} \qquad \text{as} \quad p \to p_c, \tag{7.20}$$

and we therefore conjecture that such an asymptotic relation is valid. As before, there are two aspects to this conjecture: that $\xi(p)$ approaches $\xi(p_c) = \infty$ in the manner of a negative power of $|p - p_c|$, and that the power in question does not depend on whether p approaches p_c from above or beneath.

There is a scaling relation involving the correlation length exponent ν:

$$\gamma = \nu(2 - \eta). \tag{7.21}$$

Before moving on to the hyperscaling relations, we note that (7.13), (7.14), and (7.21) provide four equations for seven quantities α, β, γ, δ, Δ, η, ν.

The *hyperscaling relations* involve the number d of dimensions, and assert that

$$d\rho = \delta + 1 \tag{7.22}$$

and

$$d\nu = 2 - \alpha, \tag{7.23}$$

so that all the terms in (7.13) equal $d\nu$. The scaling and hyperscaling relations combine to form six equations in eight unknowns. It is believed that the hyperscaling relations are valid only for values of d satisfying $d \le d_c$, for some d_c called the (*upper*) *critical dimension*. For percolation on $\mathbb{L}^d$ when $d \ge d_c$, it is believed that the process behaves roughly in the same manner as percolation on an infinite regular tree (that is, a lattice without circuits). More precisely, many believe that the critical exponents take on the corresponding values from percolation on a regular tree when $d \ge d_c$; we shall see in Section 8.1 that $\alpha = -1$, $\beta = 1$, $\gamma = 1$, $\delta = 2$, $\Delta = 2$, $\rho = \frac{1}{2}$, $\eta = 0$, and $\nu = \frac{1}{2}$ for such a percolation process. If these values are attained by percolation on $\mathbb{L}^d$ and the hyperscaling relation $d\nu = 2 - \alpha$ is valid, then $d = 6$. This calculation provides some evidence that the hyperscaling relations are valid only when $d \le 6$ and that, in six or more dimensions, the critical exponents take on the values corresponding to percolation on a regular tree.

The critical exponents α, β, γ, δ, Δ, η, ρ, ν depend of course upon the number d of dimensions, but there are physical reasons to suppose that they do not depend on the particular lattice structure. That is to say, it is believed that the critical exponents of bond percolation on any d-dimensional lattice $\mathscr{L}$ are the same as those for bond percolation on $\mathbb{L}^d$. This belief has much of its basis in the hypothesis that only over large 'length scales' do critical phenomena manifest themselves, and the local lattice structure is not easily observed over such distances. There is considerable numerical evidence to support this conjecture of 'universality', which may be phrased as follows. Consider the class of all percolation processes on d-dimensional lattices having finite vertex degrees. It is believed that this is a 'universality class' of processes, in the sense that the natures of the phase transitions thereof are similar; in particular, all processes in this class are believed to have equal critical exponents.

We may summarize the ideas of this section as follows.

(a) Macroscopic quantities such as θ, χ^f, and κ''' behave in the manner of powers of $|p - p_c|$ when p is near to p_c.
(b) The distributions of the radius and size of open clusters have tails which decay as negative powers when $p = p_c$.
(c) There is a fundamental minimal scale $\xi(p)$ over which percolation with edge-probability p is distinguishable from percolation with $p = p_c$, and this 'correlation length' behaves like a negative power of $|p - p_c|$ when p is near to p_c.
(d) The critical exponents satisfy the scaling relations.
(e) The critical exponents satisfy the hyperscaling relations if $d \leq 6$ and take on the values corresponding to percolation on a regular tree when $d \geq 6$.
(f) The critical exponents depend only on the number d of dimensions and not otherwise upon the individual structure of the underlying lattice.

In the next section, we present a more detailed account of scaling theory, leading to derivations of the scaling and hyperscaling relations. Underlying scaling theory is the theory of renormalization, and Section 7.3 is a superficial essay on this topic.

The asymptotic analysis of this chapter will appear rather shaky to some, and not without good reason. We shall use three types of notation to denote asymptotic equivalence. We write $a_n \approx b_n$ (respectively $a(p) \approx b(p)$) if $\log a_n/\log b_n \to 1$ as $n \to \infty$ (respectively $\log a(p)/\log b(p) \to 1$ as $p \to p_c$). We write $a_n \sim b_n$ if $a_n/b_n \to 1$ as $n \to \infty$, and similarly for functions of p as $p \to p_c$. The notation $\simeq$ is reserved for use in a mathematically imprecise manner to relate pairs of quantities whose asymptotic behaviours are believed to be rather similar.

7.2 Scaling Theory

The principal hypothesis of scaling theory is that quantities such as $P_p(|C| = n)$ have certain specific asymptotic expansions which are valid for all p near to p_c and for all large n. Starting from such hypotheses, we may use scaling theory to arrive at the scaling relations (7.13), (7.14), and (7.21). Other hypotheses yield the hyperscaling relations. We describe such an approach in this section; in the next section, we shall describe briefly how the theory of renormalization may be used to provide a heuristic basis for scaling theory.

We assume henceforth that all the critical exponents α, β, γ, δ, Δ, η, ρ, v are defined as in the last section. Our first scaling hypothesis is that

(7.24) $$P_p(|C| = n) \sim \begin{cases} n^{-\sigma} f_-(n/\xi(p)^\tau) & \text{if} \quad p \le p_c, \\ n^{-\sigma} f_+(n/\xi(p)^\tau) & \text{if} \quad p \ge p_c, \end{cases}$$

where σ and τ are positive constants, f_- and f_+ are smooth functions (infinitely differentiable, say) on $[0, \infty)$, and the relation $\sim$ is understood in the usual asymptotic sense to mean that the ratio of the left-hand side to the right-hand side converges to 1 as $n \to \infty$ and $p \to p_c$; note that the notation σ, τ is not standard. Following Theorems (5.78) and (6.91), it is not unreasonable to guess that $f_-(m) \approx \exp(-Am)$ and $f_+(m) \approx \exp(-Bm^{(d-1)/d})$ for some A and B as $m \to \infty$, but we shall not assume this here. Instead, we shall suppose merely that $f_-(0) = f_+(0) > 0$, and that $f_-(m) \to 0$ and $f_+(m) \to 0$ faster than any power of m^{-1} as $m \to \infty$. The first of these assumptions guarantees that the two asymptotic relations in (7.24) coincide when $p = p_c$, and the second implies that the behaviour of $P_p(|C| = n)$ is dominated when $p \neq p_c$ by a quantity which depends only on the ratio of n to a power of the correlation length.

In the matter of terminology, we note that physicists commonly refer to such a hypothesis as (7.24) as being an 'ansatz'. There are certain differences between an ansatz and a hypothesis, the former being often a guessed formula which may be used as a basis for calculations.

The asymptotic analysis which follows is somewhat speculative. Consequently, the correct form of scaling hypothesis is not entirely obvious. It is not uncommon to encounter in the literature a version of (7.24) which is assumed to be valid in the limit when $n \to \infty$ and $p \to p_c$ in such a way that $n/\xi(p)^\tau$ approaches a positive constant.

This scaling hypothesis (or ansatz) introduces the new exponents σ and τ, but these are readily expressed in terms of those exponents already to hand. We remember that $P_{p_c}(|C| = n) \approx n^{-1-1/\delta}$, and we deduce that

(7.25) $$\sigma = 1 + \delta^{-1}.$$

We next cross our fingers and perform the following calculation, typical of scaling theory. In the following, f represents f_- if $p < p_c$ and f_+ if

$p > p_c$. We have that

(7.26)
$$\begin{aligned}\chi^f(p) &= \sum_n nP_p(|C| = n)\\ &\simeq \sum_n n^{1-\sigma} f(n\xi(p)^{-\tau})\\ &\simeq \int_0^\infty n^{1-\sigma} f(n\xi(p)^{-\tau})\, dn\\ &= \xi(p)^{\tau(1-\sigma)+\tau} \int_0^\infty u^{1-\sigma} f(u)\, du\end{aligned}$$

by the substitution $u = n\xi(p)^{-\tau}$. Now $\sigma = 1 + \delta^{-1}$ and $\delta \geq 1$, giving that $1 - \sigma = -\delta^{-1} \geq -1$. If $\delta > 1$, then $1 - \sigma > -1$ and the last integral converges, but we have not eliminated the case $\delta = 1$, for which $1 - \sigma = -1$ and the integral diverges. We shall see in Section 8.2 that $\delta \geq 2$ whenever δ exists, so that this discussion is academic; an alternative argument uses the assumption that $\chi^f(p) < \infty$ when $p \neq p_c$. We now make the assumptions that $\chi^f(p) \approx |p - p_c|^{-\gamma}$ and $\xi(p) \approx |p - p_c|^{-\nu}$ to find that

$$\gamma = \nu\tau(2 - \sigma)$$

which, in conjunction with (7.25), yields

(7.27)
$$\tau = \frac{\gamma}{\nu(1 - \delta^{-1})},$$

and we have expressed σ and τ in terms of γ, δ, ν.

We move on to consider the percolation probability $\theta(p)$, and we suppose now that $p > p_c$. Then

(7.28)
$$\begin{aligned}\theta(p) &= 1 - \sum_n P_p(|C| = n)\\ &= \sum_n \{P_{p_c}(|C| = n) - P_p(|C| = n)\} \qquad \text{if} \quad \theta(p_c) = 0\\ &\simeq \sum_n n^{-\sigma}\{f_+(0) - f_+(n\xi(p)^{-\tau})\}\\ &\simeq \int_0^\infty n^{-\sigma}\{f_+(0) - f_+(n\xi(p)^{-\tau})\}\, dn\\ &= \xi(p)^{\tau(1-\sigma)} \int_0^\infty u^{-\sigma}\{f_+(0) - f_+(u)\}\, du\end{aligned}$$

as before; such asymptotic relations should not be interpreted too strictly. The last integral converges since the integrand behaves like $-u^{1-\sigma} f'_+(0)$ near $u = 0$. From the assumption $\theta(p) \approx (p - p_c)^\beta$, we have that

$$\beta = -\nu\tau(1 - \sigma),$$

giving from (7.25) and (7.27) that

$$\gamma + 2\beta = \beta(\delta + 1), \tag{7.29}$$

our first scaling relation.

Next we apply the same argument to the number $\kappa(p)$ of open clusters per vertex. As before, f represents f_- if $p < p_c$ and f_+ if $p > p_c$. We have that

$$\begin{aligned}\kappa(p) &= \sum_n \frac{1}{n} P_p(|C| = n) \\ &\simeq \sum_{n=1}^{\infty} n^{-1-\sigma} f(n\xi(p)^{-\tau}) \\ &\simeq \xi(p)^{-\tau\sigma} \int_{\xi(p)^{-\tau}}^{\infty} u^{-1-\sigma} f(u)\, du\end{aligned} \tag{7.30}$$

as before. Now, if ε is small and positive,

$$\begin{aligned}\int_{\xi(p)^{-\tau}}^{\infty} u^{-1-\sigma} f(u)\, du &= \int_{\xi(p)^{-\tau}}^{\varepsilon} u^{-1-\sigma} f(u)\, du + \int_{\varepsilon}^{\infty} u^{-1-\sigma} f(u)\, du \\ &\simeq f(0)\left[-\frac{1}{\sigma} u^{-\sigma}\right]_{\xi(p)^{-\tau}}^{\varepsilon} + \int_{\varepsilon}^{\infty} u^{-1-\sigma} f(u)\, du \\ &= A\xi(p)^{\tau\sigma} + B,\end{aligned} \tag{7.31}$$

where A and B are constants which depend upon ε. We substitute this into (7.30) to find that

$$\begin{aligned}\kappa(p) &\simeq A + B\xi(p)^{-\tau\sigma} \\ &\simeq A + B|p - p_c|^{\nu\tau\sigma}.\end{aligned} \tag{7.32}$$

We compare this with the assumption $\kappa'''(p) \approx |p - p_c|^{-1-\alpha}$, and we perform a sufficiently large leap of the imagination to arrive at

$$-1 - \alpha = \nu\tau\sigma - 3.$$

We now use (7.25), (7.27), and (7.29) to deduce the second scaling relation:

$$2 - \alpha = \gamma + 2\beta. \tag{7.33}$$

We argue similarly to obtain

$$E_p(|C|^k; |C| < \infty) \simeq D_k \xi(p)^{\tau(k+1-\sigma)}$$

if $k \geq 1$, where D_k is a constant which depends on k. Comparison with (7.7) yields a value for the gap exponent:

$$\begin{aligned}\Delta &= \tau\nu \\ &= \delta\beta \qquad \text{by (7.27) and (7.29).}\end{aligned} \tag{7.34}$$

This is just about as far as we may progress from the scaling hypothesis (7.24) for the cluster size distribution. The easiest way to obtain the relation (7.21), $\gamma = \nu(2-\eta)$, is to introduce a new scaling ansatz; we hypothesize that

$$\tau_p^f(0, x) \simeq \begin{cases} |x|^{2-d-\eta} g_-(|x|/\xi(p)) & \text{as } p \uparrow p_c \text{ and } |x| \to \infty, \\ |x|^{2-d-\eta} g_+(|x|/\xi(p)) & \text{as } p \downarrow p_c \text{ and } |x| \to \infty, \end{cases} \tag{7.35}$$

for some smooth functions g_-, g_+ (infinitely differentiable, say) satisfying $g_-(0) = g_+(0) > 0$ and $g_-(m) \to 0$, $g_+(m) \to 0$ faster than any power of m^{-1} as $m \to \infty$. Taking (7.15) into account, it is reasonable to suppose that $g_-(m) \approx e^{-Am}$ and $g_+(m) \approx e^{-Bm}$ as $m \to \infty$ for constants A and B, but we shall not assume this here. We have used the undefined asymptotic relation $\simeq$ in (7.35) for the following reason. Presumably $\tau_p^f(0, x)$ depends not only on the distance $|x|$ from the origin to x, but also on the geometry of the position of x. However, it is likely that we may re-scale $|x|$ in an appropriate way depending on this geometry, with the effect that equation (7.35), or something very similar, becomes valid with $\simeq$ replaced by $\sim$ and $|x|$ replaced by the re-scaled distance function; we do not attempt this here, but remind the reader of the discussion leading to Proposition (5.47).

Summing (7.35) over all x, we obtain

$$\chi^f(p) \simeq \sum_x |x|^{2-d-\eta} g(|x|/\xi(p)),$$

where g is interpreted as g_- or g_+ depending on whether $p < p_c$ or $p > p_c$. The number of vertices x for which $|x| = n$ has order n^{d-1}, so that

$$\begin{aligned} \chi^f(p) &\simeq A \sum_n n^{d-1} n^{2-d-\eta} g(n/\xi(p)) \\ &\simeq A \int_0^\infty n^{1-\eta} g(n/\xi(p))\, dn \\ &\simeq A\xi(p)^{2-\eta} \int_0^\infty u^{1-\eta} g(u)\, du \end{aligned}$$

for some constant A. It cannot be the case that $\eta \geq 2$, since $\chi^f(p)$ and $\xi(p)$ are assumed to diverge at $p = p_c$ in the manner of negative powers of $|p - p_c|$. Thus $\eta < 2$ and the integral converges, implying that

$$\chi^f(p) \simeq B\xi(p)^{2-\eta}$$

for some constant B; we set the exponents of $|p - p_c|$ to be equal, and obtain

$$\gamma = \nu(2-\eta), \tag{7.36}$$

our last scaling relation.

We turn now to the hyperscaling relations, beginning with the relation (7.22), $d\rho = \delta + 1$, involving the critical exponent ρ for the cluster radius. Let $p = p_c$, and let $\text{rad}(C) = \max\{|x|: x \in C\}$ be the radius of the open cluster at the origin. If $\text{rad}(C) \geq k$ say, then how large is the volume $|C|$ of C? The following is a rather crude approach to this question. Suppose that $\text{rad}(C) \geq k$ and let us consider vertices x of $\mathbb{L}^d$ satisfying $|x| \leq k$. It may not be too unreasonable to guess that the proportion of such vertices x which are joined to the origin is roughly equal to the probability that such a vertex is in an open cluster having radius at least k; this probability is approximately $k^{-1/\rho}$, so that

$$(7.37) \qquad |C| \geq k^d k^{-1/\rho} \quad \text{if and only if} \quad \text{rad}(C) \geq k,$$

to within a few orders of magnitude and some error probability. Now $P_{p_c}(n \leq |C| < \infty) \approx n^{-1/\delta}$ and $P_{p_c}(k \leq \text{rad}(C) < \infty) \approx k^{-1/\rho}$, so that (7.37) implies

$$\left(d - \frac{1}{\rho}\right)\frac{1}{\delta} = \frac{1}{\rho},$$

or

$$(7.38) \qquad d\rho = \delta + 1$$

as required.

The following argument yields the second hyperscaling relation (7.23), which asserts that $d\nu = 2 - \alpha$. Let $T(n) = \{x \in \mathbb{Z}^d: 1 \leq x_i \leq n \text{ for all } i\}$. In the calculations (7.30)–(7.32) we saw that the 'singular' part of $\kappa(p)$ is the contribution from open clusters with sizes having orders of $\xi(p)^\tau$ or greater. Let $K_n(\xi(p)^\tau)$ be the number of such open clusters in the box $T(n)$; this is not a particularly rigorous definition, but its style is in keeping with the rest of the argument. If we set $n = \xi(p)$, the correct 'length scale' for percolation with edge-probability p, then the mean number $E_p\{K_n(\xi(p)^\tau)\}$ of such clusters should not vary greatly for values of p near to p_c. Suppose then that

$$E_p\{K_{\xi(p)}(\xi(p)^\tau)\} \simeq A \qquad \text{as} \quad p \to p_c,$$

for some constant A. For any positive integer m, we may think of $T(m\xi(p))$ as comprising m^d copies of $T(\xi(p))$. Only few of the open clusters of $T(m\xi(p))$ with size $\xi(p)^\tau$ or greater intersect two or more of these copies, so that

$$(7.39) \qquad \begin{aligned} E_p\{K_{m\xi(p)}(\xi(p)^\tau)\} &\simeq m^d E_p\{K_{\xi(p)}(\xi(p)^\tau)\} \\ &\simeq Am^d. \end{aligned}$$

The 'singular' part of $\kappa(p)$ is given therefore by

$$\lim_{m\to\infty} \frac{1}{(m\xi(p))^d} E_p\{K_{m\xi(p)}(\xi(p)^\tau)\} \simeq A\xi(p)^{-d}$$

by (7.39). We deduce by another leap of the imagination that

$$2 - \alpha = d\nu, \tag{7.40}$$

the controversial hyperscaling relation.

We summarize the discussion of this section as follows.

(a) There are two attractive scaling hypotheses concerning the behaviour of $P_p(|C| = n)$ and $\tau_p^f(0, x)$ when n and $|x|$ are large, and when p is near to p_c. These scaling hypotheses provide strong evidence for the scaling relations.
(b) Two further heuristic arguments lead to the hyperscaling relations.

7.3 Renormalization

Renormalization techniques are of great value in statistical mechanics and have been applied with some success to percolation models. They provide a unified basis for scaling theory which is attractive in its simplicity and coherence. However, with the exception of a handful of special cases, we know of no major application of renormalization theory to the critical percolation phenomenon in two or more dimensions which is both valuable and completely rigorous. This notwithstanding, the intuitive appeal of 'real-space renormalization' is considerable and, for many, such arguments are the cornerstone of the theory of critical phenomena. We shall attempt to sketch this approach.

We consider bond percolation on $\mathbb{L}^d$ where $d \geq 2$, and we let b be a positive integer satisfying $b \geq 2$. We partition $\mathbb{Z}^d$ into cubes of size b^d in a natural way, the sets of this partition having the form

$$B(j_1, j_2, \ldots, j_d) = \prod_{i=1}^{d} [j_i b, (j_i + 1)b - 1]$$

as $j = (j_1, j_2, \ldots, j_d)$ ranges over the set of all d-vectors of integers; see Figure 7.1 for a two-dimensional sketch. We call the B's *blocks*, and we say that the block $B(j_1, \ldots, j_d)$ is *traversible in the rth direction* if there is an open path in the block joining some vertex x with $x_r = j_r b$ to some vertex y with $y_r = (j_r + 1)b - 1$, and also the edge leading out of y in the direction of the positive rth coordinate axis is open.

We now 'renormalize' the lattice, by replacing each block $B(j_1, \ldots, j_d)$ by a single vertex which we label $j = (j_1, \ldots, j_d)$. In this new lattice, we place edges between all pairs j, k of vertices for which

$$\sum_i |j_i - k_i| = 1,$$

so that the new lattice is a copy of $\mathbb{L}^d$; we denote the new lattice by $\mathscr{R}_b(\mathbb{L}^d)$. We declare the edges of $\mathscr{R}_b(\mathbb{L}^d)$ to be open or closed according to the following rule: the edge joining $(j_1, \ldots, j_{r-1}, j_r, j_{r+1}, \ldots, j_d)$ to

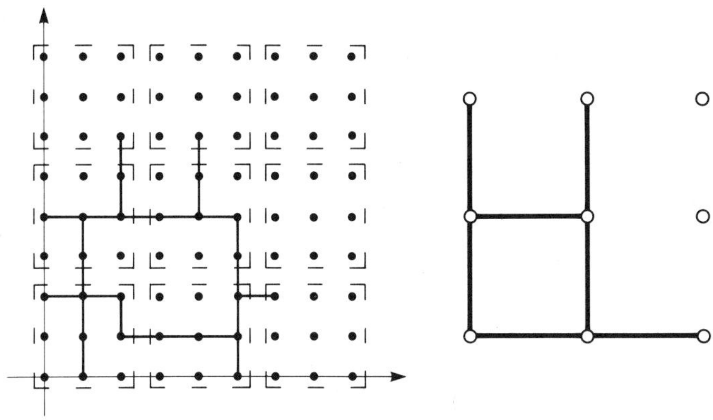

Figure 7.1. Renormalizing the square lattice with $b = 3$. The lines in the left picture represent open edges, whilst the right picture contains the 'renormalized' vertices together with edges which describe the traversibility of the 3×3 blocks of the original lattice.

$(j_1, \dots, j_{r-1}, j_r + 1, j_{r+1}, \dots, j_d)$ is declared to be open if and only if the block $B(j_1, \dots, j_d)$ is traversible in the rth direction. Thus each edge of $\mathscr{R}_b(\mathbb{L}^d)$ is open with probability $p^* = R_b(p)$, where $R_b(p)$ is the probability that $B(0, \dots, 0)$ is traversible in the first direction. We illustrate the construction in Figure 7.1, and we sketch the function R_b in Figure 7.2.

In the original lattice, the states of different edges were independent of each other. This is not true in the new lattice, since the state of each edge contains information about the states of certain other edges having

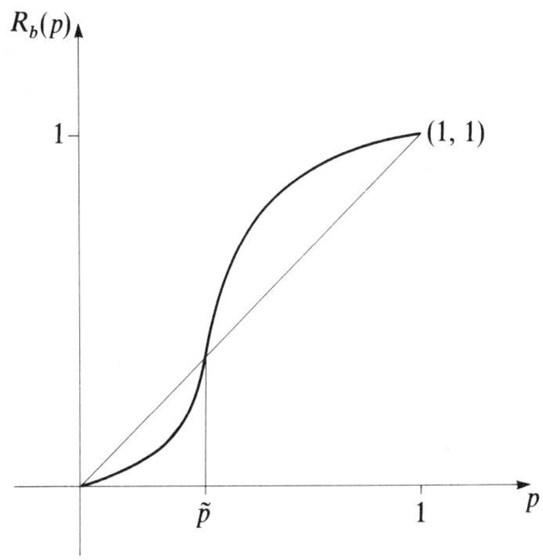

Figure 7.2. A sketch of the function R_b with emphasis on its non-trivial fixed point $\tilde{p}$. Note the characteristic 'S-shape' of the function, and the fact that the derivative $\lambda = R_b'(\tilde{p})$ at $\tilde{p}$ satisfies $\lambda > 1$.

a vertex in common. *Let us suppose however that the new process resembles closely a realization of bond percolation with edge-probability* p^*. If this were the case, then the process of renormalization would replace a bond percolation process with edge-probability p by a bond percolation process on a re-scaled lattice with edge-probability $p^* = R_b(p)$. *Let us suppose next that the large-scale connectivities of these two processes are very similar.*

Whither are we led by these two assumptions? We consider first the correlation length $\xi(p)$ of the first lattice; this is the fundamental 'length scale' of the first process, being the shortest scale over which the process can be distinguished from a critical percolation process with edge-probability p_c. In the new process, the correlation length is $\xi(p^*)$. However, each unit of length of the new lattice $\mathcal{R}_b(\mathbb{L}^d)$ corresponds to b units of length of the first, and we have supposed that the two processes are similar over long ranges. This suggests the equation

$$\xi(p) = b\xi(p^*) \tag{7.41}$$

where

$$p^* = R_b(p). \tag{7.42}$$

We now cross our fingers, and let $\tilde{p}$ be a fixed point of the mapping R_b. If $\tilde{p} = R_b(\tilde{p})$ then $\xi(\tilde{p}) = b\xi(\tilde{p})$, so that $\xi(\tilde{p})$ equals 0 or ∞. However, $\xi(\tilde{p}) = 0$ if and only if $\tilde{p}$ equals 0 or 1, and these fixed points are of little interest. It follows that the remaining fixed point of R_b is the critical probability p_c, since this is the unique value of $\tilde{p}$ for which $\xi(\tilde{p}) = \infty$. We deduce therefore that p_c is a root of the equation $R_b(p) = p$. When $b = 2$ say, the function R_b is easily calculated in closed form. Needless to say, it does not turn out in practice that $p_c = R_b(p_c)$, and the reason for this is that the two italicized hypotheses above can be considered only to be rough guidelines about the behaviour of the new process on the renormalized lattice. The process on the lattice $\mathcal{R}_b(\mathbb{L}^d)$ is *not* bond percolation, and the large-scale properties of the two lattices will generally differ. There are at least two directions which may be pursued from this point. The first is to guess that, as b grows, the fixed point of R_b approximates better and better to p_c; this argument has met with some numerical success, although current technology allows the estimation of R_b by Monte Carlo techniques only, when b is large. The second direction is theoretical rather than practical, and this is to attempt to describe and analyse with greater care the random process on the renormalized lattice. We do not pursue this here, but refer the reader to Fisher (1983) and Kesten (1987e, Section 2.3).

We return to the equations (7.41) and (7.42) which describe the effect of renormalization upon the correlation length $\xi(p)$, and we suppose that p_c is the unique non-trivial fixed point of R_b. The behaviour of $\xi(p)$ when p is near to p_c is related to the corresponding behaviour of $R_b(p)$ near to its fixed point p_c, and so we note that

(7.43) $$R_b(p) - R_b(p_c) = \lambda(p - p_c) + o(|p - p_c|)$$

where $\lambda = R_b'(p_c)$. Now, $R_b(p)$ is the probability of an increasing event which depends on only finitely many edges. We may therefore apply Theorem (2.36), and specifically inequality (2.37), to deduce that

$$\lambda = R_b'(p_c) \geq \frac{R_b(p_c)(1 - R_b(p_c))}{p_c(1 - p_c)} = 1.$$

It is not difficult to see from the proof of (2.41) that strict inequality holds here, which is to say that $\lambda > 1$. We express (7.43) in the form

(7.44) $$p^* - p_c \sim \lambda(p - p_c) \qquad \text{as} \quad p \to p_c$$

to see that the process of renormalization pushes the percolation process away from the critical probability, in the sense that $|p^* - p_c| \geq \lambda(1 + o(1))|p - p_c|$ when $|p - p_c|$ is small. We substitute (7.44) into (7.41) to find that $\psi(p - p_c) = \xi(p)$ satisfies

(7.45) $$\psi(p - p_c) \simeq b\psi(\lambda(p - p_c)) \qquad \text{when } |p - p_c| \text{ is small.}$$

Let A be a positive constant and suppose that $|p - p_c|$ is very small compared with A. We iterate (7.45) m times to obtain

(7.46) $$\psi(p - p_c) \simeq b^m \psi(\lambda^m(p - p_c))$$

and we choose m such that $\lambda^m|p - p_c| \simeq A$. With this value of m, (7.46) becomes

(7.47) $$\begin{aligned} \psi(p - p_c) &\simeq \psi(A) \exp\left(\frac{\log b}{\log \lambda}(\log A - \log|p - p_c|)\right) \\ &= D_1|p - p_c|^{-\nu}, \end{aligned}$$

where D_1 is a constant which depends on b, λ, and A, and

(7.48) $$\nu = \frac{\log b}{\log \lambda}.$$

We have shown that the equation $\xi(p) = b\xi(p^*)$, where $p^* = R_b(p)$, implies that $\xi(p)$ behaves in the manner of a power of $|p - p_c|$ when p is near to p_c, and furthermore it specifies the numerical value of the exponent ν. This 'exact calculation' of ν is only an approximation to the actual value of ν, but the error is expected to be small in the limit as $b \to \infty$.

We have based the argument so far upon the effect of renormalization upon the correlation length. Parallel analyses are possible for such quantities as the connectivity functions and the number of open clusters per vertex. For example, if we concentrate on the latter function $\kappa(p)$, we obtain

(7.49) $$\kappa(p) \simeq b^{-d}\kappa(p^*)$$

in place of $\xi(p) = b\xi(p^*)$, and this relation yields

$$\kappa(p) \simeq D_2|p - p_c|^{\nu d}, \tag{7.50}$$

where D_2 is a constant and ν is given by (7.48). As before, we have deduced that κ behaves roughly as a power of $|p - p_c|$ when p is near to p_c; combined with the hypothesis (7.6) that $\kappa'''(p) \approx |p - p_c|^{-1-\alpha}$, we find that $2 - \alpha = d\nu$, in agreement with the hyperscaling relation (7.23). We may use the 'exact value' (7.48) of ν to find the 'exact value' of α. Care is advisable in interpreting the asymptotic relation in (7.50); for example, contrary to appearances, it is not the case that $\kappa(p_c) = 0$.

The theory of renormalization lends support to the hypothesis of universality. We may think of renormalization as an operator acting on a class of processes including d-dimensional percolation models. The values of critical exponents should depend only on the re-scaling factor b and the behaviour of this operator near to its fixed point—see (7.48), for example. These values should thus depend upon the number d of dimensions but not upon other features of individual processes within a large class of 'similar' processes.

It is not difficult to pursue such arguments further, and we refer the reader to Essam (1980), Stauffer (1979, 1985), Fisher (1983), Reynolds, Stanley, and Klein (1980), Stanley, Reynolds, Redner, and Family (1982), and the references therein.

7.4 The Incipient Infinite Cluster

The term 'incipient infinite cluster' has been used in the physics literature in an attempt to describe the infinite open cluster when $p = p_c$. Unfortunately, no such cluster exists when $d = 2$, and it is widely believed that this is the case also when $d > 2$. How then may we make mathematical sense of the physical idea of the infinite open cluster at the critical point p_c? Kesten (1986b) has made the following proposal in the case of two dimensions, and parallel developments may be valid in higher dimensions. Suppose that $d = 2$. The event $\{|C| = \infty\}$, that the origin is in an infinite open cluster, has probability 0 when $p = p_c$. We may however force this event to occur by suitable conditioning, and this may be done in two ways.

First we take $p > p_c$, and we define the probability measure ν_p by

$$\nu_p(A) = P_p(A \,|\, |C| = \infty) \tag{7.51}$$

for cylinder events A. Kesten (1986b) has proved that the limit of $\nu_p(A)$ exists as $p \downarrow p_c$: denoting the limit by $\nu(A)$, we have that

$$\nu(A) = \lim_{p \downarrow p_c} \nu_p(A) \tag{7.52}$$

for cylinder events A.

Secondly, we take $p = p_c$ and let $B(n)$ be the usual box with side-length $2n$ and centre at the origin. For cylinder events A, we write

$$(7.53) \qquad \nu_n(A) = P_{p_c}(A \mid 0 \leftrightarrow \partial B(n))$$

for the probability of A conditional on there being an open path from the origin to the surface of $B(n)$. It turns out that the limit of $\nu_n(A)$ exists as $n \to \infty$, and that this limit coincides with the limit in (7.52). That is to say

$$(7.54) \qquad \lim_{n \to \infty} \nu_n(A) = \nu(A)$$

for cylinder events A. The limit function ν is defined on cylinder events only, but it has a unique extension to the σ-field $\mathscr{F}$.

It is clear by the nature of the conditioning in (7.51) and (7.53) that the origin has ν-probability 1 of being in an infinite open cluster $\mathscr{I}$, and it turns out that this cluster is (ν-a.s.) the unique infinite open cluster of the lattice. We call this infinite cluster the *incipient infinite cluster.*

It is an open problem both to prove that $\theta(p_c) = 0$ and to give a rigorous justification of the above argument in three or more dimensions under the hypothesis that $\theta(p_c) = 0$.

Physicists are interested in the geometry of the incipient infinite cluster, and particularly in its properties of statistical self-similarity over different length-scales. Very little rigorous progress has been made in this direction, even in the case of two dimensions. Presumably the incipient infinite cluster permeates the space which contains it, but is rather sparse within this space. It is believed to have zero density but to be extremely 'ramified', which is to say that it contains long dendritic connections across space but relatively few interconnections; see Figure 7.3 for a sketch of such a network in two dimensions.

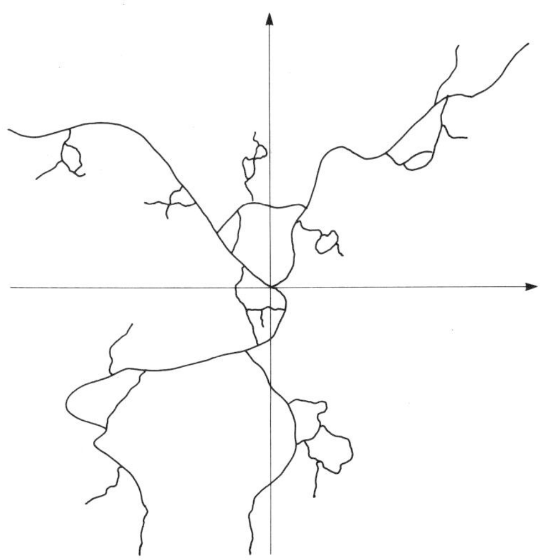

Figure 7.3. An artist's impression of a sparse open cluster having connections over long ranges but few interconnections between distant parts.

A first step towards understanding the geometry of $\mathscr{I}$ in two dimensions is to estimate the density of $\mathscr{I}$. Towards this end, we ask for an estimate of the probability that a vertex x of $\mathbb{Z}^2$ lies in $\mathscr{I}$. Let m be a positive integer and suppose that x is a vertex with $\|x\| = m$. It may not be unreasonable to guess that, for large n and 'most' such vertices x,

$$\begin{aligned} v_n(0 \leftrightarrow x) &= P_{p_c}(0 \leftrightarrow x \mid 0 \leftrightarrow \partial B(n)) \\ &\simeq m^{-1/\rho}, \end{aligned}$$

since, conditional on there being an open path from the origin to a vertex of $\partial B(n)$, the vertex x is required to be connected to vertices in the open cluster at the origin, and some of these vertices are approximately distance m from x. We let n tend to infinity to obtain

(7.55) $$v(x \in \mathscr{I}) \simeq \|x\|^{-1/\rho} \qquad \text{for} \quad x \in \mathbb{Z}^2.$$

The number of vertices of $B(n)$ which lie in $\mathscr{I}$ is approximately

(7.56) $$\begin{aligned} E_v(|B(n) \cap \mathscr{I}|) &= \sum_{x \in B(n)} v(x \in \mathscr{I}) \\ &\simeq \sum_{x \in B(n)} \|x\|^{-1/\rho} \\ &\simeq D_1 \sum_{m=1}^{n} m \cdot m^{-1/\rho} \\ &\simeq D_2 n^{2-1/\rho} \end{aligned}$$

for some constants D_1 and D_2, where E_v denotes the expectation operator associated with the measure v. This rough argument indicates that the density of $\mathscr{I}$ in $B(n)$ behaves in the manner of $n^{-1/\rho}$ as $n \to \infty$, and furthermore we may think of $\mathscr{I}$ as being a subset of $\mathbb{L}^2$ having dimension $2 - 1/\rho$; this last number is sometimes referred to as the 'fractal dimension' of the incipient infinite cluster. Kesten (1986b) has shown how to make this argument rigorous in two dimensions, in as much as this is possible without assuming the existence of the critical exponent ρ. The corresponding argument for d dimensions indicates a value of $d - 1/\rho$ for the hypothetical 'fractal dimension' of the hypothetical incipient infinite cluster.

It has been suggested by de Gennes (1976) that the asymptotic properties of a random walk on the incipient infinite cluster $\mathscr{I}$ may give insight into the geometry of $\mathscr{I}$ and particularly its conductivity when it is viewed as an infinite random electrical network. Thinking of the random walker as an 'ant in a labyrinth', we may be interested in such quantities as the mean square displacement of the ant after it has taken n steps. Let X_n be the position of such an ant after n steps, and suppose that $X_0 = 0$. Kesten (1986c) has shown that the random walk is *subdiffusive* in two dimensions, in the sense that

(7.57) $$E(X_n^2) \leq An^{1-\varepsilon}$$

for some constants A and $\varepsilon > 0$, and for all n; we recall that symmetric random walk on $\mathbb{Z}^2$ has mean square displacement equal to n, so that (7.57) asserts that the ant diffuses at a rate which is of a lesser order of magnitude than that of an unrestricted random walk. To see the intuitive reasoning behind this, we introduce the idea of the *backbone* $\mathscr{B}$ of $\mathscr{I}$, being the set of vertices x of $\mathscr{I}$ from which there exist both an infinite open path and an open path to the origin which is edge-disjoint from the first path (note that this definition of the backbone is somewhat non-standard, other authors defining the backbone as the set of vertices from which there exist two or more edge-disjoint infinite open paths). The idea is that the ant divides its time between occupying vertices in the backbone $\mathscr{B}$ and occupying other vertices. However, time spent outside $\mathscr{B}$ is wasted since the ant must return to the backbone in order to progress substantially in its march about the lattice (see Figure 7.3). It turns out that the number of vertices of $B(n)$ which belong to the backbone is approximately $n^2 P_{p_c}(A_n)$ where A_n is the event that there are two or more edge-disjoint open paths from the origin to the surface of $B(n)$. We have from the BK inequality that

$$\begin{aligned} P_{p_c}(A_n) &\leq P_{p_c}(0 \leftrightarrow \partial B(n))^2 \\ &\approx n^{-2/\rho}, \end{aligned}$$

if ρ exists. Thus the volume of $\mathscr{B} \cap B(n)$ has at most order $n^{2-2/\rho}$, which is smaller than the order $n^{2-1/\rho}$ of $\mathscr{I} \cap B(n)$. This difference in orders of magnitude is the principal reason for the subdiffusive behaviour (7.57) of the ant. See Kesten (1986b, c) for more details.

7.5 Notes

Sections 7.1 and 7.2. The principal reviews of scaling theory are those of Essam (1980) and Stauffer (1979, 1985). We shall not attempt an accurate historical bibliography here, but refer the reader to the references contained in these extensive reviews. We defer a discussion of rigorous results until Chapters 8 and 9, but note that Kesten (1987b) has proved the validity of all scaling relations in two dimensions which do not involve α, subject to the assumption that ρ and ν exist. In related work, Kesten (1987c) has proved the hyperscaling relation $d\rho = \delta + 1$ in the case $d = 2$, subject to a similar assumption of existence.

The recent reviews of Aizenman (1987), Kesten (1987e), and Newman (1987a) contain interesting discussions concerning critical exponents and scaling theory in the context of percolation.

The predictions of theoretical physics do not stop at the scaling and hyperscaling relations. Certain exact values are conjectured for critical exponents in two dimensions: $\alpha = -2/3$, $\beta = 5/36$, $\gamma = 43/18$, $\delta = 91/5$, and so on. See Kesten (1987b), Stauffer (1981, 1985), and the references therein.

It was Toulouse (1974) who discovered the argument that the (upper) critical dimension is 6. For more recent work concerning the critical dimension, see Tasaki (1987a, b, 1988) and Chayes and Chayes (1988).

For a general introduction to the theory of phase transitions, we refer the reader to the book by Stanley (1971).

Section 7.3. Fisher (1983) has written a beautiful account of renormalization, together with a discussion of the hypothesis of universality and the failure of the hyperscaling laws in large dimensions. See also Kesten (1987e) and the references at the end of the section.

Section 7.4. Other approaches to the 'incipient infinite cluster' are contained in Chayes and Chayes (1986a), Chayes, Chayes, and Durrett (1987), Coniglio (1985), Ben-Avraham and Havlin (1982), Kapitulnik, Aharony, Deutscher, and Stauffer (1983), and elsewhere. The rigorous results for two dimensions may be found in Kesten (1986b, c), who considers also the asymptotic properties of a random walk on the family tree of a critical branching process conditioned on non-extinction. See De Masi, Ferrari, Goldstein, and Wick (1985) for an account of random walk on the infinite open cluster of the percolation model when $p > p_c$.

Chayes, Chayes, and Durrett (1987) have studied the possible existence of an infinite open cluster in two dimensions when the probability that the edge e is open is $p_c + f(e)$ where $f(e) \approx |e|^{-\lambda}$ and $|e|$ is the distance from the origin to the midpoint of e. They show that the existence or not of an infinite open cluster depends on whether $\lambda < \nu^{-1}$ or $\lambda > \nu^{-1}$.

See Chayes and Chayes (1987), Menshikov (1987b), and the references therein for further discussion of the backbone of the infinite open cluster.

8

Near the Critical Point: Rigorous Results

8.1 Percolation on a Tree

It is the presence of circuits in $\mathbb{L}^d$ which causes difficulties in exact calculations: there are many different paths joining two specified vertices, and pairs of such paths generally have edges in common. In attempting to understand critical phenomena, it is usual to tackle first the corresponding problem on a rather special lattice, being a lattice which is devoid of circuits; we do this in the hope that the ensuing calculation will be simple but will help in the development of insight into more general situations. Such a lattice is called a *tree*, and it is to percolation on trees that this section is devoted.

Let T be the labelled binary tree, drawn in Figure 8.1. We label the vertices of T with sequences of 1's and 2's in the following way. We label the *root* of T by $\varnothing$, the empty sequence. We label the two children of the vertex labelled $\lambda_1\lambda_2 \dots \lambda_n$ with the labels $\lambda_1\lambda_2 \dots \lambda_n 1$ and $\lambda_1\lambda_2 \dots \lambda_n 2$. Any vertex whose label is a sequence of length n is said to belong to the nth generation of T.

Let p satisfy $0 < p < 1$. We declare each edge of T to be *open* with probability p and *closed* otherwise, independently of all other edges. We call the resulting process (*bond*) *percolation on* T, and it is on this process that we shall concentrate.

We note that T is not a regular tree, since all vertices have degree 3 except the root which has degree 2. It is customary to add another branch from the root, isomorphic to the two which are already present, in order to equalize the degrees of the vertices. Such an addition has no effect on the critical exponents of the percolation process on T; since it is primarily in these exponents that we shall be interested, we shall not bother to do this. The critical exponents are similarly insensitive to the

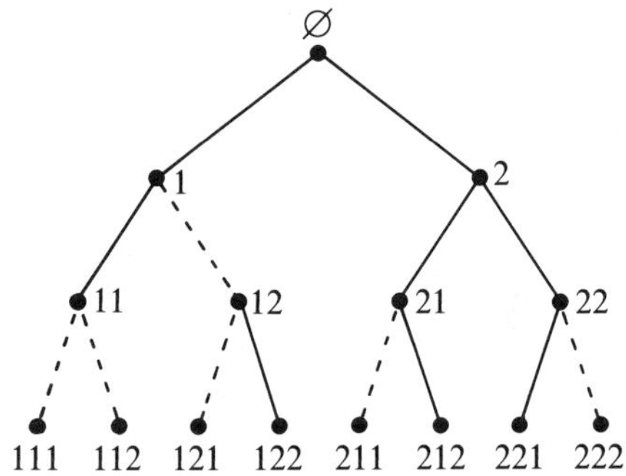

Figure 8.1. Four generations of the labelled binary tree, the unbroken edges being open and the broken edges being closed.

common degree of the vertices of T; thus we lose no substantial degree of generality by restricting our attention to the *binary* tree rather than considering the 'r-ary' tree in which each vertex has r children.

Such trees are sometimes called *Bethe lattices*, after Hans Bethe. Mathematical physicists sometimes use the term 'mean field theory' to describe the statistical mechanics of interaction processes on trees; this is not the only use of the term.

There is a sense in which percolation on T corresponds to percolation on $\mathbb{L}^d$ in the limit as $d \to \infty$, and such an interpretation contributes to our understanding of percolation in high dimensions. We may think of T as being embedded in $\mathbb{L}^\infty$, with each edge joining an nth generation vertex to an $(n+1)$th generation vertex of T lying parallel to the nth coordinate axis of $\mathbb{L}^\infty$. In this embedding, the vertex of T with label $\lambda_1\lambda_2\ldots\lambda_n$ corresponds to the vertex $(x_1, x_2, \ldots, x_n, 0, 0, \ldots)$ where $x_i = -1$ if $\lambda_i = 1$ and $x_i = 1$ if $\lambda_i = 2$. In performing such an embedding we need to be careful when calculating distances between vertices. The natural distance function on T is defined as follows: the distance $\delta(x, y)$ is the number of edges in the unique path of T joining x to y. Thus $\delta(\varnothing, x)$ equals the generation number of x. With T embedded in $\mathbb{L}^\infty$, it is not unnatural to consider instead the euclidean distance

$$(8.1) \qquad d(\varnothing, x) = \sqrt{\sum_{i=1}^{\infty} x_i^2},$$

where $(x_1, x_2, \ldots)$ is the vertex of $\mathbb{L}^\infty$ which corresponds to x. It is not difficult to see that $d(\varnothing, x) = \sqrt{\delta(\varnothing, x)}$, and this difference between d and δ should be held in our minds.

Exact calculations are possible for percolation on a tree. We shall show how to perform these, but in advance of doing this we present a preview of our conclusions. It is easy to see that the critical probability of percolation on T is given by $p_c = \frac{1}{2}$. We may calculate the critical exponents of the process exactly, and we shall see that $\alpha = -1$, $\beta = 1$,

$\gamma = 1$, $\delta = 2$, $\Delta = 2$, and $\rho = \frac{1}{2}$. The definitions in Section 7.1 of η and the correlation length $\xi(p)$ are not suitable for percolation on a tree. At the end of this section, we discuss alternative definitions of these quantities and deduce that $\eta = 0$ and $\nu = \frac{1}{2}$. The numerical values of ρ, η, and ν depend on which distance function is used in their definitions. The values above are calculated using the 'euclidean' distance function $d(\varnothing, x)$; had we used $\delta(\varnothing, x)$ instead, they would each have been 1.

We note that the numerical values of these critical exponents satisfy the scaling relations

$$2 - \alpha = \gamma + 2\beta = \beta(\delta + 1),$$
$$\Delta = \beta\delta.$$

As we have remarked earlier, substitution of these values into the hyperscaling relations $d\nu = 2 - \alpha$, $d\rho = \delta + 1$, yields $d = 6$, and this may be seen as evidence that the critical exponents for percolation on $\mathbb{L}^d$ take on the corresponding values for percolation on a tree when $d \geq 6$.

We write C for the open cluster of T containing the root $\varnothing$. Clearly, C is the family tree of a Galton–Watson branching process with a single progenitor and family sizes having the binomial distribution with parameters 2 and p. The cluster C is finite if and only if this branching process becomes extinct, and the probability of this is the smallest nonnegative root of the equation $s = G(s)$, where $G(s) = (1 - p + ps)^2$ is the probability generating function of a typical family size. It is easy to check that this root equals 1 if $p \leq \frac{1}{2}$ and equals $\{p^{-1}(1-p)\}^2$ if $p \geq \frac{1}{2}$. We have proved that $\theta(p) = P_p(|C| = \infty)$ is given by

$$\theta(p) = \begin{cases} 0 & \text{if } p \leq \frac{1}{2}, \\ 1 - \{p^{-1}(1-p)\}^2 & \text{if } p \geq \frac{1}{2}. \end{cases} \tag{8.2}$$

Thus the critical probability p_c of percolation on T equals $\frac{1}{2}$, and the percolation probability $\theta(p)$ may be expressed in the closed form (8.2). We are interested here in critical exponents, and so we differentiate the function $1 - \{p^{-1}(1-p)\}^2$ at $p = \frac{1}{2}$ to obtain

$$\theta(p) \sim 8(p - \tfrac{1}{2}) \quad \text{as} \quad p \downarrow \tfrac{1}{2}. \tag{8.3}$$

We compare this with the hypothesis that $\theta(p) \approx (p - p_c)^\beta$ for some β as $p \downarrow p_c$, to arrive at the following proposition.

(8.4) Proposition. *For percolation on the binary tree, the critical exponent β equals* 1.

We shall see that all critical exponents may be calculated exactly and rigorously in the case of percolation on T. We consider γ next. Suppose that $p < p_c$ $(= \frac{1}{2})$ and write $\chi(p) = E_p|C|$ as usual. The vertex x of T lies in C if and only if every edge is open in the unique path of the tree joining $\varnothing$ to x. There are 2^n vertices in the nth generation, so that

$$\begin{aligned}(8.5)\qquad \chi(p) &= \sum_{n=0}^{\infty} 2^n p^n \\ &= (1-2p)^{-1} \qquad \text{if} \quad p < \tfrac{1}{2} \\ &= \tfrac{1}{2}(\tfrac{1}{2}-p)^{-1},\end{aligned}$$

which we compare with the hypothesis that $\chi(p) \approx (p_c - p)^{-\gamma}$ as $p \uparrow p_c$, to deduce that $\gamma = 1$. The complete hypothesis involving γ asserts that $\chi^f(p) \approx |p - p_c|^{-\gamma}$ as $p \to p_c$, and we have verified this only in the case $p < p_c$. We therefore turn our attention to the case $p > p_c$, and recall that

$$\chi^f(p) = E_p(|C|; |C| < \infty),$$

the mean size of the family tree of the branching process on the event that the process becomes extinct. Let X_n be the number of nth generation descendants of $\varnothing$ in the branching process. The sequence $X = (X_n : n \geq 0)$ is a Markov chain with transition probabilities

$$(8.6)\qquad P_p(X_{n+1} = j \mid X_n = i) = \binom{2i}{j} p^j (1-p)^{2i-j}.$$

Let $\bar{X} = (\bar{X}_n : n \geq 0)$ be the sequence of generation sizes of the branching process conditioned on the event E that the process becomes extinct. It is not difficult to check that $\bar{X}$ is a Markov chain whose transition probabilities are given by the following calculation (in which 'P_p' is used to denote the appropriate probability measure):

$$\begin{aligned}P_p(\bar{X}_{n+1} = j \mid \bar{X}_n = i) &= \frac{P_p(\bar{X}_{n+1} = j, \bar{X}_n = i)}{P_p(\bar{X}_n = i)} \\ &= \frac{P_p(X_{n+1} = j, X_n = i, E)/P_p(E)}{P_p(X_n = i, E)/P_p(E)} \\ &= \frac{P_p(X_{n+1} = j, E \mid X_n = i) P_p(X_n = i)}{P_p(E \mid X_n = i) P_p(X_n = i)} \\ &= \frac{P_p(X_{n+1} = j \mid X_n = i) P_p(E)^j}{P_p(E)^i} \qquad \text{for} \quad i, j \geq 0.\end{aligned}$$

We substitute from (8.6) and use the fact that $P_p(E) = 1 - \theta(p) = \{p^{-1}(1-p)\}^2$ to find that

$$\begin{aligned}P_p(\bar{X}_{n+1} = j \mid \bar{X}_n = i) &= \binom{2i}{j} p^j (1-p)^{2i-j} \left(\frac{1-p}{p}\right)^{2(j-i)} \\ &= \binom{2i}{j} (1-p)^j p^{2i-j},\end{aligned}$$

which has the same form as (8.6) with p replaced by $1 - p$. We have thus shown that the distribution of the branching process, conditional on extinction, is identical to that of a subcritical branching process with parameter $1 - p$. In particular

$$\begin{aligned}\chi^f(p) &= E_p(|C|; |C| < \infty) \\ &= E_p(|C| \,|\, |C| < \infty)P_p(|C| < \infty) \\ &= \chi(1 - p)(1 - \theta(p)) \\ &= \tfrac{1}{2}(p - \tfrac{1}{2})^{-1}\left(\frac{1 - p}{p}\right)^2\end{aligned} \tag{8.7}$$

by (8.2) and (8.5). Thus $\chi^f(p) \approx (p - p_c)^{-1}$ when $p > p_c$, and we have proved the next proposition.

(8.8) Proposition. *For percolation on the binary tree, the critical exponent γ equals* 1.

Whilst on the subject of the mean size of C, we recall the gap exponent Δ given by

$$\frac{E_p(|C|^{k+1}; |C| < \infty)}{E_p(|C|^k; |C| < \infty)} \approx |p - p_c|^{-\Delta} \qquad \text{as} \quad p \to p_c \tag{8.9}$$

for $k \geq 1$. We may suppose that $p < p_c \ (= \frac{1}{2})$, since if $p > \frac{1}{2}$ then the discussion prior to (8.7) gives that

$$E_p(|C|^k; |C| < \infty) = E_{1-p}(|C|^k)P_p(|C| < \infty). \tag{8.10}$$

We suppose therefore that $p < \frac{1}{2}$ and $k \geq 1$. Then

$$\begin{aligned}E_p(|C|^k; |C| < \infty) &= E_p(|C|^k) \\ &= \sum_{n=1}^{\infty} n^k P_p(|C| = n).\end{aligned} \tag{8.11}$$

There is more than one way to discover how this behaves as p approaches $\frac{1}{2}$ from beneath, and here is such a way. It is not difficult to calculate $P_p(|C| = n)$ exactly: either use probability generating functions and the natural recursion, or read Durrett (1985a, p. 429) and the references therein, to find that

$$P_p(|C| = n) = \frac{1}{n}\binom{2n}{n-1}p^{n-1}(1 - p)^{n+1}. \tag{8.12}$$

We write $\varepsilon = \frac{1}{2} - p$ and substitute (8.12) into (8.11) to find that

$$E_p(|C|^k) = \sum_{n=1}^{\infty} n^{k-1}\binom{2n}{n-1}(\tfrac{1}{2} - \varepsilon)^{n-1}(\tfrac{1}{2} + \varepsilon)^{n+1}. \tag{8.13}$$

We have from Stirling's formula that

$$\binom{2n}{n-1} = \frac{n}{n+1}\binom{2n}{n}$$

$$\sim \frac{4^n}{\sqrt{\pi n}} \quad \text{as} \quad n \to \infty.$$

We approximate to the right-hand side of (8.13) by an integral and we make the substitution $x = -n\log(1 - 4\varepsilon^2)$ to obtain

(8.14)

$$A\int_1^\infty n^{k-1}\frac{(1-4\varepsilon^2)^n}{\sqrt{n}}\,dn \sim A\{-\log(1-4\varepsilon^2)\}^{-k+(1/2)}\int_0^\infty x^{k-(3/2)}e^{-x}\,dx$$

$$\sim A(4\varepsilon^2)^{-k+(1/2)}\Gamma(k - \tfrac{1}{2}),$$

as $\varepsilon \downarrow 0$, where A is a constant and Γ is the gamma function. This asymptotic analysis is rather approximate, but it is not difficult to make it rigorous when $k \geq 1$. We compare the last formula with (8.9) to obtain the next result.

(8.15) Proposition. *For percolation on the binary tree, the critical exponent* Δ *equals* 2.

The nature of the above calculation suggests looking for more powerful asymptotic results as $p \to \frac{1}{2}$ than merely those concerning the moments of $|C|$. One way to do this is to define a random variable Y_p with mass function

$$P(Y_p = n) = \frac{nP_p(|C| = n)}{\sum_m mP_p(|C| = m)} = \frac{nP_p(|C| = n)}{\chi^f(p)}. \tag{8.16}$$

Some people think of Y_p as having approximately the distribution of a 'typical' finite open cluster, but the intuition behind this is not entirely convincing, and we omit it.

It follows from the discussion leading to (8.7) that Y_p and Y_{1-p} have the same distribution. It is a reasonably elementary exercise in branching processes and asymptotic analysis to show that $|p - \frac{1}{2}|^2 Y_p$ converges in distribution as $p \to \frac{1}{2}$, with the limit distribution being gamma with parameters $\frac{1}{2}$ and 4. That is to say

$$P(|p - \tfrac{1}{2}|^2 Y_p \leq y) \to \int_0^y \frac{2}{\sqrt{\pi}} x^{-1/2} e^{-4x}\,dx \tag{8.17}$$

as $p \to \frac{1}{2}$. A similar calculation shows that, for $k \geq 1$,

$$\begin{aligned}(8.18)\qquad E_p(|C|^k; |C| < \infty) &= E(Y_p^{k-1})\chi^f(p) && \text{by (8.16)}\\ &\sim A_k|p - \tfrac{1}{2}|^{-2(k-1)}\chi^f(p) && \text{as in (8.17)}\\ &\sim \tfrac{1}{2}A_k|p - \tfrac{1}{2}|^{1-2k} && \text{by (8.7)}\end{aligned}$$

for some constant A_k depending on k, in agreement with (8.14).

The critical exponent α is defined by $\kappa'''(p) \approx |p - p_c|^{-1-\alpha}$ as $p \to p_c$, where $\kappa(p) = E_p(|C|^{-1})$. We substitute $k = -1$ into (8.13) to find that

$$(8.19)\qquad \kappa(p) = \frac{1 + 2\varepsilon}{1 - 2\varepsilon}\sum_{n=1}^{\infty}\frac{1}{n^2}\binom{2n}{n-1}\frac{1}{4^n}(1 - 4\varepsilon^2)^n,$$

where $\varepsilon = \frac{1}{2} - p$. Differentiating with respect to ε, we discover that the infinite series in (8.19) is infinitely differentiable term by term, except possibly at $\varepsilon = 0$. It is now a minor calculation somewhat similar to (8.14) to show that κ''' is bounded on a set of the form $(\frac{1}{2} - \zeta, \frac{1}{2}) \cup (\frac{1}{2}, \frac{1}{2} + \zeta)$ for some $\zeta > 0$, and thus $\alpha = -1$. It turns out that $\kappa'''(p)$ has a jump discontinuity at $p = \frac{1}{2}$.

(8.20) Proposition. *For percolation on the binary tree, the critical exponent α equals -1.*

Little extra work is needed to calculate the critical exponent δ. We have from (8.12) and Stirling's formula that

$$P_{1/2}(|C| = n) \sim \frac{1}{\sqrt{\pi n^3}},$$

giving that $\delta = 2$. For the critical exponent ρ, we argue as follows. If $p = \frac{1}{2}$, the corresponding branching process is critical, and the classical result of Kolmogorov (see Harris (1963, p. 21)) implies that

$$(8.21)\qquad P_{1/2}(\mathrm{rad}(C) \geq n) \sim \frac{4}{n} \quad \text{as} \quad n \to \infty,$$

where $\mathrm{rad}(C) = \max\{\delta(\varnothing, x): x \in C\}$; it is an interesting exercise to derive this from first principles. Mathematical physicists prefer to use the 'euclidean' distance $d(\varnothing, x) = \sqrt{\delta(\varnothing, x)}$. We rewrite (8.21) as

$$P_{1/2}(\sqrt{\mathrm{rad}(C)} \geq n) \sim \frac{4}{n^2} \quad \text{as} \quad n \to \infty,$$

and compare this with the hypothesis that $P_{p_c}(\sqrt{\mathrm{rad}(C)} = n) \approx n^{-1-1/\rho}$, to obtain $\rho = \frac{1}{2}$.

(8.22) Proposition. *For bond percolation on the binary tree, the critical exponents δ and ρ satisfy $\delta = 2$, $\rho = \frac{1}{2}$.*

There remain the exponents η and ν, defined in Section 7.1 by

$$(8.23) \qquad P_{p_c}(0 \leftrightarrow x) \approx |x|^{2-d-\eta}, \qquad \xi(p) \approx |p - p_c|^{-\nu}.$$

This definition of η is not suitable for percolation on a tree, since

$$P_{1/2}(\varnothing \leftrightarrow x) = (\tfrac{1}{2})^{\delta(\varnothing, x)},$$

which does not decay as a power of $\delta(\varnothing, x)$. A similar problem arises for ν, since the definition of the correlation length $\xi(p)$ is not suitable for percolation on trees. We complete this section with a discussion of alternative definitions of η and $\xi(p)$ suitable for the binary tree, and we begin with η. Rather than basing our approach upon the connectivity function $\tau_{p_c}(0, x)$, we work with the number of vertices within distance n of the origin which are joined to the origin by open paths. Let $S(n)$ be the ball with radius n and centre at the origin. A rough calculation for percolation on $\mathbb{L}^d$ suggests

$$(8.24) \qquad \begin{aligned} E_{p_c}|C \cap S(n)| &= \sum_{x:|x| \le n} \tau_{p_c}(0, x) \\ &\simeq \sum_{m=0}^{n} m^{d-1} m^{2-d-\eta} \qquad \text{by (8.23)} \\ &\simeq n^{2-\eta}, \end{aligned}$$

and we take this formula as our template for the corresponding calculation for the binary tree. In the latter case, we have that

$$\begin{aligned} E_{p_c}|C \cap S(n)| &= \sum_{x:\delta(\varnothing, x) \le n} P_{1/2}(\varnothing \leftrightarrow x) \\ &= \sum_{m=0}^{n} 2^m (\tfrac{1}{2})^m \\ &\sim n, \end{aligned}$$

implying that

$$\begin{aligned} E_{p_c}|\{x: \varnothing \leftrightarrow x \text{ and } d(\varnothing, x) \le n\}| &= E_{p_c}|C \cap S(n^2)| \\ &\sim n^2, \end{aligned}$$

and thence $\eta = 0$ by comparison with (8.24).

As remarked in Section 7.1, another possible definition of correlation length is the quantity $\varphi(p)$ given by

$$\varphi(p) = \sqrt{\frac{1}{\chi^f(p)} \sum_x |x|^2 \tau_p^f(0, x)}.$$

Once again there is ambiguity over the choice of the distance function $|x|$. Following the preference of mathematical physicists, we substitute $\sqrt{\delta(\varnothing, x)}$ for $|x|$ and obtain

$$\varphi(p) = \sqrt{\frac{1}{\chi^f(p)} \sum_{n=0}^{\infty} n 2^n \zeta_p(n)},$$

where $\zeta_p(n)$ is the probability that there is an open path in T joining the root $\varnothing$ to a given vertex in the nth generation, but that the branching process with progenitor $\varnothing$ is extinct. It is not difficult to show that

$$\zeta_p(n) = (1 - \theta(p))\zeta_{1-p}(n) \quad \text{if} \quad p > \tfrac{1}{2}$$

as in the discussion prior to (8.7), and furthermore

$$\chi^f(p) = (1 - \theta(p))\chi(1 - p) \qquad \text{if} \quad p > \tfrac{1}{2}.$$

Thus $\varphi(p) = \varphi(1 - p)$, so that the behaviour of $\varphi(p)$ when p is near to $\frac{1}{2}$ is determined by its behaviour for $p < \frac{1}{2}$. Suppose then that $p < \frac{1}{2}$. Then $\zeta_n(p) = p^n$, giving that

$$\text{(8.25)} \qquad \begin{aligned} \varphi(p) &= \sqrt{\frac{1}{\chi^f(p)} \sum_{n=0}^{\infty} n(2p)^n} \\ &= \sqrt{\frac{p}{\frac{1}{2} - p}} \qquad \text{by (8.5)} \\ &\approx (\tfrac{1}{2} - p)^{-1/2} \qquad \text{as} \quad p \uparrow \tfrac{1}{2}. \end{aligned}$$

(8.26) Proposition. *For percolation on the binary tree, the critical exponents η and ν satisfy $\eta = 0$ and $\nu = \frac{1}{2}$.*

8.2 Inequalities for Critical Exponents

We return to the muddier waters of percolation on $\mathbb{L}^d$ when $d \geq 2$. As we have recounted earlier, it is believed that the critical exponents take on the corresponding values for percolation on a tree when d is sufficiently large, and there is reason to suppose that this occurs first when $d = 6$. Whereas no exact formulae are known for critical exponents on $\mathbb{L}^d$ (indeed, the very existence of critical exponents remains unproven), we have at our disposal certain rigorous inequalities which are of value. Consider for example the exponent β for the percolation probability $\theta(p)$. A byproduct of the proof of the uniqueness of the critical probability (in particular, Theorem (3.8) or Theorem (3.48)) was the observation that $\theta(p) - \theta(p_c) \geq a(p - p_c)$ for some a (> 0) and for all small positive values of $p - p_c$. It follows that, if $\theta(p) \approx (p - p_c)^\beta$ for $p > p_c$, then $\beta \leq 1$.

(8.27) Proposition. *If the critical exponent β exists, then $\beta \leq \beta_T$, where $\beta_T = 1$ is the corresponding exponent for percolation on the binary tree.*

Thus β is bounded on one side by its value for a tree; it is conjectured that equality is valid here if $d \geq 6$. Similar inequalities are expected for the other critical exponents, and we describe some such results here. We assume throughout this section that $\theta(p_c) = 0$, although this is not essential for all our conclusions.

(8.28) Proposition. *If the critical exponent γ exists, then $\gamma \geq \gamma_T$, where $\gamma_T = 1$ is the corresponding exponent for percolation on the binary tree.*

(8.29) Proposition. *If the critical exponent δ exists, then $\delta \geq \delta_T$, where $\delta_T = 2$ is the corresponding exponent for percolation on the binary tree. If, further, the critical exponent β exists, then $\beta \geq 2/\delta$.*

We note that $\beta_T = 2/\delta_T$; many mathematical physicists express this by saying that 'the inequality $\beta \geq 2/\delta$ saturates in mean field'. This inequality is a strengthening of the fact that $\delta \geq 2$, since $\beta \leq 1$.

Many other inequalities for critical exponents are known, and we shall not list them here. For references, see the notes at the end of this chapter.

Proof of Proposition (8.28). We follow Aizenman and Newman (1984). First we express $\chi(p)$ as the sum of the two-point connectivities:

$$\chi(p) = \sum_{x \in \mathbb{Z}^d} \tau_p(0, x) = \sum_{x \in \mathbb{Z}^d} P_p(0 \leftrightarrow x). \tag{8.30}$$

We should like to differentiate with respect to p and use Russo's formula, but there are two obstacles in that the summations are infinite and the events $\{0 \leftrightarrow x\}$ depend on the states of infinitely many edges. In order to overcome this difficulty, we restrict ourselves for the moment to percolation on the finite box $B(n)$, and later we shall take the limit as $n \to \infty$. We write $\chi_n(p, v)$ for the mean number of vertices in the box $B(n)$ which are joined to the vertex v by open paths of $B(n)$, and we define

$$\hat{\chi}_n(p) = \max\{\chi_n(p, v) : v \in B(n)\}. \tag{8.31}$$

Now, $\chi_n(p, v) \leq \chi(p)$ for all $v \in B(n)$, giving that $\hat{\chi}_n(p) \leq \chi(p)$. On the other hand,

$$\begin{aligned} \hat{\chi}_n(p) &\geq \chi_n(p, 0) \\ &= \sum_{x \in B(n)} P_p(0 \leftrightarrow x \text{ in } B(n)). \end{aligned} \tag{8.32}$$

However, $P_p(0 \leftrightarrow x \text{ in } B(n)) \to P_p(0 \leftrightarrow x)$ as $n \to \infty$, so that

$$\begin{aligned} \chi(p) &\geq \hat{\chi}_n(p) \\ &\geq \sum_{x \in B(n)} P_p(0 \leftrightarrow x \text{ in } B(n)) \\ &\to \sum_{x \in \mathbb{Z}^d} P_p(0 \leftrightarrow x) \qquad \text{as} \quad n \to \infty \\ &= \chi(p), \end{aligned}$$

by the bounded convergence theorem. Thus

$$\hat{\chi}_n(p) \to \chi(p) \qquad \text{as} \quad n \to \infty, \tag{8.33}$$

whether or not $\chi(p)$ is finite.

Let v be a vertex of $B(n)$. We shall concentrate upon $\chi_n(p, v)$, which we write in the form

$$\chi_n(p, v) = \sum_{x \in B(n)} P_p(v \leftrightarrow x \text{ in } B(n)).$$

By Russo's formula,

$$\frac{d}{dp}\chi_n(p, v) = \sum_{x \in B(n)} \sum_{e \in B(n)} P_p(e \text{ is pivotal for } A_n(v, x)), \tag{8.34}$$

where $A_n(v, x)$ is the event that there is an open path from v to x in $B(n)$, and the second summation is over all edges e in $B(n)$. Writing $e = \langle a, b\rangle$, we have that, if e is pivotal for $A_n(v, x)$, then there is an open path in $B(n)$ from v to either a or b, and there is a disjoint open path in $B(n)$ from x to the other endvertex of e. See Figure 8.2 for a sketch of the situation. Therefore

(8.35)

$$\begin{aligned} P_p(e \text{ pivotal for } A_n(v, x)) &\le P_p(A_n(v, a) \circ A_n(x, b)) + P_p(A_n(v, b) \circ A_n(x, a)) \\ &\le P_p(A_n(v, a))P_p(A_n(x, b)) \\ &\quad + P_p(A_n(v, b))P_p(A_n(x, a)) \end{aligned}$$

by the BK inequality. We substitute this into (8.34) and sum over all vertices x and edges $e = \langle a, b\rangle$ to obtain

$$\begin{aligned} \frac{d}{dp}\chi_n(p, v) &\le \sum_{e=\langle a,b\rangle} \{P_p(A_n(v, a))\chi_n(p, b) + P_p(A_n(v, b))\chi_n(p, a)\} \\ &\le \hat{\chi}_n(p) \sum_{e=\langle a,b\rangle} \{P_p(A_n(v, a)) + P_p(A_n(v, b))\} \\ &\le 2d\hat{\chi}_n(p)\chi_n(p, v) \\ &\le 2d\hat{\chi}_n(p)^2. \end{aligned} \tag{8.36}$$

It would be convenient to take the limit as $n \to \infty$ in order to deduce that

$$\frac{d}{dp}\chi(p) \le 2d\chi(p)^2, \tag{8.37}$$

but certain difficulties arise in justifying the step $\chi_n'(p, v) \to \chi'(p)$. These difficulties may be overcome directly, but we shall argue differently here. We note that $\hat{\chi}_n(p)$ is the maximum of a finite number of polynomial functions $\chi_n(p, v)$. Therefore $\hat{\chi}_n(p)$ is differentiable except possibly for finitely many values of p. Also, whenever $\hat{\chi}_n$ is differentiable, we have that

$$\begin{aligned} \frac{d}{dp}\hat{\chi}_n(p) &\le \max_{v \in B(n)} \left\{\frac{d}{dp}\chi_n(p, v)\right\} \\ &\le 2d\hat{\chi}_n(p)^2 \qquad \text{by (8.36).} \end{aligned} \tag{8.38}$$

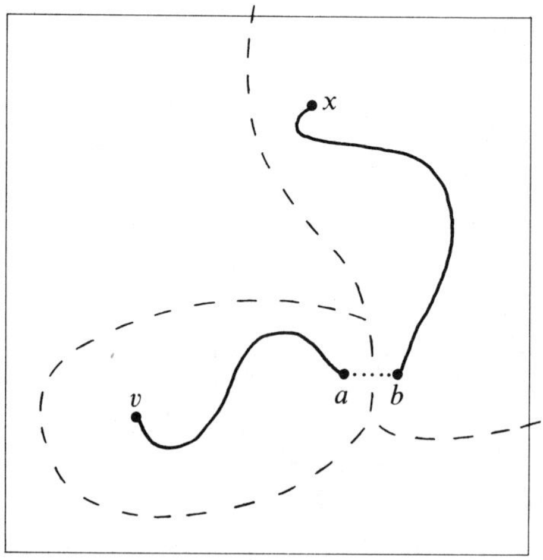

Figure 8.2. A sketch of the event that $e = \langle a, b\rangle$ is pivotal for $A_n(v, x)$. If e is open then $v \leftrightarrow x$ in $B(n)$, but not otherwise.

To recapitulate, we have that $\hat{\chi}_n(p)$ is continuous on [0, 1] and satisfies

$$\hat{\chi}_n(p)^{-2}\frac{d}{dp}\hat{\chi}_n(p) \leq 2d \tag{8.39}$$

except possibly for a finite set of values of p. We integrate (8.39) over the interval $[p, p_c]$ to obtain

$$[-\hat{\chi}_n(\pi)^{-1}]_p^{p_c} \leq 2d(p_c - p),$$

or

$$\frac{1}{\hat{\chi}_n(p)} - \frac{1}{\hat{\chi}_n(p_c)} \leq 2d(p_c - p). \tag{8.40}$$

We may at last allow the limit $n \to \infty$. As $n \to \infty$, it is the case that $\hat{\chi}_n(p) \to \chi(p)$ by (8.33), and $\hat{\chi}_n(p_c) \to \chi(p_c) = \infty$ by (5.52). Inequality (8.40) becomes

$$\chi(p) \geq \frac{1}{2d(p_c - p)} \qquad \text{for} \quad p < p_c. \tag{8.41}$$

Comparison with the hypothesis $\chi(p) \approx (p_c - p)^{-\gamma}$ for $p < p_c$ yields $\gamma \geq 1$ as required. Note that we have used the fact that $\chi(p_c) = \infty$, proved in (5.52). This reference to earlier work is easily avoided by noting that (8.39) is valid for all $p \in [0, 1]$ at which $\hat{\chi}_n$ is differentiable. We integrate (8.39) over the interval $[p, p_c + \varepsilon]$ where ε is small, to obtain

$$\frac{1}{\hat{\chi}_n(p)} - \frac{1}{\hat{\chi}_n(p_c + \varepsilon)} \leq 2d(p_c + \varepsilon - p).$$

Now $\hat{\chi}_n(p_c + \varepsilon) \to \chi(p_c + \varepsilon) = \infty$ as $n \to \infty$, and ε was arbitrary, so that (8.41) is valid as before. ■

Proof of Proposition (8.29). We give two proofs that $\delta \geq 2$. For the first, we follow Aizenman and Barsky (1987). The ground was prepared for this in Chapter 3, whence we extract Proposition (3.49): if p is such that $\chi^f(p) = \infty$, then

$$\begin{aligned}\theta(p, \psi) &= 1 - \sum_{n=1}^{\infty} (1 - \psi)^n P_p(|C| = n) \\ &\geq a\psi^{1/2}\end{aligned}$$

for some $a = a(p) > 0$ and all small positive values of ψ. We apply this with $p = p_c$, noting that $\chi^f(p_c) = \chi(p_c) = \infty$ if $\theta(p_c) = 0$, to obtain

$$\theta(p_c, \psi) \geq a\psi^{1/2} \tag{8.42}$$

for small positive ψ.

The inequality $\delta \geq 2$ follows by a standard application of the following Tauberian theorem for power series (see Feller (1971, p. 447)). Let $(q_m: m \geq 1)$ be a monotonic sequence of positive real numbers such that

$$Q(\psi) = \sum_{m=1}^{\infty} (1 - \psi)^m q_m$$

converges for $0 < \psi \leq 1$. If L is a slowly varying function and $0 < \zeta < 1$ then the two relations

$$Q(\psi) \sim \psi^{-\zeta} L(\psi^{-1}) \qquad \text{as} \quad \psi \downarrow 0$$

and

$$q_m \sim \frac{1}{\Gamma(\zeta)} m^{\zeta - 1} L(m) \qquad \text{as} \quad m \to \infty$$

are equivalent.

We may apply this theorem as follows. First, we note that

$$\begin{aligned}\sum_{m=1}^{\infty} (1 - \psi)^m P_{p_c}(|C| \geq m) &= \sum_{m=1}^{\infty} \sum_{n=m}^{\infty} (1 - \psi)^m P_{p_c}(|C| = n) \\ &= \frac{1 - \psi}{\psi} \sum_{n=1}^{\infty} \{1 - (1 - \psi)^n\} P_{p_c}(|C| = n) \\ &= \frac{1 - \psi}{\psi} \theta(p_c, \psi),\end{aligned} \tag{8.43}$$

where we have used the assumption that $\theta(p_c) = 0$, giving that $\sum_n P_{p_c}(|C| = n) = 1$. Let us assume that

$$P_{p_c}(|C| = m) \approx m^{-1-1/\delta} \qquad \text{as} \quad m \to \infty \tag{8.44}$$

where $1 \le \delta < \infty$. For $0 < \varepsilon < \delta^{-1}$, there exists a constant A_1 such that

$$P_{p_c}(|C| \ge m) \le A_1 m^{\varepsilon - 1/\delta} \qquad \text{for} \quad m \ge 1.$$

Applying the Tauberian theorem with $q_m = A_1 m^{\varepsilon - 1/\delta}$, we find that

$$\sum_{m=1}^{\infty} (1-\psi)^m P_{p_c}(|C| \ge m) \le A_2 \psi^{-(1+\varepsilon-1/\delta)}$$

for some constant A_2 and all small positive ψ. By (8.43),

$$\theta(p_c, \psi) \le \frac{A_2}{1-\psi} \psi^{-\varepsilon+1/\delta} \tag{8.45}$$

for all small positive ψ, yielding in the light of (8.42) that $\delta^{-1} \le \frac{1}{2} + \varepsilon$. We now take the limit as $\varepsilon \downarrow 0$ to obtain that $\delta \ge 2$.

It is easy to see by the foregoing arguments that (8.44) together with $2 \le \delta < \infty$ implies that

$$\theta(p_c, \psi) \approx \psi^{1/\delta} \qquad \text{as} \quad \psi \downarrow 0, \tag{8.46}$$

an expression which has been used by some as the definition of the critical exponent δ.

We follow Newman (1987c) for the rest of this proof. Let us write $\Pi_{nmb}(p)$ for the probability

$$\Pi_{nmb}(p) = a_{nmb} p^m (1-p)^b \tag{8.47}$$

that the origin is an open cluster with n vertices, m edges, and b boundary edges; as usual, a_{nmb} is the number of such animals. It is easy to see that

$$\Pi_{nmb}(p-\varepsilon) = \left(1 - \frac{\varepsilon}{p}\right)^m \left(1 + \frac{\varepsilon}{1-p}\right)^b \Pi_{nmb}(p) \tag{8.48}$$

for any small ε, positive or negative. We wish to relate β to δ, which is to say that we wish to relate $\theta(p_c + \varepsilon)$ to $\theta(p_c, \psi)$, for $\varepsilon, \psi > 0$. We argue as follows. Suppose that $\varepsilon > 0$. Then

$$P_{p_c - \varepsilon}(|C| < \infty) = \sum_{n,m,b} \Pi_{nmb}(p_c - \varepsilon) = 1,$$

so that

$$\begin{aligned}
\theta(p_c + \varepsilon) &= 1 - \sum_{n,m,b} \Pi_{nmb}(p_c + \varepsilon) \\
&= 1 - \left(\sum_{n,m,b} \Pi_{nmb}(p_c - \varepsilon)\right)\left(\sum_{n,m,b} \Pi_{nmb}(p_c + \varepsilon)\right) \\
&= 1 - E_{p_c}\left\{\left(1 - \frac{\varepsilon}{p_c}\right)^{|C_e|}\left(1 + \frac{\varepsilon}{1-p_c}\right)^{|\Delta C|}; |C| < \infty\right\} \\
&\qquad \times E_{p_c}\left\{\left(1 + \frac{\varepsilon}{p_c}\right)^{|C_e|}\left(1 - \frac{\varepsilon}{1-p_c}\right)^{|\Delta C|}; |C| < \infty\right\}
\end{aligned}$$

by (8.48), where C_e is the set of edges of C and ΔC is its external boundary. We use the Cauchy–Schwarz inequality to deduce that

$$\theta(p_c + \varepsilon) \le 1 - \left[E_{p_c}\left\{ \left(1 - \frac{\varepsilon^2}{p_c^2}\right)^{|C_e|/2} \left(1 - \frac{\varepsilon^2}{(1-p_c)^2}\right)^{|\Delta C|/2} ; |C| < \infty \right\} \right]^2.$$

However, $|C_e| \le 2d|C|$ and $|\Delta C| \le 2d|C|$ from (4.14) and (4.15), and it follows from the definition of $\theta(p, \psi)$ that

(8.49) $$\theta(p_c + \varepsilon) \le 1 - \{1 - \theta(p_c, \psi(\varepsilon))\}^2,$$

where

(8.50) $$\psi(\varepsilon) = 1 - \left\{ \left(1 - \frac{\varepsilon^2}{p_c^2}\right)\left(1 - \frac{\varepsilon^2}{(1-p_c)^2}\right) \right\}^d$$

$$= A_1 \varepsilon^2 + o(\varepsilon^2) \quad \text{as} \quad \varepsilon \downarrow 0$$

for some positive constant A_1. Now $\psi(\varepsilon) \to 0$ as $\varepsilon \downarrow 0$, and therefore $\theta(p_c, \psi(\varepsilon)) \to 0$ also, giving from (8.49) that

(8.51) $$\theta(p_c + \varepsilon) \le 2(1 + o(1))\theta(p_c, \psi(\varepsilon)) \quad \text{as} \quad \varepsilon \downarrow 0.$$

We have from Theorem (3.8) that $\theta(p_c + \varepsilon) \ge a\varepsilon$ for some a (> 0) and all small ε, and thus (8.50) and (8.51) imply that

$$\theta(p_c, \psi) \ge A_2 \psi^{1/2}$$

for some positive constant A_2 and all small positive values of ψ. We compare this with (8.46) and deduce that $\delta \ge 2$, if δ exists.

The inequality $\beta \ge 2/\delta$ follows similarly: if $\theta(p_c, \psi) \approx \psi^{1/\delta}$ and $\theta(p_c + \varepsilon) \approx \varepsilon^\beta$ then (8.50) and (8.51) yield the required inequality. ■

8.3 Notes

Section 8.1. Exact calculations for processes on trees provide a favoured activity for statistical physicists. Their principal tactics are the standard methods of branching processes, for which the main references are Harris (1963), Athreya and Ney (1972), and Asmussen and Hering (1983). We have borrowed heavily from the treatment presented by Durrett (1985a); see also Section 5 of Kesten (1987b). The idea of studying the asymptotic properties of the moments of the cluster size by way of the random variable Y_p appears to be due to Aizenman and Newman (1984).

Section 8.2. The inequality $\beta \le 1$ is due to Chayes and Chayes (1987); see also Menshikov (1987b) and the discussion in the notes for Section 3.2. Aizenman and Newman (1984) proved $\gamma \ge 1$, and Aizenman and Barsky (1987) proved $\delta \ge 2$ and also $\beta \ge (\delta - 1)^{-1}$. The inequality $\beta \ge 2/\delta$ appears in Newman (1987c), along with a similar proof of the inequalities $\gamma, \gamma' \ge 2(1 - \delta^{-1})$, where γ and γ' are the critical exponents for $\chi^f(p)$ when $p < p_c$ and $p > p_c$, respectively; Newman does not assume that $\gamma = \gamma'$. See also Newman (1986, 1987b), and Aizenman, Kesten, and Newman (1987).

Certain inequalities of debatable interest may be obtained directly from inequality (3.22); assuming only the existence of the critical exponents in question, we may find that $\gamma \le d$ and $\nu \le 1$ if $\rho < 1$, whereas $\gamma \le \rho d$ and $1 \le \nu \le \rho$ if $\rho > 1$. See the proof of Theorem (9.93) also.

In recent work, Tasaki (1987a, b) has shown that, if the critical exponents of percolation on $\mathbb{L}^d$ equal those of percolation on a binary tree, then $d \ge 6$. Thus the (upper) critical dimension d_c cannot be less than 6. Chayes and Chayes (1988) have reached the same conclusion using different arguments.

Other relevant references are Durrett (1985a), Durrett and Nguyen (1985), Nguyen (1985, 1987a, b), Chayes, Chayes, Fisher, and Spencer (1986), and Newman (1987a).

Rather more is known in the special case when $d = 2$, and we shall discuss such results in the next chapter. Briefly, Kesten (1987b, c) has proved the validity of the scaling and hyperscaling relations $2\nu = \gamma + 2\beta = \beta(\delta + 1)$, $2\rho = \delta + 1$, $\gamma = \nu(2 - \eta)$, under the assumption that δ and ρ exist. He has no such exact result for the exponent α, and indeed it remains an open problem to show that $\kappa(p)$ is not thrice differentiable at p_c. Kesten (1981, 1982) has found bounds for quantities such as θ and χ of the form $\theta(p) \le (p - p_c)^\sigma$ for $p > p_c$ and $\chi(p) \le (p_c - p)^{-\tau}$ for $p < p_c$. Such inequalities imply that $\beta > 0$ and $\gamma < \infty$, together with certain other rather basic facts which remain unsettled in three and higher dimensions. Finally, Kesten (1987b) discusses various bounds for critical exponents in two dimensions which imply that many of these exponents cannot equal the corresponding values for percolation on the binary tree. We reproduce his table.

Rigorous bounds for exponents for $d = 2$	'Exact values'	Exponents for binary tree
$\alpha < 0$	$\alpha = -2/3$	$\alpha = -1$
$\beta < 1$	$\beta = 5/36$	$\beta = 1$
$\gamma \ge 8/5$	$\gamma = 43/18$	$\gamma = 1$
$\delta \ge 5$	$\delta = 91/5$	$\delta = 2$
$\nu > 1$	$\nu = 4/3$	$\nu = \frac{1}{2}$

The so-called 'exact values' are those predicted in the work of den Nijs (1979), Nienhuis, Riedel, and Schick (1980), and Pearson (1980); see also Stauffer (1981). We leave this subject now, with the remark that the inequality $\beta < 1$ for two dimensions was proved by Kesten and Zhang (1987). They prove rigorously that $\theta(p) \ge A(p - p_c)^b$ for $p > p_c$ and some constants A and $b < 1$, this implies that the right-hand derivative of $\theta(p)$ at p_c is infinite, and justifies our sketch of the function θ in Figure 1.4.

How may we show that the critical exponents take on the corresponding values from percolation on a tree? Aizenman and Newman (1984) have introduced the 'triangle condition' in an attempt to answer this question. We define

$$\nabla(p) = \sum_{x, y \in \mathbb{Z}^d} \tau_p(0, x)\tau_p(x, y)\tau_p(y, 0). \tag{8.52}$$

It is known that, if $\nabla(p_c) < \infty$, then β, γ, and δ exist and take the values 1, 1, and 2, respectively. It is sufficient for $\nabla(p_c) < \infty$ that $d > 6$ and

$$\tau_{p_c}(0, x) = O(|x|^{2-d}) \qquad \text{as} \quad |x| \to \infty. \tag{8.53}$$

See Aizenman and Barsky (1988), and the related discussions in Aizenman (1987), Newman (1987a), and Nguyen (1987b). (*Added in proof.*) Hara and Slade (1988) have shown that $\nabla(p_c) < \infty$ if the number d of dimensions is sufficiently large ($d \geq 209$ is certainly enough). This implies that β, γ, δ, and Δ exist and take their 'mean field' values in high dimensions, as well as $\theta(p_c(d)) = 0$.

9

Bond Percolation in Two Dimensions

9.1 Introduction

Until recently, percolation was a game that was played largely on the plane. There is a special reason why percolation in two dimensions is more approachable than percolation in higher dimensions. To every planar two-dimensional lattice $\mathscr{L}$ there corresponds a 'dual' planar lattice $\mathscr{L}_{\mathrm{d}}$ whose edges are in one–one correspondence with the edges of $\mathscr{L}$; furthermore, in a natural embedding of these lattices in the plane, every finite connected subgraph of $\mathscr{L}$ is surrounded by a circuit of $\mathscr{L}_{\mathrm{d}}$. Each edge of $\mathscr{L}$ corresponds to a unique edge of $\mathscr{L}_{\mathrm{d}}$, so that the percolation process on $\mathscr{L}$ generates a percolation process on $\mathscr{L}_{\mathrm{d}}$. In this dual pair of processes, the origin of $\mathscr{L}$ is in an infinite open cluster if and only if it is in the interior of no closed circuit of $\mathscr{L}_{\mathrm{d}}$; such observations may be used to show that, in certain circumstances, $\mathscr{L}$ contains an infinite open cluster if and only if $\mathscr{L}_{\mathrm{d}}$ contains no infinite closed cluster (almost surely), which is to say that

$$p_c(\mathscr{L}) + p_c(\mathscr{L}_{\mathrm{d}}) = 1, \tag{9.1}$$

where $p_c(\mathscr{L})$ and $p_c(\mathscr{L}_{\mathrm{d}})$ are the associated critical probabilities. We saw a similar argument in the proof of Theorem (1.10), where it was shown that the square lattice is self-dual in the sense that the dual lattice of $\mathbb{L}^2$ is isomorphic to $\mathbb{L}^2$. Equation (9.1) implies immediately in this case that $p_c(\mathbb{L}^2) = \frac{1}{2}$, the celebrated exact calculation proved by Kesten (1980a) using arguments based on work of Harris, Russo, Seymour, and Welsh.

We explore the consequences of planar duality in this chapter. We shall see that the picture is rather more complete in two dimensions than in three and more; it is known, for example, that there is no infinite open cluster when $p = p_c$, and that the tail $P_p(n \leq |C| < \infty)$ of the sizes of finite open clusters decays at least as fast (as $n \to \infty$) as $\exp(-\eta(p)n^{1/2})$ for some $\eta(p) > 0$, throughout the supercritical phase $p > p_c$. In addition, duality allows rather greater progress in the direction of describing the geometry of open clusters.

We begin this chapter with a discussion of the technique of planar duality. We shall then apply this technique in proving that $p_c(\mathbb{L}^2) = \frac{1}{2}$; one of the steps of the proof is to show that $\theta(\frac{1}{2}) = 0$, so that there is (almost surely) no infinite open cluster *at* the critical point. In Section 9.4 we study the supercritical percolation process and we show how duality may be used to extend the results of Chapter 6 to the whole of the supercritical phase; in particular, we show the exponential decay of the truncated connectivity functions, and we study the distribution of the size of a large finite open cluster. There follows a discussion of percolation on subsets of $\mathbb{L}^2$, and particularly an account of the problem of ascertaining how the critical probability of percolation on the subset S of the lattice depends on the geometry of S. Section 9.6 contains a brief survey of central limit theorems for the contents of large finite regions of the lattice; we deal there with such quantities as the number of open clusters contained within a large contour γ, and the number of vertices inside γ which are joined to vertices of γ by open paths. The chapter terminates with two sections devoted to open crossings of large rectangles and rigorous results for power law behaviour of $\theta(p)$, $\chi(p)$, and related quantities, when p is near to p_c.

We adopt the following convention throughout this chapter: the letter 'd' stands for 'dual' rather than 'dimension' unless the contrary is clear from the context.

9.2 Planar Duality

Let G be a planar graph, drawn in the plane in the manner of Figure 9.1. With G, we may associate another graph G_d called the (*planar*) *dual*

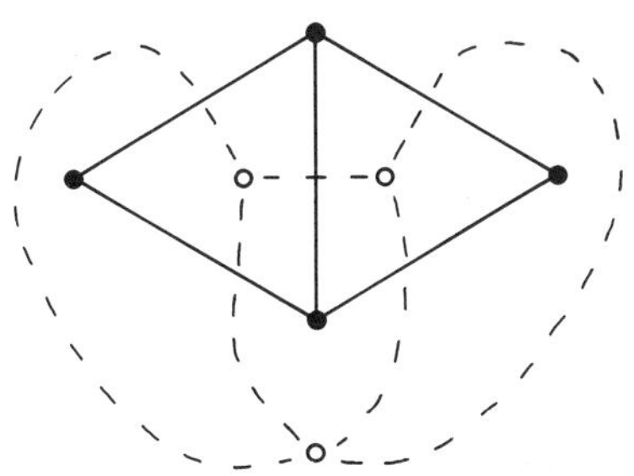

Figure 9.1. A graph and its dual graph.

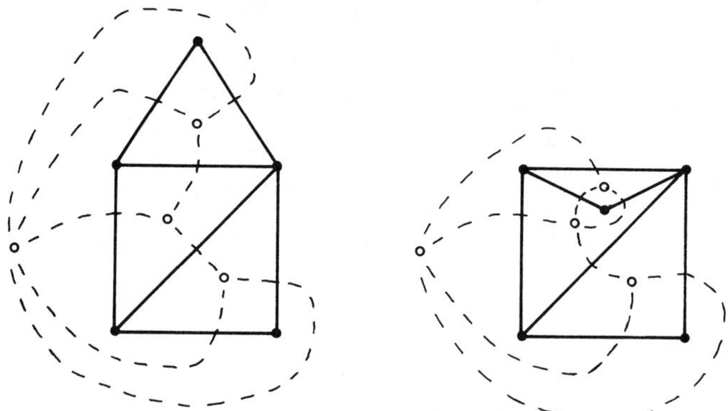

Figure 9.2. Two planar embeddings of the same graph. Note that the duals are not isomorphic, the first containing a vertex with degree five and the second containing only vertices with degrees three and four.

of G, and we do this in the following way. In each face of G (including the infinite face, if it exists) we place a vertex of G_d; for each edge e of G, we place a corresponding edge joining those two vertices of G_d which lie in the two faces of G abutting e. See Figure 9.1 again for an illustration of this construction. Some care is needed in constructing dual graphs since the dual of a planar graph G depends on the way in which G is embedded in the plane; see Figure 9.2.

We are concerned here with the dual graphs of lattices. A glance at Figure 9.3 indicates that the dual of the square lattice $\mathbb{L}^2$ is isomorphic to $\mathbb{L}^2$; for the sake of being concrete, we shall take the dual of $\mathbb{L}^2$ to be

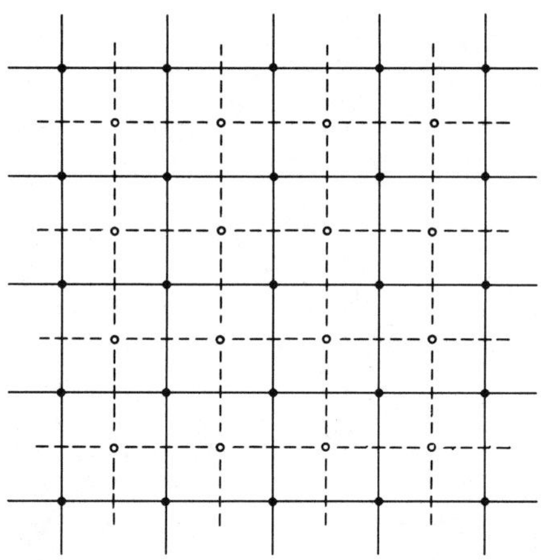

Figure 9.3. Part of the square lattice and its dual. Note that the dual lattice is isomorphic to the original lattice.

the lattice $\mathbb{L}_d^2$ with vertex set $\{x + (\frac{1}{2}, \frac{1}{2}): x \in \mathbb{Z}^2\}$ and edges joining all pairs of vertices which are unit distance apart. Thus each edge of $\mathbb{L}^2$ is bisected by an edge of $\mathbb{L}_d^2$ and vice versa. We denote by O_d the *origin* $(\frac{1}{2}, \frac{1}{2})$ of $\mathbb{L}_d^2$.

Let G be a connected subgraph of $\mathbb{L}^2$, and let $\triangle G$ be the edge boundary of G, defined to be the set of edges of $\mathbb{L}^2$ which do not lie in G but which are incident to at least one vertex of G. The following proposition is crucial to the study of percolation on $\mathbb{L}^2$.

(9.2) Proposition. *Let G be a finite connected subgraph of $\mathbb{L}^2$. There exists a unique circuit $\Sigma(G)$ of $\mathbb{L}_d^2$ containing G in its interior and with the property that every edge of $\Sigma(G)$ crosses an edge of $\triangle G$.*

We do not prove this proposition in all its topological glory, but prefer to call it 'obvious' in the light of the picture in Figure 9.4. Kesten (1982, p. 386) presents a more rigorous account of this result. Elementary topological considerations, such as the isoperimetric inequality, may be used to estimate the length of the circuit $\Sigma(G)$ referred to in the proposition. First, if G has n vertices then it must be the case that $\Sigma(G)$ has at least $\lambda\sqrt{n}$ vertices, for some absolute positive constant λ; similarly, if $\Sigma(G)$ has m vertices then G must have at least m/ζ vertices for some absolute positive constant ζ. We suppose henceforth that λ and ζ are positive constants such that

$$\lambda\sqrt{|G|} \leq |\Sigma(G)| \leq \zeta|G| \tag{9.3}$$

for all finite connected subgraphs G of $\mathbb{L}^2$.

Let $0 \leq p \leq 1$ and consider percolation on $\mathbb{L}^2$ with edge-probability p. We declare each edge of $\mathbb{L}_d^2$ to be *open* (respectively *closed*) if it crosses an open (respectively closed) edge of $\mathbb{L}^2$. This gives rise to a percolation process on $\mathbb{L}_d^2$, which we call the *dual percolation process*. It follows from

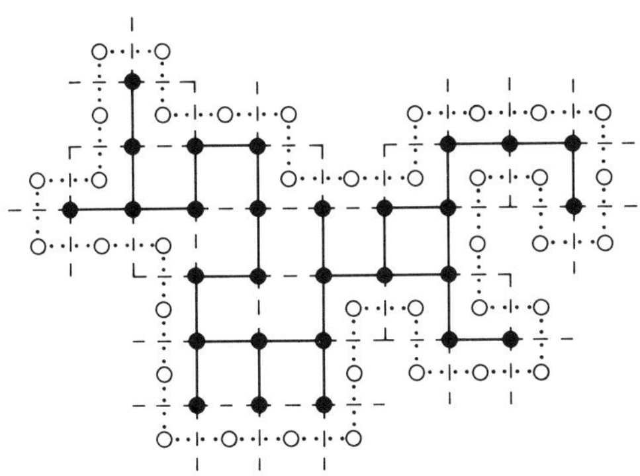

Figure 9.4. A finite connected subgraph of $\mathbb{L}^2$, surrounded by a circuit in the dual each of whose edges traverses an edge in the edge boundary of the subgraph.

the self-duality of the square lattice that the statistical properties of the dual process are identical to those of the original process.

Let C be the open cluster of $\mathbb{L}^2$ at the origin. It is a consequence of Proposition (9.2) that C is finite if and only if the origin of $\mathbb{L}^2$ is contained in the interior of a closed circuit of the dual lattice $\mathbb{L}^2_{\mathrm{d}}$, and it was this argument which we used in the proof of Theorem (1.10).

In an early 'exact calculation' of the critical probabilities of certain two-dimensional lattices, Sykes and Essam (1964) presented an argument which is simple, beautiful, and highly plausible. It remains an open problem to make their argument completely rigorous, even though their 'exact conclusions' are now known to be valid. Their principal observation is the following theorem, the proof of which is an exercise in the use of duality.

(9.4) Theorem. *For percolation on the square lattice $\mathbb{L}^2$, the number $\kappa(p)$ of open clusters per vertex satisfies*

(9.5) $$\kappa(p) = \kappa(1 - p) + 1 - 2p.$$

It is but a small non-rigorous step to deduce from this theorem that $p_c = \frac{1}{2}$, and we argue as follows. Let us suppose that κ is an infinitely differentiable function of p except at p_c, where some derivative of κ fails to exist. A glance at (9.5) indicates that $p_c = 1 - p_c$, and so $p_c = \frac{1}{2}$. A similar argument applied to a dual pair $(\mathscr{L}, \mathscr{L}_{\mathrm{d}})$ of lattices yields (9.1). Even assuming the fact that $p_c = \frac{1}{2}$, can we justify the hypothesis of this argument? We know from Theorem (5.108) that κ is analytic on $[0, p_c)$, and it follows from (9.5) that κ is analytic on $(1 - p_c, 1]$ also. We shall prove in the next two sections that $p_c = \frac{1}{2}$, so that κ is analytic except possibly at p_c. It is an open problem to show that κ is not infinitely differentiable at p_c. Certainly κ is twice differentiable at p_c (see the discussion after Theorem (9.93)) but it is believed that κ is not thrice differentiable at this point; this is related to the conjecture that the critical exponent α in (7.6) satisfies $-1 \leq \alpha < 0$.

Proof of Theorem (9.4). Of principal value here is Euler's formula (see Bondy and Murty (1976, p. 14) or Wilson (1979, p. 66)). Let G be a finite planar graph, drawn in the plane with $v(G)$ vertices, $e(G)$ edges, $f(G)$ finite faces, and $c(G)$ connected components. Euler's formula states that

(9.6) $$c(G) = v(G) - e(G) + f(G).$$

Suppose that $0 \leq p \leq 1$, and consider percolation on $\mathbb{L}^2$ with edge-probability p. Let $B(n)$ be the box with side-length $2n$ and centre at the origin. Let G_n be the subgraph of $\mathbb{L}^2$ with vertex set $B(n)$ together with all open edges joining such vertices. We apply (9.6) and take expectations to obtain

$$E_p(c(G_n)) = |B(n)| - E_p(e(G_n)) + E_p(f(G_n)). \tag{9.7}$$

We divide by $|B(n)|$ and let $n \to \infty$. From the definition of the function $\kappa(p)$, and particularly Theorem (4.2), we have that

$$\frac{1}{|B(n)|} E_p(c(G_n)) \to \kappa(p) \qquad \text{as} \quad n \to \infty. \tag{9.8}$$

Also, the mean number of edges in G_n is $2p|B(n)| + o(|B(n)|)$, so that

$$\frac{1}{|B(n)|} E_p(e(G_n)) \to 2p \qquad \text{as} \quad n \to \infty. \tag{9.9}$$

It remains to deal with the mean number of faces of G_n, and it is here that we make use of duality. Let $G_{n,\mathrm{d}}$ be the following subgraph of the dual lattice $\mathbb{L}^2_\mathrm{d}$: $G_{n,\mathrm{d}}$ has vertex set $B(n)_\mathrm{d} = \{x + (\frac{1}{2}, \frac{1}{2})\colon -n \le x_1, x_2 < n\}$ together with all closed edges joining two such vertices in the dual lattice. We claim that every finite face of G_n contains a unique connected component of $G_{n,\mathrm{d}}$, and we argue as follows. First note that no edge of $G_{n,\mathrm{d}}$ crosses an edge of G_n, since every edge of $G_{n,\mathrm{d}}$ is closed and every edge of G_n is open. Every finite face of G_n contains some vertex of $G_{n,\mathrm{d}}$, and hence some component of $G_{n,\mathrm{d}}$ also. On the other hand, no finite face of G_n can contain two or more such components since, if u and v are vertices of $G_{n,\mathrm{d}}$ which are contained in the same finite face of G_n, then there exists a closed path of $B(n)_\mathrm{d}$ joining u to v. This fact requires more careful proof than that given here; the reader is referred to Figure 9.5 for a pictorial version of the argument and to Kesten (1982, p. 244) for more analytical details. Thus each finite face of G_n contains a unique component of $G_{n,\mathrm{d}}$. How many components of $G_{n,\mathrm{d}}$ do not lie in the finite faces of G_n? Any such component lies in the infinite face of G_n and therefore contains a vertex u of $G_{n,\mathrm{d}}$ lying on its boundary $\partial B(n)_\mathrm{d}$. There are $o(|B(n)|)$ such vertices, and therefore $o(|B(n)|)$ such components of $G_{n,\mathrm{d}}$. We have shown that the number of finite faces of G_n differs from the number of components of $G_{n,\mathrm{d}}$ by at most $o(|B(n)|)$. Hence

$$\frac{1}{|B(n)|} E_p(f(G_n)) \to \kappa(1-p) \qquad \text{as} \quad n \to \infty, \tag{9.10}$$

since the mean number of closed components of $G_{n,\mathrm{d}}$ is $(1 + o(1)) \times |B(n)_\mathrm{d}|\kappa(1-p)$ by Theorem (4.2), and $|B(n)_\mathrm{d}|/|B(n)| \to 1$ as $n \to \infty$.

We divide through (9.7) by $|B(n)|$ and let $n \to \infty$ to obtain from (9.8)–(9.10) that

$$\kappa(p) = 1 - 2p + \kappa(1-p)$$

as required. ∎

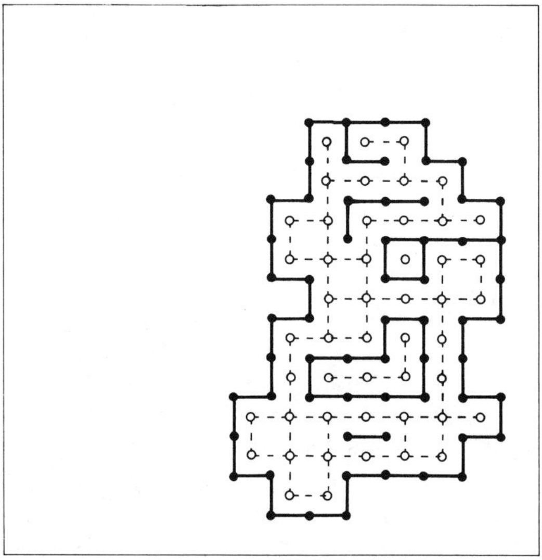

Figure 9.5. The part of G_n sketched here contains three faces, two of which are in the interior of the third. Each such face contains a unique component of the dual.

9.3 The Critical Probability Equals $\frac{1}{2}$

The principal purpose of this section is to prove the following famous exact calculation.

(9.11) Theorem. *The critical probability of bond percolation on* $\mathbb{L}^2$ *equals* $\frac{1}{2}$.

This result is one of the most important milestones in the history of percolation. Here is some intuitive reasoning to justify the value $\frac{1}{2}$. First, if $\theta(\frac{1}{2}) > 0$ then, when $p = \frac{1}{2}$, there exists (almost surely) both an infinite open cluster in $\mathbb{L}^2$ and an infinite closed cluster in $\mathbb{L}^2_{\mathrm{d}}$. The plane is a somewhat confined space for two non-intersecting infinite clusters, and so their co-existence seems unlikely. This is an indication that $\theta(\frac{1}{2}) = 0$, giving that $p_c \geq \frac{1}{2}$. On the other hand, if $p < p_c$ then we may accept a picture of many finite open clusters adrift in an ocean of closed edges of the dual. Presumably this ocean contains an infinite closed cluster, so that $1 - p \geq p_c$; thus $p \leq 1 - p_c$ whenever $p < p_c$, which implies that $p_c \leq \frac{1}{2}$.

The two main ingredients of the streamlined proof which follows are the fact that $\chi(p) < \infty$ when $p < p_c$ (Theorem (3.2)), and the uniqueness of the infinite cluster whenever such a cluster exists (Theorem (6.1)); taken in conjunction with the self-duality of the square lattice, these two ingredients guarantee the result. Approached from this point of view, the result is very natural and its proof rather short and accessible, in considerable contrast to the proof which appeared first in the literature. The latter proof was the crowning achievement of four papers published over a period of 21 years. Harris (1960) proved that $\theta(\frac{1}{2}) = 0$, and therefore $p_c \geq \frac{1}{2}$, using arguments of some geometric complexity and ingenuity; as a byproduct he obtained the uniqueness of the infinite cluster in two dimensions. Attention was drawn to the role of the mean cluster size by Russo (1978) and by Seymour and Welsh (1978) in independent but essentially equivalent work. Finally, Kesten (1980a) showed how to build on their arguments to obtain the full result. We shall take a quite different route here, using the uniqueness of the infinite cluster in place of Harris's arguments, and appealing to the uniqueness of critical points (Theorem (3.2)) in place of Kesten's argument and the difficult parts of the Russo–Seymour–Welsh technology.

We divide the theorem into two parts, and shall prove first the following lemma.

(9.12) Lemma. *It is the case that* $\theta(\frac{1}{2}) = 0$, *and therefore the critical probability* p_c *satisfies* $p_c \geq \frac{1}{2}$.

Once we have proved that $p_c = \frac{1}{2}$ then this lemma states that $\theta(p_c) = 0$, so that there is (almost surely) no infinite open cluster *at* the critical point. As already noted several times, this implies that θ is continuous on the whole interval $[0, 1]$, a statement which is believed to be valid in all dimensions d (≥ 2). A further consequence of this lemma is the following, which is stated here largely for historical reasons.

(9.13) Lemma. Existence of open circuits above p_c. *Suppose that* p *is such that* $\theta(p) > 0$, *and let* n *be a positive integer. With probability* 1, *there exists an open circuit of* $\mathbb{L}^2$ *containing the box* $B(n)$ *in its interior.*

We recall that $B(n)$ is the box with side-length $2n$ and centre at the origin. By the *interior* of a circuit, we mean the bounded region of $\mathbb{R}^2$ contained strictly within the circuit.

Lemmas (9.12) and (9.13) are essentially equivalent to each other. Harris proved (9.13) directly, and deduced (9.12) together with the uniqueness of the infinite open cluster. On the other hand, it is easy to deduce (9.13) from (9.12), and here is a sketch of the argument; it is not difficult to fill in the gaps. Suppose that $\theta(p) > 0$ but there exists a positive integer n such that the probability that $B(n)$ is contained in the interior of an open circuit equals $1 - \delta$ where $\delta > 0$. We have from Lemma (9.12) that $p > \frac{1}{2}$. The box $B(n)$ is in the interior of no open circuit if and only if some vertex of the dual lattice $\mathbb{L}_d^2$ lying 'just out-

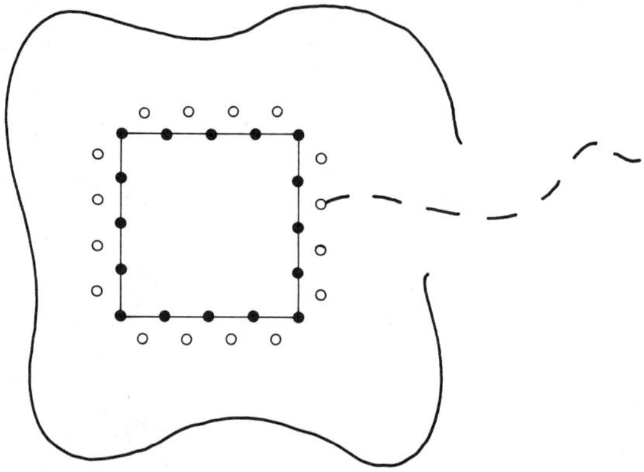

Figure 9.6. If $B(n)$ is not in the interior of an open circuit, then some vertex of the dual is in an infinite closed cluster. The broken line represents a closed path in the dual.

side' $B(n)$ is in an infinite closed cluster of the dual (see Figure 9.6). This observation is easily checked by applying Proposition (9.2) to the connected subgraph G of $\mathbb{L}_{\mathrm{d}}^2$ containing the 'dual box' $B_{\mathrm{d}} = \{x + (\frac{1}{2}, \frac{1}{2})$: $-n-1 \leq x_1, x_2 \leq n\}$, together with all edges of the dual joining pairs of such vertices, and all vertices and closed edges of the dual joined to B_{d} by closed paths; by the proposition, G is finite if and only if it is in the interior of some open circuit of the original lattice $\mathbb{L}^2$. Thus the probability that there exists an infinite closed cluster in the dual is at least δ, and therefore the density $1-p$ of closed edges exceeds the critical density p_c. That is to say, $1 - p \geq p_c$, and so $p \leq 1 - p_c$. However $p_c \geq \frac{1}{2}$ and therefore $p \leq \frac{1}{2}$, a contradiction.

We turn now to the proof of Lemma (9.12). In this proof we shall make use of an elementary device which is of value later in this chapter: if $A_1, A_2, \ldots, A_m$ are increasing events with equal probability, then

(9.14) $$P_p(A_1) \geq 1 - \left\{1 - P_p\left(\bigcup_{i=1}^m A_i\right)\right\}^{1/m}.$$

To see this, note that

$$\begin{aligned} 1 - P_p\left(\bigcup_{i=1}^m A_i\right) &= P_p\left(\bigcap_{i=1}^m A_i^c\right) \\ &\geq \prod_{i=1}^m P_p(A_i^c) \qquad \text{by the FKG inequality} \\ &= (1 - P_p(A_1))^m. \end{aligned}$$

We call this the 'square root trick' following Cox and Durrett (1988).

Proof of Lemma (9.12). We follow an argument of Y. Zhang (1988, unpublished). Consider bond percolation on $\mathbb{L}^2$ with edge-probability $\frac{1}{2}$, and suppose that $\theta(\frac{1}{2}) > 0$. For each positive integer n, let $A^{\mathrm{l}}(n)$ (respectively $A^{\mathrm{r}}(n)$, $A^{\mathrm{t}}(n)$, $A^{\mathrm{b}}(n)$) be the event that some vertex on the left (respectively right, top, bottom) side of the square $T(n) = [0, n]^2$ is in an infinite open path of $\mathbb{L}^2$ which uses no other vertex of $T(n)$. Clearly $A^{\mathrm{l}}(n)$, $A^{\mathrm{r}}(n)$, $A^{\mathrm{t}}(n)$, and $A^{\mathrm{b}}(n)$ are increasing events having equal probability and whose union is the event that some vertex in $T(n)$ is in an infinite open cluster. There exists an infinite open cluster with probability 1, so that

$$P_{1/2}(A^{\mathrm{l}}(n) \cup A^{\mathrm{r}}(n) \cup A^{\mathrm{t}}(n) \cup A^{\mathrm{b}}(n)) \to 1 \qquad \text{as} \quad n \to \infty.$$

It follows by the square root trick (9.14) that

$$(9.15) \qquad P_{1/2}(A^u(n)) \to 1 \qquad \text{as} \quad n \to \infty \quad \text{for} \quad u = \mathrm{l, r, t, b}.$$

We choose N such that

$$(9.16) \qquad P_{1/2}(A^u(N-1)) > \tfrac{7}{8} \qquad \text{for} \quad u = \mathrm{l, r, t, b},$$

and furthermore

$$(9.17) \qquad P_{1/2}(A^u(N)) > \tfrac{7}{8} \qquad \text{for} \quad u = \mathrm{l, r, t, b}.$$

Moving to the dual lattice, we define the dual box $T(n)_{\mathrm{d}} = \{x + (\frac{1}{2}, \frac{1}{2}): 0 \le x_1, x_2 < n\}$. Let $A^{\mathrm{l}}_{\mathrm{d}}(n)$ (respectively $A^{\mathrm{r}}_{\mathrm{d}}(n)$, $A^{\mathrm{t}}_{\mathrm{d}}(n)$, $A^{\mathrm{b}}_{\mathrm{d}}(n)$) be the event that some vertex on the left (respectively right, top, bottom) side of $T(n)_{\mathrm{d}}$ is in an infinite closed path of $\mathbb{L}^2_{\mathrm{d}}$ which uses no other vertex of $T(n)_{\mathrm{d}}$. Each edge of $\mathbb{L}^2_{\mathrm{d}}$ is closed with probability $\frac{1}{2}$, so that

$$(9.18) \qquad P_{1/2}(A^u_{\mathrm{d}}(N)) = P_{1/2}(A^u(N-1)) > \tfrac{7}{8} \qquad \text{for} \quad u = \mathrm{l, r, t, b},$$

by (9.16). Consider now the event $A = A^{\mathrm{l}}(N) \cap A^{\mathrm{r}}(N) \cap A^{\mathrm{t}}_{\mathrm{d}}(N) \cap A^{\mathrm{b}}_{\mathrm{d}}(N)$, that there are infinite open paths of $\mathbb{L}^2 \backslash T(N)$ touching the left and right sides of $T(N)$, and infinite closed paths of $\mathbb{L}^2_{\mathrm{d}} \backslash T(N)_{\mathrm{d}}$ touching the top and bottom sides of $T(N)_{\mathrm{d}}$; see Figure 9.7. The probability that A does not occur satisfies

$$\begin{aligned} P_{1/2}(A^c) &\le P_{1/2}(A^{\mathrm{l}}(N)^c) + P_{1/2}(A^{\mathrm{r}}(N)^c) + P_{1/2}(A^{\mathrm{t}}_{\mathrm{d}}(N)^c) + P_{1/2}(A^{\mathrm{b}}_{\mathrm{d}}(N)^c) \\ &\le \tfrac{1}{2} \qquad \text{by (9.17) and (9.18),} \end{aligned}$$

giving that $P_{1/2}(A) > \frac{1}{2}$. If A occurs then $\mathbb{L}^2 \backslash T(N)$ contains two disjoint infinite open clusters, since the clusters in question are separated physically by infinite closed paths of the dual; any open path of $\mathbb{L}^2 \backslash T(N)$ joining these two clusters would contain an edge which crosses a closed edge of the dual, and no such edge can exist. Similarly, on A, the graph $\mathbb{L}^2_{\mathrm{d}} \backslash T(N)_{\mathrm{d}}$ contains two disjoint infinite closed clusters, separated physically by infinite open paths of $\mathbb{L}^2 \backslash T(N)$. Now, the whole lattice $\mathbb{L}^2$ contains (almost surely) a *unique* infinite open cluster, and it follows

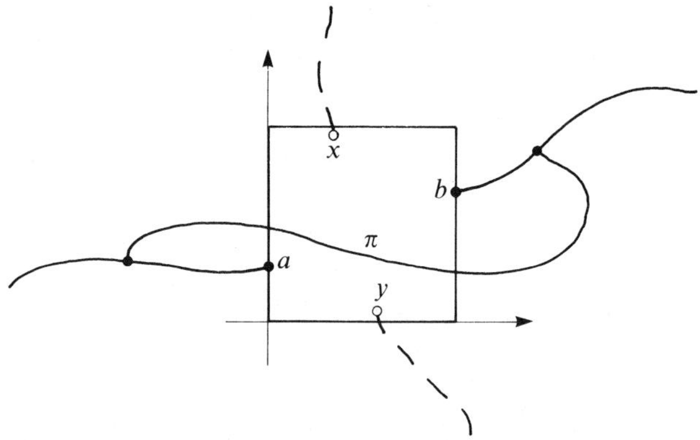

Figure 9.7. Vertices a and b are in infinite open clusters of $\mathbb{L}^2\backslash T(N)$, and vertices x and y are in infinite closed clusters of $\mathbb{L}_d^2\backslash T(N)_d$. If there is a unique infinite open cluster, then there exists an open path π joining a to b, and thus the infinite closed clusters at x and y are disjoint.

that there exists (almost surely on A) an open connection of the lattice between the fore-mentioned infinite open clusters. By the geometry of the situation (see Figure 9.7 again), this connection traverses $T(N)$ and forms a barrier to possible closed connections of the dual joining the two infinite closed clusters. Hence, almost surely on A, the dual lattice contains two or more infinite closed clusters, an event having probability 0. Thus $P_{1/2}(A) = 0$ in contradiction of the deduction that $P_{1/2}(A) > \frac{1}{2}$. The initial hypothesis that $\theta(\frac{1}{2}) > 0$ is therefore incorrect, and the proof is complete. ■

Proof of Theorem (9.11). It remains only to show that $p_c \leq \frac{1}{2}$. There is more than one way of going about this, and we shall give two proofs, both of which originate in the work of Russo (1978) and Seymour and Welsh (1978).

Here is our first proof. We shall show that, if $p < p_c$, then there is strictly positive probability that the origin of the dual lattice is in an infinite closed cluster. Such a cluster exists with positive probability only if the probability $1 - p$, that a given edge is closed, is at least as large as p_c. Thus $1 - p \geq p_c$ whenever p satisfies $p < p_c$, and this implies that $p_c \leq \frac{1}{2}$ as required.

Suppose that $p < p_c$, so that the mean cluster size $\chi(p)$ satisfies

(9.19) $$\chi(p) = \sum_{n=1}^{\infty} P_p(|C| \geq n) < \infty.$$

Let M be a positive integer. We denote by A_M the event that there is an open path π in $\mathbb{L}^2$ joining a vertex of the form $(k, 0)$ with $k < 0$ to a vertex of the form $(l, 0)$ with $l \geq M$, and having the property that all the

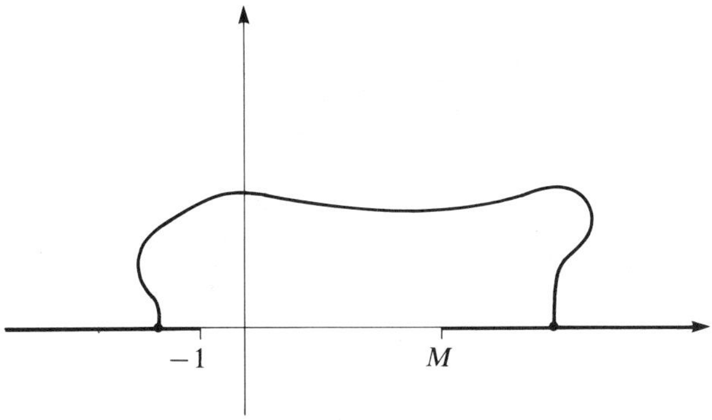

Figure 9.8. A sketch of the event A_M.

vertices of π other than its endvertices lie strictly above the horizontal axis. A sketch of the event A_M appears in Figure 9.8. We may see that

$$P_p(A_M) \le P_p\left(\bigcup_{l=M}^{\infty} \{(l, 0) \leftrightarrow (k, 0) \text{ for some } k < 0\}\right)$$

$$\le \sum_{l=M}^{\infty} P_p(|C| \ge l)$$

since, if $(l, 0)$ is joined to some vertex $(k, 0)$ with $k < 0$, then $(l, 0)$ is in an open cluster of size at least l. From (9.19), we may choose M large enough to ensure that $P_p(A_M) \le \frac{1}{2}$. We claim that if A_M does not occur then there exists an infinite closed cluster in the dual lattice. Let L be the set of vertices $\{(m + \frac{1}{2}, \frac{1}{2})\colon 0 \le m < M\}$ in the dual, and let $C(L)$ be the set of vertices in the dual which are connected to some vertex in L by a closed path. If $|C(L)| < \infty$, then we apply Proposition (9.2) to find that $C(L)$ is surrounded in the dual of $\mathbb{L}_d^2$ by an open circuit Σ (this is seen easiest by applying the proposition to the graph with vertex set $C(L)$ and all edges of $\mathbb{L}_d^2$ joining pairs of such vertices). However, the dual of $\mathbb{L}_d^2$ is the original lattice $\mathbb{L}^2$, and such an open circuit Σ must contain an open path joining some vertex $(k, 0)$ with $k < 0$ to some vertex $(l, 0)$ where $l \ge M$, and otherwise lying entirely in the upper half-plane. Thus

$$P_p(|C(L)| < \infty) \le P_p(A_M) \le \tfrac{1}{2},$$

giving that $P_p(|C(L)| = \infty) > \frac{1}{2}$. Now, if $|C(L)| = \infty$ then one of the M vertices of L is in an infinite closed cluster of $\mathbb{L}_d^2$, giving that

$$P_p(0_d \text{ in infinite closed cluster}) \ge \frac{1}{M} P_p(|C(L)| = \infty)$$

$$> \frac{1}{2M},$$

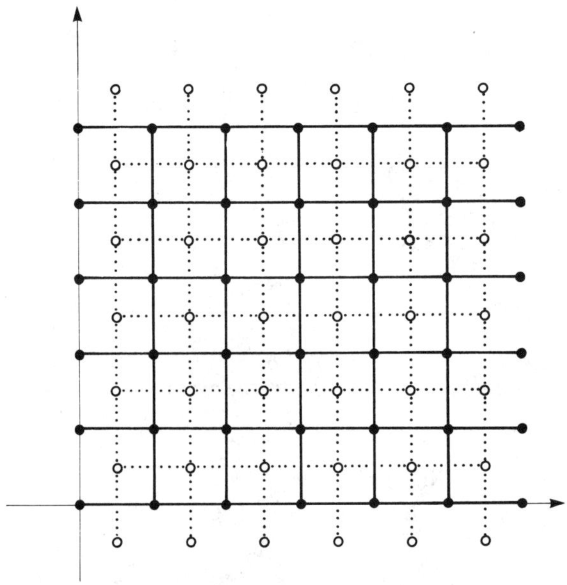

Figure 9.9. A picture of $S(5)$ and its dual $S(5)_d$.

and we have shown that there is strictly positive probability that the origin of $\mathbb{L}_d^2$ is in an infinite closed cluster, as required. The first proof that $p_c \le \frac{1}{2}$ is now complete.

Our second proof of the inequality $p_c \le \frac{1}{2}$ makes use of a beautiful application of duality. Let $S(n)$ be the subgraph of $\mathbb{L}^2$ with vertex set $\{x \in \mathbb{Z}^2: 0 \le x_1 \le n+1, 0 \le x_2 \le n\}$ and edge set comprising all edges joining pairs of vertices in $S(n)$ *except* those joining pairs x, y with either $x_1 = y_1 = 0$ or $x_1 = y_1 = n+1$; see Figure 9.9 for a drawing of $S(n)$. Let $S(n)_d$ be the subgraph of $\mathbb{L}_d^2$ with vertex set $\{x + (\frac{1}{2}, \frac{1}{2}): 0 \le x_1 \le n, -1 \le x_2 \le n\}$ and edge set comprising all edges of $\mathbb{L}_d^2$ joining pairs of vertices in $S(n)_d$ *except* those joining pairs u, v with either $u_2 = v_2 = -\frac{1}{2}$ or $u_2 = v_2 = n + \frac{1}{2}$. A glance at Figure 9.9 will bring this definition to life; note that $S(n)_d$ may be obtained by rotating $S(n)$ anticlockwise through a right angle and relocating the vertex labelled (0, 0) at the point $(n + \frac{1}{2}, -\frac{1}{2})$.

We denote by A_n the event that there is an open path of $S(n)$ joining a vertex on the left side of $S(n)$ to a vertex on its right side. Similarly, we denote by B_n the event that there is a closed path of $S(n)_d$ joining a vertex on the top side of $S(n)_d$ to a vertex on its bottom side. It is easily seen that $A_n \cap B_n = \varnothing$, the empty event, since if both A_n and B_n occur then there is an open path in $S(n)$ which crosses a closed path in $S(n)_d$; see Figure 9.9. Where these two paths cross, there is an open edge of $\mathbb{L}^2$

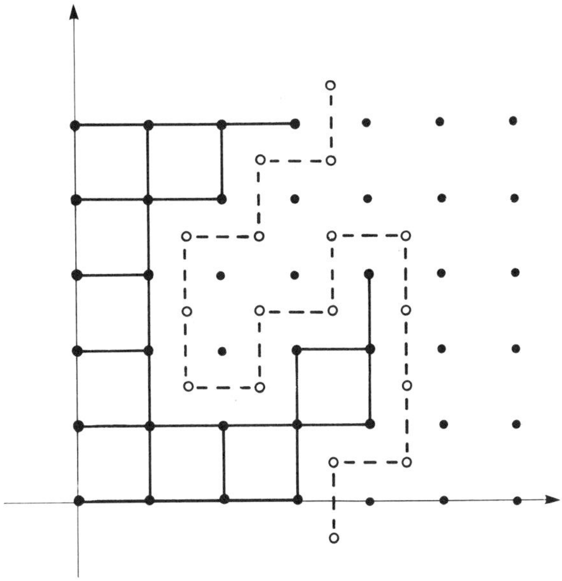

Figure 9.10. If there is no open path traversing $S(n)$ from left to right then there exists a closed path crossing $S(n)_d$ from top to bottom.

which is crossed by a closed edge of $\mathbb{L}_d^2$, and this is impossible. Thus $A_n \cap B_n = \varnothing$. On the other hand, either A_n or B_n must occur. To see this, suppose that A_n does not occur, and let D be the set of all vertices of $S(n)$ which are attainable from the left side of $S(n)$ along open paths; we turn D into a graph by adding all open edges of $S(n)$ joining pairs of vertices in D. By a minor variant of Proposition (9.2) (see Figure 9.10) there is a closed path of $\mathbb{L}_d^2$ crossing $S(n)_d$ from top to bottom which crosses only edges of $S(n)$ contained in the edge boundary of D. Thus B_n occurs whenever A_n does not occur. We have proved that A_n and B_n are disjoint events which partition the sample space, and so

$$P_p(A_n) + P_p(B_n) = 1.$$

On the other hand, $P_p(B_n) = P_{1-p}(A_n)$, since $S(n)_d$ is isomorphic to $S(n)$ and each edge of $S(n)_d$ is closed with probability $1 - p$. Hence

$$P_p(A_n) + P_{1-p}(A_n) = 1,$$

and in particular $P_{1/2}(A_n) = \frac{1}{2}$. We state this conclusion as a lemma, to facilitate its availability for future use.

(9.20) Lemma. *Let A_n be the event that there exists an open path in the rectangle $[0, n + 1] \times [0, n]$ joining some vertex on its left side to some vertex on its right side. Then $P_{1/2}(A_n) = \frac{1}{2}$.*

This latter definition of A_n may appear at first sight to differ from the original definition; however, it is easily seen that the two definitions are equivalent, since the existence of such an open crossing does not depend on the states of edges in the left and right sides of the rectangle.

Suppose now that $p_c > \frac{1}{2}$. Then $p = \frac{1}{2}$ is in the subcritical phase, for which the probability of an open path from the origin to a vertex on the line $L_n = \{(n, k): k \in \mathbb{Z}\}$ decays exponentially as $n \to \infty$. That is to say, if $p_c > \frac{1}{2}$ then there exists $\sigma > 0$ such that

(9.21) $$P_{1/2}(0 \leftrightarrow L_n) \leq e^{-\sigma n} \qquad \text{for all } n,$$

by either Theorem (3.4) or Theorem (5.1). In this case

$$\begin{aligned} P_{1/2}(A_n) &\leq \sum_{k=0}^{n} P_{1/2}((0, k) \leftrightarrow L_n) \\ &\leq (n+1)e^{-\sigma n} \\ &\to 0 \qquad \text{as} \quad n \to \infty, \end{aligned}$$

in contradiction of the fact that $P_{1/2}(A_n) = \frac{1}{2}$ for all n. It follows that $p_c \leq \frac{1}{2}$, and the second proof is complete. ∎

9.4 Tail Estimates in the Supercritical Phase

There are various unresolved conjectures concerning the sizes of finite open clusters when $p > p_c(d)$ and the number d of dimensions satisfies $d \geq 3$. We may use duality to resolve most such questions in the case of two dimensions. There are two principal results in this direction. The first concerns the probability $\tau_p^f(0, e_n)$ that the origin is in a finite open cluster containing the vertex $e_n = (n, 0)$, and the second concerns the rate of decay of the cluster size distribution as $n \to \infty$.

In preparation, we recall from Section 6.6 that, for two or more dimensions, the correlation length $\xi(p)$ is defined by the asymptotic relation

(9.22) $$\tau_p^f(0, e_n) \approx e^{-n/\xi(p)} \qquad \text{as} \quad n \to \infty.$$

It is known that $\xi(p) < \infty$ if $p \in (0, p_c(d))$ and $d \geq 2$, but it is an open question to prove $\xi(p) < \infty$ throughout the supercritical phase when $d \geq 2$. In the case of two dimensions, the finiteness of $\xi(p)$ above $p_c(2)$ is part of the following result.

(9.23) Theorem. Exponential decay of the truncated connectivity function. *Suppose that* $d = 2$ *and* $1 > p > p_c$ $(=\frac{1}{2})$. *Then* $\xi(p) = \frac{1}{2}\xi(1-p)$, *and therefore* $0 < \xi(p) < \infty$.

Our second result concerns the asymptotic behaviour of $P_p(|C| = n)$ when $p \in (p_c(d), 1)$. We saw in Section 6.7 that this probability is at least as large as $\exp(-\gamma(p)n^{(d-1)/d})$ for some $\gamma(p) > 0$ where d is the number of dimensions, and it is conjectured that the actual asymptotic behaviour of $P_p(|C| = n)$ is of this form.

(9.24) Theorem. *Suppose that $d = 2$ and $1 > p > p_c$ $(=\frac{1}{2})$. There exists $\eta(p) > 0$ such that*

$$P_p(|C| = n) \le \exp(-\eta(p)n^{1/2}) \qquad \text{for all } n. \tag{9.25}$$

This provides an upper bound of the correct form which is valid throughout the supercritical phase in two dimensions. It remains an open problem to show the existence of the limit

$$\lim_{n\to\infty} \left\{-\frac{1}{\sqrt{n}} \log P_p(|C| = n)\right\}$$

when $p \in (p_c, 1)$.

We turn now to the proofs of these theorems. In our proof of Theorem (9.23) we shall need a lemma which is of some interest in its own right. Let k be a positive integer and let T_k be the tube $\{x \in \mathbb{Z}^2: |x_2| \le k\}$. We write '$0 \leftrightarrow e_n$ in T_k' for the event that there is an open path of T_k joining the origin to the vertex e_n. It is clear that

$$\lim_{k\to\infty} P_p(0 \leftrightarrow e_n \text{ in } T_k) = P_p(0 \leftrightarrow e_n) = \tau_p(0, e_n). \tag{9.26}$$

(9.27) Lemma. *Suppose that $0 < p < 1$. The limit*

$$\varphi_k(p) = \lim_{n\to\infty} \left\{-\frac{1}{n} \log P_p(0 \leftrightarrow e_n \text{ in } T_k)\right\} \tag{9.28}$$

exists for each $k \ge 1$. Furthermore

$$P_p(0 \leftrightarrow e_n \text{ in } T_k) \le \exp(-n\varphi_k(p)) \qquad \text{for all } n \text{ and } k. \tag{9.29}$$

The functions $\varphi_k(p)$ satisfy

$$0 < \varphi_k(p) < \infty \qquad \text{for all} \quad k \ge 1 \quad \text{and} \quad 0 < p < 1, \tag{9.30}$$

$$\varphi_k(p) \downarrow \varphi(p) \qquad \text{as} \quad k \to \infty, \tag{9.31}$$

where

$$\varphi(p) = \lim_{n\to\infty} \left\{-\frac{1}{n} \log \tau_p(0, e_n)\right\}. \tag{9.32}$$

Thus, for each tube T_k, the two-point connectivity function decays in the manner of an exponential function of the distance between the two points, and furthermore the constant φ_k in the exponent converges to the unrestricted constant φ as the width $2k$ of the tube tends to ∞. If $p < \frac{1}{2}$ then $\varphi(p) = \xi(p)^{-1}$, the reciprocal of the correlation length. See Theorem (5.44) and the discussion around (5.54) for the relevant results in the subcritical phase. It is easy to adapt Lemma (9.27) to obtain a result valid in all dimensions.

Proof of Lemma (9.27). This proof is implicit in Chayes, Chayes, Grimmett, Kesten, and Schonmann (1988) and appears explicitly in Durrett and Schonmann (1988b). It is easy to adapt the proof of Theorem (5.44) to see that the limit in (9.28) exists and that

$$\varphi_k(p) = \inf_n \left\{ -\frac{1}{n} \log P_p(0 \leftrightarrow e_n \text{ in } T_k) \right\}; \tag{9.33}$$

hence (9.29) holds. Certainly $\varphi_k(p) < \infty$ since $P_p(0 \leftrightarrow e_n \text{ in } T_k) \geq p^n$, the probability that each of the n edges in the shortest path from the origin to e_n is open. Here is a direct argument which proves that $\varphi_k(p) > 0$. Let A_i be the event that all edges of T_k joining two vertices of the form (i, r) and $(i + 1, r)$ for some r are closed. The A_i's are independent events and $P_p(A_i) = (1 - p)^{2k+1}$. Also

$$\begin{aligned} P_p(0 \leftrightarrow e_n \text{ in } T_k) &\leq P_p(A_i \text{ does not happen for } 0 \leq i < n) \\ &= \{1 - (1 - p)^{2k+1}\}^n, \end{aligned}$$

giving that

$$\varphi_k(p) \geq -\log\{1 - (1 - p)^{2k+1}\}.$$

Finally we prove (9.31). It is the case that $T_k \subseteq T_{k+1} \subseteq \mathbb{Z}^2$, and so

$$\varphi_k(p) \geq \varphi_{k+1}(p) \geq \varphi(p) \qquad \text{for all } k; \tag{9.34}$$

hence the limit $\lim_{k\to\infty} \varphi_k(p)$ exists and satisfies

$$\lim_{k\to\infty} \varphi_k(p) \geq \varphi(p). \tag{9.35}$$

On the other hand, we combine (9.26) with (9.29) to obtain

$$\begin{aligned} \tau_p(0, e_n) &\leq \lim_{k\to\infty} \exp(-n\varphi_k(p)) \\ &= \exp\left(-n \lim_{k\to\infty} \varphi_k(p)\right). \end{aligned}$$

We take logarithms of this inequality, divide by n, let n tend to infinity, and find that

$$\varphi(p) \geq \lim_{k\to\infty} \varphi_k(p)$$

by (9.32). The result follows from (9.35). ■

Proof of Theorem (9.23). We follow Chayes, Chayes, Grimmett, Kesten, and Schonmann (1988). We first derive an upper bound for

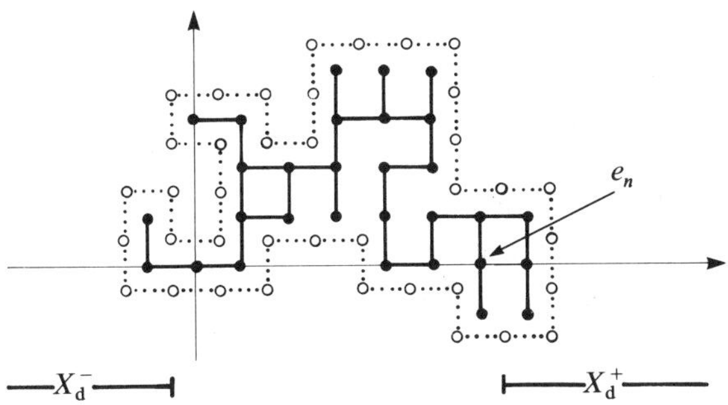

Figure 9.11. Note the existence of two edge-disjoint closed paths joining X_d^- to X_d^+, each being a sub-path of a closed circuit surrounding the open cluster at the origin.

$\tau_p^f(0, e_n)$, and then we show that this bound is sharp to the 'logarithmic' order.

Suppose that $p_c < p < 1$. If the origin and $e_n = (n, 0)$ are in the same finite open cluster, then there exists a closed circuit in the dual lattice containing both these vertices in its interior (see Proposition (9.2) and the discussion thereafter). Writing X_d^- for the half-axis $\{(k + \frac{1}{2}, \frac{1}{2}): k < 0\}$ and X_d^+ for the half-line $\{(l + \frac{1}{2}, \frac{1}{2}): l \geq n\}$ in the dual lattice, we may see from Figure 9.11 that such a circuit contains two edge-disjoint closed paths joining vertices in X_d^- to vertices in X_d^+. Writing A_n for the event that some vertex in X_d^- is joined by a closed path in the dual lattice to some vertex in X_d^+, we have that

$$\tau_p^f(0, e_n) \leq P_p(A_n \circ A_n). \tag{9.36}$$

The event A_n is increasing in the set-theoretic lattice of subsets of the edge-set of $\mathbb{L}_d^2$, and we may therefore apply the BK inequality (particularly inequality (2.17)) to find that

$$\tau_p^f(0, e_n) \leq P_p(A_n)^2. \tag{9.37}$$

Now,

$$\begin{aligned} P_p(A_n) &\leq \sum_{k=-\infty}^{-1} \sum_{l=n}^{\infty} P_p\big((k + \tfrac{1}{2}, \tfrac{1}{2}) \text{ joined to } (l + \tfrac{1}{2}, \tfrac{1}{2}) \text{ by a closed path of the dual}\big) \\ &= \sum_{k=-\infty}^{-1} \sum_{l=n}^{\infty} P_{1-p}\big((k, 0) \leftrightarrow (l, 0)\big). \end{aligned} \tag{9.38}$$

However, $1 - p < p_c$ since $p > p_c$ $(=\frac{1}{2})$, and so

$$P_{1-p}\big((k, 0) \leftrightarrow (l, 0)\big) \approx e^{-(l-k)/\xi(1-p)}$$

as $l - k \to \infty$, by (9.22). Furthermore $\xi(1 - p) < \infty$. Therefore

$$\sum_{k=-\infty}^{-1} \sum_{l=n}^{\infty} P_{1-p}\big((k, 0) \leftrightarrow (l, 0)\big) \approx e^{-n/\xi(1-p)},$$

and it follows from (9.37) and (9.38) that

$$\liminf_{n\to\infty}\left\{-\frac{1}{n}\log\tau_p^f(0,e_n)\right\}\geq\frac{2}{\xi(1-p)},\tag{9.39}$$

which is one half of the conclusion of the theorem.

For the other half, we need to show that the above approximation to $\tau_p^f(0, e_n)$ is not too bad, and it will follow that

$$\limsup_{n\to\infty}\left\{-\frac{1}{n}\log\tau_p^f(0,e_n)\right\}\leq\frac{2}{\xi(1-p)};\tag{9.40}$$

the remainder of the proof is devoted to showing this, and it is here that we make use of Lemma (9.27). Let n be a positive integer and let $m = \lfloor\sqrt{n}\rfloor$; later we shall take the limit as $n\to\infty$. We denote by R_{mn} the rectangle $[-m, n+m]\times[-m, m]$. We write $O(R_{mn})$ for the event that R_{mn} is contained in the interior of a closed circuit of the dual, and $\{0\leftrightarrow e_n \text{ in } R_{mn}\}$ for the event that there is an open path in R_{mn} joining the origin to e_n. These two events depend on disjoint sets of edges and are therefore independent. If they both occur then the origin and e_n are in the same finite open cluster of $\mathbb{L}^2$, and therefore

$$\begin{aligned}\tau_p^f(0,e_n)&\geq P_p(O(R_{mn}), \text{ and } 0\leftrightarrow e_n \text{ in } R_{mn})\\&=P_p(O(R_{mn}))P_p(0\leftrightarrow e_n \text{ in } R_{mn}).\end{aligned}\tag{9.41}$$

We need to estimate the last two probabilities, and we begin with $P_p(O(R_{mn}))$. Let k be a positive integer and define the events

$A_1 = \{$there exists a closed path in the dual from $(-m-k-\frac{1}{2}, m+k+\frac{1}{2})$ to $(n+m+k+\frac{1}{2}, m+k+\frac{1}{2})$ lying entirely in the tube $T_1 = \{(u+\frac{1}{2}, v+\frac{1}{2}): u\in\mathbb{Z}, m\leq v\leq m+2k\}\}$,

$A_2 = \{$there exists a closed path in the dual from $(n+m+k+\frac{1}{2}, m+k+\frac{1}{2})$ to $(n+m+k+\frac{1}{2}, -m-k-\frac{1}{2})$ lying entirely in the tube $T_2 = \{(u+\frac{1}{2}, v+\frac{1}{2}): n+m\leq u\leq n+m+2k, v\in\mathbb{Z}\}\}$,

$A_3 = \{$there exists a closed path in the dual from $(n+m+k+\frac{1}{2}, -m-k-\frac{1}{2})$ to $(-m-k-\frac{1}{2}, -m-k-\frac{1}{2})$ lying entirely in the tube $T_3 = \{(u-\frac{1}{2}, v-\frac{1}{2}): u\in\mathbb{Z}, -m-2k\leq v\leq -m\}\}$,

$A_4 = \{$there exists a closed path in the dual from $(-m-k-\frac{1}{2}, -m-k-\frac{1}{2})$ to $(-m-k-\frac{1}{2}, m+k+\frac{1}{2})$ lying entirely in the tube $T_4 = \{(u-\frac{1}{2}, v-\frac{1}{2}): -m-2k\leq u\leq -m, v\in\mathbb{Z}\}\}$.

See Figure 9.12 for a drawing of these events. It is clear that R_{mn} is surrounded by a closed circuit of the dual whenever $A_1\cap A_2\cap A_3\cap A_4$

occurs. The A_i's are decreasing events, and thus we may apply the FKG inequality (and particularly (2.7)) to find that

$$\begin{aligned}(9.42)\qquad P_p(O(R_{mn})) &\geq P_p(A_1 \cap A_2 \cap A_3 \cap A_4)\\ &\geq P_p(A_1)P_p(A_2)P_p(A_3)P_p(A_4).\end{aligned}$$

The first and third terms, $P_p(A_1)$ and $P_p(A_3)$, equal the probability that two points, distance $n + 2m + 2k + 1$ apart, are joined by a closed path in a tube of width $2k$; the second and fourth terms equal the probability that two points, distance $2m + 2k + 1$ apart, are joined by a closed path in a tube of width $2k$. It follows from Lemma (9.27) applied to closed paths that

$$(9.43)\qquad \frac{1}{n}\log P_p(A_i) \to -\varphi_k(1-p) \quad \text{as} \quad n\to\infty, \quad \text{for} \quad i = 1, 3,$$

$$(9.44)\qquad \frac{1}{n}\log P_p(A_i) \to 0 \quad \text{as} \quad n\to\infty, \quad \text{for} \quad i = 2, 4,$$

where φ_k is given in (9.28) and we have used the fact that $m/n \to 0$ as $n \to \infty$. Hence

$$\liminf_{n\to\infty}\left\{\frac{1}{n}\log P_p(O(R_{mn}))\right\} \geq -2\varphi_k(1-p).$$

This holds for every $k \geq 1$, and so we may take the limit as $k \to \infty$ to obtain

$$\begin{aligned}(9.45)\qquad \liminf_{n\to\infty}\left\{\frac{1}{n}\log P_p(O(R_{mn}))\right\} &\geq -2\varphi(1-p)\\ &= -\frac{2}{\xi(1-p)},\end{aligned}$$

where $\xi(1-p) = \varphi(1-p)^{-1}$ is the correlation length.

We turn next to the last term in (9.41). We write $B(m)$ and $B(m, e_n)$ for the boxes with side-length $2m$ and centres at the origin and e_n, respectively, and we define the following events:

$D_1 = \{$there exists an open path in R_{mn} joining a vertex on its left side to a vertex on its right side$\}$,

$D_2 = \{$there exist open paths in $B(m)$ joining 0 to the upper side and to the lower side of $B(m)\}$,

$D_3 = \{$there exist open paths in $B(m, e_n)$ joining e_n to the upper side and to the lower side of $B(m, e_n)\}$.

See Figure 9.13 for a drawing of these events. It is clear that there exists an open path in R_{mn} from the origin to e_n whenever $D_1 \cap D_2 \cap D_3$ occurs. The D_i's are increasing events, and thus we may use the FKG inequality

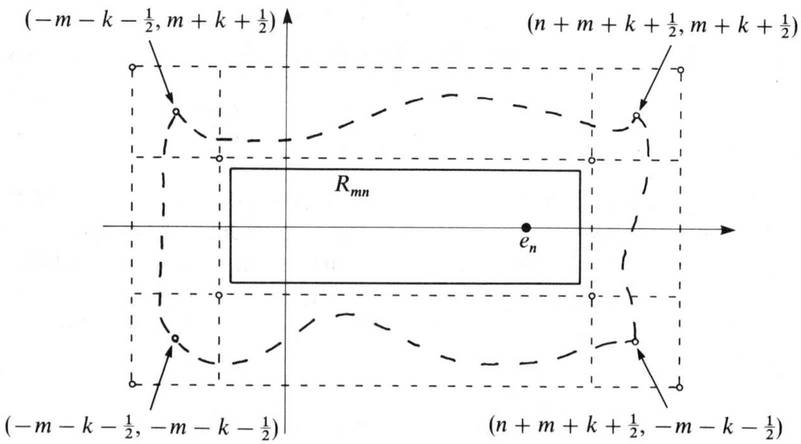

Figure 9.12. The four heavy dashed lines are the four closed paths referred to in the definitions of A_1, A_2, A_3, A_4. If such paths exist then R_{mn} is in the interior of a closed circuit of the dual.

to find that

$$P_p(0 \leftrightarrow e_n \text{ in } R_{mn}) \geq P_p(D_1)P_p(D_2)P_p(D_3). \tag{9.46}$$

If D_1 does not occur then there exists a closed path in the dual of R_{mn} which separates the left side of R_{mn} from its right side. For this to occur, one of the dual vertices $(r+\frac{1}{2}, m+\frac{1}{2})$ for $-m \leq r < n+m$ must be joined by a closed path in the dual to the line $\{(u+\frac{1}{2}, -m-\frac{1}{2}): -\infty < u < \infty\}$, and the chance of this is at most

$$\sum_{r=-m}^{n+m-1} P_{1-p}(0 \leftrightarrow \partial B(2m)) \leq (n+2m)e^{-2m\sigma}$$

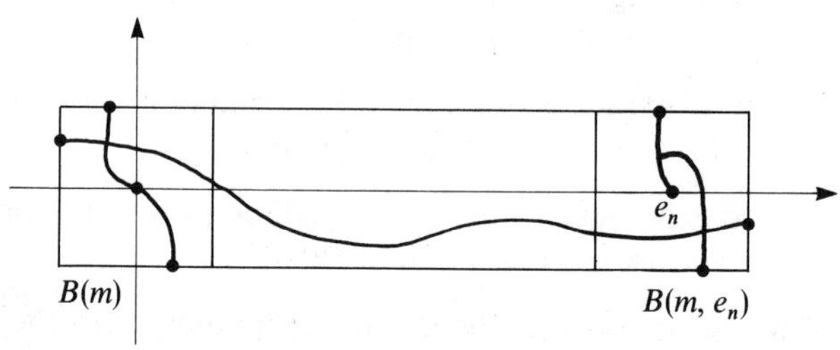

Figure 9.13. If 0 is joined to the top and bottom sides of $B(m)$, e_n is similarly joined in $B(m, e_n)$, and R_{mn} is traversed from left to right, then 0 is joined to e_n in R_{mn}.

for some $\sigma > 0$, by Theorem (5.1). Hence

(9.47) $$P_p(D_1) \to 1 \quad \text{as} \quad n \to \infty.$$

Finally, $P_p(D_2) = P_p(D_3)$, and so we consider $P_p(D_2)$ only. Let D_4 be the event that there is an open path of $B(m)$ joining the origin to the upper side of $B(m)$, and let D_5 be the event that a similar path exists to the lower side of $B(m)$. The events D_4 and D_5 are increasing, and their intersection equals D_2. Hence

$$\begin{aligned} P_p(D_2) &= P_p(D_4 \cap D_5) \\ &\geq P_p(D_4)P_p(D_5) \end{aligned}$$

by the FKG inequality. However $P_p(D_4) = P_p(D_5)$ by symmetry, and furthermore

$$P_p(D_4) \geq \tfrac{1}{4}\theta(p),$$

since if the origin is in an infinite open cluster then it is joined by an open path of $B(m)$ to at least one of the four sides of $B(m)$. We combine these last inequalities to obtain

(9.48) $$P_p(D_2) = P_p(D_3) \geq \tfrac{1}{16}\theta(p)^2,$$

which, with (9.46) and (9.47), implies that

(9.49) $$P_p(0 \leftrightarrow e_n \text{ in } R_{mn}) \geq (1 + o(1))\tfrac{1}{16}\theta(p)^2 \quad \text{as} \quad n \to \infty.$$

We return now to (9.41), take logarithms, and use (9.45) and (9.49) to find that

$$\begin{aligned} \limsup_{n\to\infty} \left\{ -\frac{1}{n} \log \tau_p^f(0, e_n) \right\} &= -\liminf_{n\to\infty} \left\{ \frac{1}{n} \log P_p(O(R_{mn})) \right\} \\ &\leq \frac{2}{\xi(1-p)} \end{aligned}$$

as required in (9.40). The proof is complete. ∎

Proof of Theorem (9.24). This is a rather straightforward application of the technique of duality. Suppose that $p \in (p_c, 1)$ and the origin is in a finite open cluster of size n. From the remarks after Proposition (9.2), there exists a closed circuit of the dual lattice $\mathbb{L}_d^2$ which contains the origin of $\mathbb{L}^2$ in its interior and which has at least $\lambda\sqrt{n}$ vertices, where λ is a positive constant. This circuit must contain some vertex of $\mathbb{L}_d^2$ of the form $(k + \frac{1}{2}, \frac{1}{2})$ where $0 \leq k \leq n - 1$, and therefore some such vertex of $\mathbb{L}_d^2$ must lie in a closed cluster of $\mathbb{L}_d^2$ of size $\lambda\sqrt{n}$ or greater. Each edge of $\mathbb{L}_d^2$ is closed with probability $1 - p$, and $1 - p < p_c$ $(=\frac{1}{2})$ since $p > p_c$ $(=\frac{1}{2})$. Thus the process of closed edges in $\mathbb{L}_d^2$ is subcritical, giving by Theorem (5.78) that there exists $\tau > 0$, depending on p, such that

$$P_p(0_d \text{ is in closed cluster of } \mathbb{L}_d^2 \text{ with size at least } \lambda\sqrt{n}) \leq e^{-\tau\lambda\sqrt{n}}.$$

Therefore

$$P_p(|C| = n) \le \sum_{k=0}^{n-1} P_p((k + \tfrac{1}{2}, \tfrac{1}{2}) \text{ in closed cluster of } \mathbb{L}^2_d \text{ with size at least } \lambda\sqrt{n})$$

$$\le ne^{-\tau\lambda\sqrt{n}}.$$

We choose $\eta(p)$ accordingly. ∎

9.5 Percolation on Subsets of the Square Lattice

Let G be a connected subgraph of $\mathbb{L}^2$. The percolation process on $\mathbb{L}^2$ induces a percolation process on G, obtained by deleting all vertices and edges of $\mathbb{L}^2$ which do not lie in G. How do the properties of this induced process depend on the geometry of G? Of particular interest is the question of determining whether or not G contains an infinite open cluster for a given value of p. Clearly, no infinite cluster exists in G when $p \le \frac{1}{2}$, since the complete lattice contains no such cluster in this case. We may guess that, if $p > \frac{1}{2}$, then all 'reasonably fat' graphs G contain an infinite open cluster. For example, it is not difficult to show that every non-trivial cone contains (almost surely) an infinite cluster whenever $p > \frac{1}{2}$. How 'thin' need the graph G be in order that it contain no infinite open cluster? There is a rather precise answer to this question.

We begin with some notation. Let G be a subset of $\mathbb{Z}^2$; we shall think of G as being a graph whose edges are those of $\mathbb{L}^2$ joining pairs of vertices in G. We shall be particularly interested in sets G of the form

(9.50) $$G(f) = \{x \in \mathbb{Z}^2 \colon 0 \le x_2 \le f(x_1), x_1 \ge 0\},$$

where f is a specified function on $[0, \infty)$ taking non-negative values.

Let p satisfy $0 \le p \le 1$, and consider percolation on $\mathbb{L}^2$ with edge-probability p. We write $\psi(p, G)$ for the probability that there is an infinite open connected subgraph of G, and we say in this case that 'G contains an infinite open cluster'. As in the case of the complete lattice $\mathbb{L}^2$ (see Theorem (1.11)), $\psi(p, G)$ is the probability of a tail event of a family of independent random variables, so that $\psi(p, G)$ equals either 0 or 1 for all p and G. We denote by $p_c(G)$ the *critical probability* of G:

(9.51) $$p_c(G) = \sup\{p \colon \psi(p, G) = 0\}.$$

An alternative definition of the critical probability $p_c(G)$ is by way of the probability that a given vertex of G is in an infinite open cluster of G. Let x be a vertex of G, and let $\theta(p, x; G)$ be the probability that x is in an infinite open cluster of G. It is not difficult to see (use the argument of Section 6.4) that, for all pairs x, y of vertices which belong to the same component of G, it is the case that $\theta(p, x; G) > 0$ if and only if

$\theta(p, y; G) > 0$. If G is connected, then we may define $p_c(G)$ by

$$(9.52) \qquad p_c(G) = \sup\{p: \theta(p, x; G) = 0\},$$

a definition which is independent of the choice of x. By the argument of Theorem (1.11), the two definitions (9.51) and (9.52) of $p_c(G)$ are equivalent whenever G is connected. We shall be concerned here with connected graphs G only, and so we shall use these definitions interchangeably.

Before stating the main result, we recall from Section 5.2 the definition of the correlation length $\xi(p)$ for $p < p_c$ $(=\frac{1}{2})$. Writing $B(n)$ for the usual box with side-length $2n$ and centre at the origin, we have that

$$(9.53) \qquad P_p(0 \leftrightarrow \partial B(n)) \approx e^{-n/\xi(p)} \quad \text{as} \quad n \to \infty \quad \text{if} \quad 0 < p < \tfrac{1}{2},$$

where ξ is a strictly increasing and continuous function on $(0, \frac{1}{2})$ satisfying $\xi(p) \to 0$ as $p \downarrow 0$ and $\xi(p) \to \infty$ as $p \uparrow \frac{1}{2}$. Furthermore, the probability that there exists an open path from the origin to the vertex $e_n = (n, 0)$ satisfies

$$(9.54) \qquad \tau_p(0, e_n) \approx e^{-n/\xi(p)} \quad \text{as} \quad n \to \infty.$$

See Figure 9.14 for a sketch of the function ξ.

(9.55) Theorem. *Let $a > 0$, and let f be a non-negative function on $[0, \infty)$ satisfying*

$$\frac{f(u)}{\log u} \to a \quad \text{as} \quad u \to \infty.$$

The critical probability of $G(f)$ is the unique number $\pi \in (\frac{1}{2}, 1)$ satisfying

$$\xi(1 - \pi) = a.$$

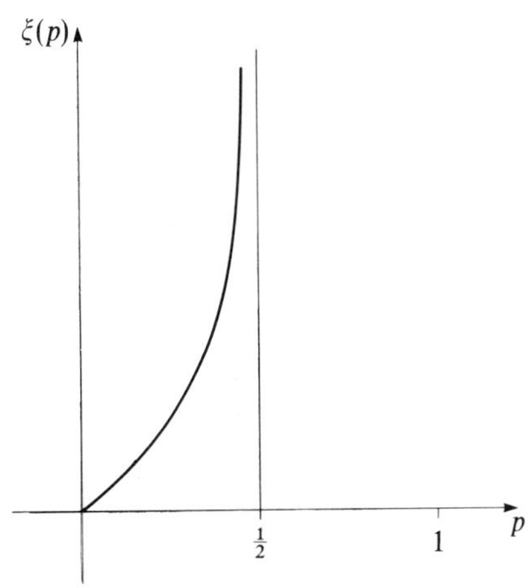

Figure 9.14. A sketch of the function $\xi(p)$ for $p < \frac{1}{2}$.

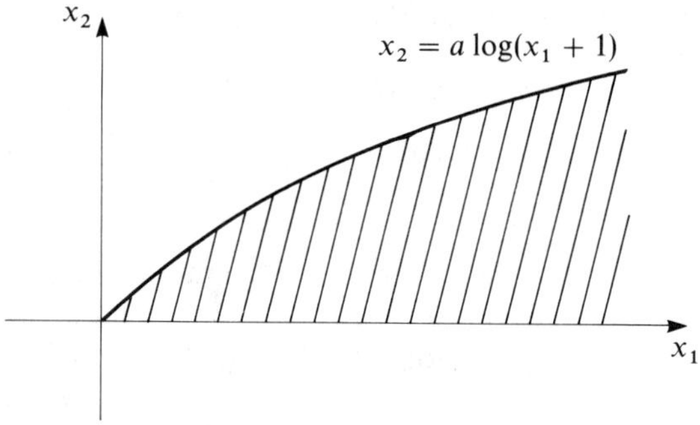

Figure 9.15. A sketch of the region G_a.

We note that ξ maps $(0, \frac{1}{2})$ continuously onto $(0, \infty)$; the strict monotonicity of ξ ensures that the equation $\xi(1-\pi) = a$ is satisfied by a unique π in $(\frac{1}{2}, 1)$, for any given $a > 0$. The assertion of the theorem amounts to saying that the existence or not of an infinite open cluster in the set

$$G_a = \{x \in \mathbb{Z}^2 : 0 \le x_2 \le a\log(x_1+1),\ x_1 \ge 0\} \tag{9.56}$$

depends on whether a is larger or smaller than $\xi(1-p)$; see Figure 9.15 for a picture of the set G_a.

We may choose the function f in such a way that there exists an infinite open cluster when p *equals* the critical probability for $G(f)$. That is to say, there exist functions f such that

$$P_{p_c(G(f))}(G(f) \text{ contains an infinite open cluster}) = 1.$$

It is not difficult to construct such functions, and we shall see a recipe at the end of the proof of Theorem (9.55): it suffices to let $a > 0$ and define f by

$$f(u) = a\log(u+1) + b\log\{1 + \log(u+1)\}$$

where $b > 2a$. In the language of mathematical physicists, the corresponding graph $G(f)$ has a 'first order' phase transition in the sense that the probability $\theta(p, x; G(f))$, that the vertex x lies in an infinite open cluster of $G(f)$, is discontinuous at the critical point $p_c(G(f))$.

Proof of Theorem (9.55). We stay close to Grimmett (1981a, 1983). We shall sometimes use real-valued quantities in places where integers are required; this is for notational convenience only and has no mathematical significance. Let $a > 0$, and suppose that f is a function on $[0, \infty)$ which takes non-negative values and satisfies $f(u)/\log u \to a$ as $u \to \infty$.

We consider percolation on $G(f)$, and we suppose first that the edge-probability p satisfies $p > \frac{1}{2}$ and $a < \xi(1-p)$. We shall show that $G(f)$

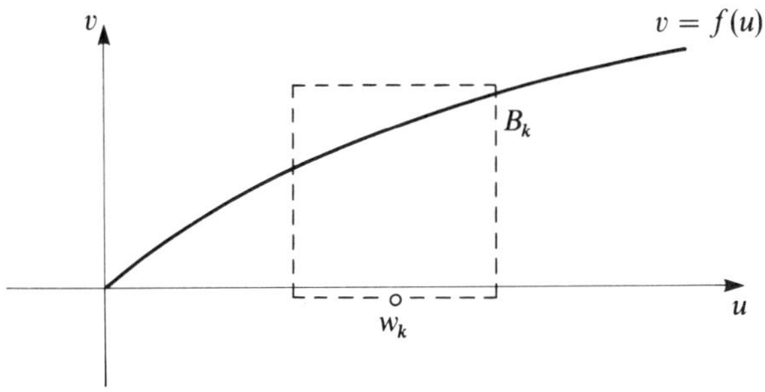

Figure 9.16. The box B_k is the smallest box in the dual having w_k in the middle of its lower side and its upper side strictly above the curve $v = f(u)$.

contains (almost surely) no infinite open cluster. We choose $\delta > 0$ such that $(1 + \delta)a < \xi(1 - p)$, and we define w_k to be the vertex $(k^{1+\delta} + \frac{1}{2}, -\frac{1}{2})$ of the dual lattice $\mathbb{L}_d^2$. For each k, we construct a box B_k of the dual in the following way. Amongst all square boxes in the dual lattice with w_k in the middle of the lower side, B_k is the smallest with the property that the upper side lies strictly above the curve $v = f(u)$; see Figure 9.16. Now, $f(u)/\log u \to a$ as $u \to \infty$, and it is a simple matter to deduce that the side-length l_k of B_k satisfies

$$l_k = a(1 + o(1)) \log k^{1+\delta} \quad \text{as} \quad k \to \infty. \tag{9.57}$$

We think of B_k as a box with side-length l_k and centre at $w_k + (0, \frac{1}{2}l_k)$. Let A_k be the event that B_k contains a closed path from a vertex on its upper side to a vertex on its lower side. The probability of A_k is no smaller than the probability that the centre of B_k is joined by closed paths of B_k to both its upper and lower sides. Thus

$$\begin{aligned} P_p(A_k) &\geq P_p(0 \text{ is joined by closed paths to top and bottom sides of } B(\tfrac{1}{2}l_k)) \\ &\geq \{\tfrac{1}{4}P_{1-p}(0 \leftrightarrow \partial B(\tfrac{1}{2}l_k))\}^2 \end{aligned} \tag{9.58}$$

by the FKG inequality, since the probability that the origin is joined to any given side of $B(\frac{1}{2}l_k)$ is at least one quarter of the probability that it is joined to the surface of $B(\frac{1}{2}l_k)$. However,

$$P_{1-p}(0 \leftrightarrow \partial B(\tfrac{1}{2}l_k)) \approx e^{-l_k(2\xi(1-p))^{-1}} \quad \text{as} \quad k \to \infty$$

by (9.53). We combine (9.58) with (9.57) and deduce that

$$\begin{aligned} P_p(A_k) &\geq \tfrac{1}{16} \exp\{-(1 + o(1))l_k/\xi(1 - p)\} \\ &= \tfrac{1}{16}k^{-(1+o(1))(1+\delta)a/\xi(1-p)} \end{aligned}$$

as $k \to \infty$, which implies that

$$\sum_k P_p(A_k) = \infty \tag{9.59}$$

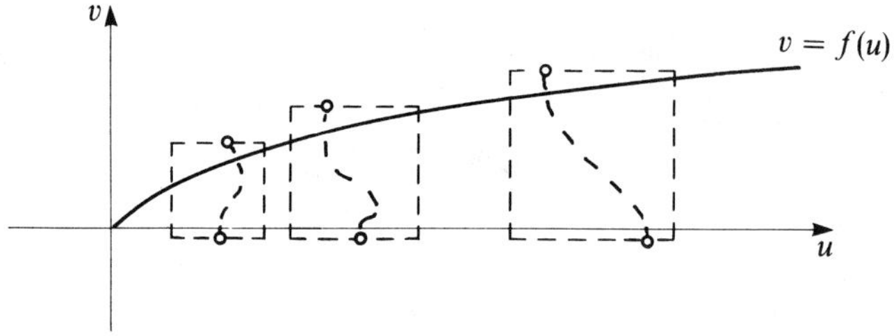

Figure 9.17. The existence of infinitely many closed paths traversing the boxes $\{B_k: k \geq K\}$ prevents the formation of an infinite open cluster in $G(f)$.

since $(1 + \delta)a < \xi(1 - p)$. On the other hand, the distance from w_k to w_{k+1} is $(k + 1)^{1+\delta} - k^{1+\delta} \sim (1 + \delta)k^{\delta}$, and the side-length of B_k is $O(\log k)$, so that there exists K such that $\{B_k: k \geq K\}$ are disjoint boxes; this implies that $\{A_k: k \geq K\}$ are independent events. Equation (9.59) implies that (almost surely) infinitely many of the events $\{A_k: k \geq K\}$ occur, which implies in turn that $G(f)$ is traversed from top to bottom in the dual by infinitely many closed paths. These closed paths partition $G(f)$ into finite regions (see Figure 9.17). Each open cluster of $G(f)$ is contained within one of these regions, since otherwise there is an open edge of $\mathbb{L}^2$ crossing a closed edge of $\mathbb{L}^2_{\mathrm{d}}$, and this is impossible. Thus all open clusters of $G(f)$ are (almost surely) finite, and the first part of the proof is complete.

Secondly, we suppose that $p > \frac{1}{2}$ and $a > \xi(1 - p)$. Let f be a non-negative function satisfying $f(u)/\log u \to a$ as $u \to \infty$. We shall show that $G(f)$ contains (almost surely) an infinite open cluster. First, we choose α such that $a > \alpha > \xi(1 - p)$ and we let D_k be the box of the dual lattice with centre $(k + \frac{1}{2}, \frac{1}{2})$ and side-length $2\alpha \log k$. For all large values of k (say $k \geq L$) it is the case that D_k lies strictly beneath the curve $v = f(u) - 2$. Let E_k be the event that $(k + \frac{1}{2}, \frac{1}{2})$ is joined by a closed path of the dual to a vertex on the surface ∂D_k of D_k. We have from (9.53) that

(9.60) $$P_p(E_k) = P_{1-p}(0 \leftrightarrow \partial B(\alpha \log k)) \approx k^{-\alpha/\xi(1-p)} \quad \text{as} \quad k \to \infty,$$

giving that

(9.61) $$\sum_k P_p(E_k) < \infty$$

from the assumption $\alpha > \xi(1 - p)$. Thus there exists $M \geq L$ such that

$$P_p\left(\bigcup_{k \geq M} E_k\right) < \tfrac{1}{2}.$$

However, if none of $\{E_k: k \geq M\}$ occurs then no vertex $(k + \frac{1}{2}, \frac{1}{2})$ with

$k \geq M$ is joined by a closed path to a vertex of the dual lying above the curve $v = f(u) - 2$. We move to the dual $\mathbb{L}^2$ of $\mathbb{L}^2_{\mathrm{d}}$ to see that there is probability at least $\frac{1}{2}$ that $G(f)$ contains an infinite open path. The probability of the last event is either 0 or 1, and we have proved that $G(f)$ contains almost surely an infinite open cluster, as required.

Finally we show that, for any $a > 0$, there exists a function f such that $f(u)/\log u \to a$ as $u \to \infty$ and

$$P_{p_c(G(f))}(G(f) \text{ contains an infinite open cluster}) > 0.$$

We consider a function f satisfying

$$f(u) = a \log u + b \log \log u \qquad \text{for all large } u,$$

where $b > 2a$. This time we let D_k be the largest box of the dual lattice with centre at $(k + \frac{1}{2}, \frac{1}{2})$ and lying strictly beneath the curve $v = f(u) - 2$. It is not difficult to show that D_k has side-length $2f(k) + O(1)$ as $k \to \infty$, so that (9.60) may be replaced by

$$\begin{aligned} P_p(E_k) &= P_{1-p}\big(0 \leftrightarrow \partial B(f(k) + O(1))\big) \\ &\leq \mu(\log k) \exp\left(-\frac{a \log k + b \log \log k}{\xi(1-p)} \right) \end{aligned}$$

for all large k and some absolute constant μ, where we have used the right-hand inequality in (5.11). The critical probability of percolation on $G(f)$ is the number $\pi \in (\frac{1}{2}, 1)$ satisfying $\xi(1 - \pi) = a$, and with this choice of π we have that

$$P_\pi(E_k) \leq \frac{\mu}{k(\log k)^\alpha},$$

where $\pi = p_c(G(f))$ and $\alpha = ba^{-1} - 1 > 1$. It follows as before that

$$P_\pi(G(f) \text{ contains an infinite open cluster}) = 1$$

as required. ∎

9.6 Central Limit Theorems

Let K_n be the number of open clusters in the box $B(n)$ when each edge is open with probability p. It was proved in Theorem (4.2) that

$$\frac{1}{|B(n)|} K_n \to \kappa(p) \quad \text{a.s.}$$

as $n \to \infty$, where $\kappa(p) = E_p(|C|^{-1})$ is the number of open clusters per vertex. We may view this limit theorem as a 'strong law' for the number of open clusters. In studying the asymptotic behaviour of sequences of random variables, the classical theory turns next to central limit theorems, and thus we ask whether the sequence $(K_n: n \geq 1)$ is such that

$$\frac{K_n - E_p(K_n)}{\sqrt{\operatorname{var}_p(K_n)}}$$

is asymptotically normally distributed as $n \to \infty$, where var_p denotes variance relative to P_p. This is indeed the case whenever $p \neq 0, \frac{1}{2}, 1$, and this fact is one of a class of central limit theorems for quantities arising out of percolation on $\mathbb{L}^2$.

The general framework is as follows. Any circuit c of the lattice $\mathbb{L}^2$ has an interior and an exterior in the plane $\mathbb{R}^2$. We denote by $\Lambda(c)$ the subgraph of $\mathbb{L}^2$ containing all vertices and edges which are in either c or its interior. We write as usual $\partial\Lambda(c)$ for the surface of $\Lambda(c)$, being the set of vertices in c which have neighbours which do not belong to $\Lambda(c)$. Suppose that $\mathbf{c} = (c_n: n \geq 1)$ is a sequence of circuits satisfying $|\Lambda(c_n)| \to \infty$ as $n \to \infty$ and that, for each n, we are provided with a function f_n mapping the set of connected subgraphs of $\mathbb{L}^2$ into the real line $\mathbb{R}$. Suppose that $0 < p < 1$, and consider bond percolation on $\mathbb{L}^2$ with edge-probability p. We define the sequence $\mathbf{Z} = (Z_n: n \geq 1)$ of random variables by

$$Z_n = \sum_{x \in \Lambda(c_n)} f_n(C(x)), \tag{9.62}$$

where the sum is over all vertices in $\Lambda(c_n)$. We say that '$\mathbf{Z}$ satisfies the central limit theorem' if

$$P_p\left(\frac{Z_n - E_p(Z_n)}{\sqrt{\operatorname{var}_p(Z_n)}} \leq u\right) \to \int_{-\infty}^{u} \frac{1}{\sqrt{2\pi}} e^{-v^2/2}\, dv$$

for all u, as $n \to \infty$.

This general framework contains many specific examples of interest. For example, let us set c_n to be the bounding circuit of the box $B(n)$ and define

$$f_n(C(x)) = \frac{1}{|C_n(x)|} \qquad \text{for} \quad x \in B(n),$$

where $C_n(x)$ is the set of vertices of $B(n)$ attainable from x along open paths of $B(n)$; then

$$Z_n = \sum_{x \in B(n)} \frac{1}{|C_n(x)|}$$

equals the number K_n of open clusters of $B(n)$, as in Section 4.1.

The aim of this section is to survey known central limit theorems for sequences of the form (9.62), and particularly to discuss the available techniques and their fields of application. We shall give no proofs, but refer the reader to the original papers.

There are three basic methods for proving the asymptotic normality of $Y_n = (Z_n - E_p(Z_n))(\operatorname{var}_p(Z_n))^{-1/2}$.

(1) *Method of moments.* We estimate the semi-invariants (otherwise known as the cumulants) of Y_n and show that they converge to those of the normal distribution. (See Malyshev (1975), Cox and Grimmett (1981, 1984), and the references therein.)

(2) *Association.* If the f_n's are increasing functions on the set-theoretic lattice of subgraphs of $\mathbb{L}^2$ (that is to say, $f_n(G) \le f_n(H)$ whenever G is a subgraph of H), then the random variables $(f_n(C(x)): x \in \Lambda(c_n))$ are positively correlated: this fact can be used to estimate the characteristic function of Y_n directly, thereby to study its asymptotic behaviour as $n \to \infty$. (See Newman (1980), Newman and Wright (1981), Cox and Grimmett (1984), and the references therein.)

(3) *Strong mixing.* We estimate the correlation coefficients of events related to sets A and B of vertices, as a function of the distance between A and B. This coefficient decays sufficiently rapidly to allow a central limit theorem. (See Neaderhouser (1978, 1981), Grimmett (1979), Bolthausen (1982), Herrndorf (1985), and the references therein.)

Before discussing the relative strengths and weaknesses of these three approaches, we state a central limit theorem. We say that a function f on the set of connected subgraphs of $\mathbb{L}^2$ is *constant on infinite graphs* if $f(G) = f(H)$ for all infinite connected subgraphs G and H.

(9.63) Theorem. Central limit theorem. *Suppose that* $\mathbf{c} = (c_n: n \ge 1)$ *is a sequence of circuits of* $\mathbb{L}^2$ *satisfying* $|\Lambda(c_n)| \to \infty$ *as* $n \to \infty$. *Let* $(f_n: n \ge 1)$ *be a sequence of real-valued functions on the set of connected subgraphs of* $\mathbb{L}^2$ *which intersect* $\Lambda(c_n)$. *Suppose that* $0 < p < 1$ *and that*

(9.64) *for all* n, f_n *is a bounded function,*

(9.65) *if* $p > \frac{1}{2}$ *then, for all* n, f_n *is constant on infinite graphs,*

$$\liminf_{n\to\infty} \frac{1}{|\Lambda(c_n)|} \operatorname{var}_p(Z_n) > 0, \tag{9.66}$$

where

$$Z_n = \sum_{x \in \Lambda(c_n)} f_n(C(x)). \tag{9.67}$$

Then $\mathbf{Z} = (Z_n: n \ge 1)$ *satisfies the central limit theorem so long as* $p \ne \frac{1}{2}$.

The conditions of this theorem are not the best possible, but this formulation is sufficient for most of our purposes. The result may be proved either by the method of moments (Cox and Grimmett (1981, 1984)) or by the theory of strong mixing (Herrndorf (1985)). It is easy to check whether or not the functions f_n are constant on infinite graphs and bounded. Condition (9.66) can be more difficult to verify, although it is satisfied automatically whenever the f_n's are either all increasing or all decreasing and there exists $\sigma^2 > 0$ such that

$$\operatorname{var}_p(f_n(C(x))) \geq \sigma^2$$

for all $x \in \Lambda(c_n)$ and all n. To see this, just note that

$$\operatorname{var}_p(Z_n) \geq \sum_{x \in \Lambda(c_n)} \operatorname{var}_p(f_n(C(x)))$$

by the FKG inequality, whenever the f_n's are either all increasing or all decreasing.

The condition that $p \neq p_c$ $(=\frac{1}{2})$ is essential for the proof of the theorem, since we rely upon the rapid decay of correlations over large distances. The condition of constantness on infinite graphs is important only when $p > \frac{1}{2}$, since if $p < \frac{1}{2}$ then all open clusters are finite with probability 1.

Each of the three techniques of proof has its limitations as well as its advantages, and we discuss some of these next.

(1) *Method of moments.* This is a flexible technique which may be applied in situations where the conditions of a general theorem such as Theorem (9.63) are not satisfied. On the other hand, it seems to require that, for all k, the kth absolute moment of $f_n(C(x))$ is bounded uniformly in k and n; this has not proved to be an important restriction in practice, since all interesting functions studied so far have been bounded.

(2) *Association.* This is a beautiful and simple argument which seems to be best suited to the case when the regions $\Lambda(c_n)$ are rectangular boxes.

(3) *Strong mixing.* Using such techniques, we reduce to a minimum the moment conditions required of the $f_n(C(x))$'s. On the other hand, these methods appear at first sight to require that the functions f_n be constant on infinite graphs; perhaps this assumption is not essential for the method.

Related central limit theorems have been proved by Brånvall (1980) and Grimmett (1981b), but we shall say nothing about these here.

Here are some applications of the central limit theorem. We assume throughout that $\mathbf{c} = (c_n: n \geq 1)$ is a sequence of circuits satisfying $|\Lambda(c_n)| \to \infty$ as $n \to \infty$. For our first example we define

$$f_n(C(x)) = \begin{cases} 1 & \text{if } |C(x)| = \infty, \\ 0 & \text{if } |C(x)| < \infty, \end{cases}$$

so that $Z_n = \sum_{x \in \Lambda(c_n)} f_n(C(x))$ is the number of vertices in $\Lambda(c_n)$ which are in infinite open clusters. It is not difficult to verify the conditions of the theorem when $\frac{1}{2} < p < 1$, and we deduce that the number of vertices of $\Lambda(c_n)$ in the infinite open cluster satisfies the central limit theorem.

For our second example we define

$$f_n(C(x)) = \begin{cases} 1 & \text{if } x \leftrightarrow \partial\Lambda(c_n), \\ 0 & \text{otherwise,} \end{cases}$$

so that Z_n is the number of vertices of $\Lambda(c_n)$ which are joined to its boundary by open paths. Once again the conditions of the theorem are easily verified when $p > \frac{1}{2}$, and the asymptotic normality of Z_n follows in this case. If $p < \frac{1}{2}$ then Z_n is a 'boundary effect' rather than a 'volume effect', and so condition (9.66) is not generally valid. Nevertheless, the method of moments may be used to show the asymptotic normality of Z_n in this case also (see Cox and Grimmett (1981)). It may be shown that there exist positive functions $A_1(p)$ and $A_2(p)$ such that

$$A_1(p)|\partial\Lambda(c_n)| \leq E_p(Z_n),\ \mathrm{var}_p(Z_n) \leq A_2(p)|\partial\Lambda(c_n)| \qquad \text{if} \quad 0 < p < \tfrac{1}{2},$$

$$A_1(p)|\Lambda(c_n)| \leq E_p(Z_n),\ \mathrm{var}_p(Z_n) \leq A_2(p)|\Lambda(c_n)| \qquad \text{if} \quad \tfrac{1}{2} < p < 1.$$

Cast in the framework of a porous stone immersed in a bucket of water, these observations amount to quantified versions of the statement that the wetting of interior vertices is a boundary effect when $p < p_c$ $(=\frac{1}{2})$ and a volume effect when $p > p_c$.

We turn next to the number of open clusters contained in a large region of the lattice. We need to be careful when counting such clusters, since there may exist open clusters of the lattice whose intersections with $\Lambda(c_n)$ are disconnected. It can even be the case that there exists an open cluster of the lattice all of whose vertices belong to $\Lambda(c_n)$ but whose intersection with $\Lambda(c_n)$ is disconnected; see Figure 9.18 for illustrations of some of these difficulties. As a starting point we define

$$(9.68) \qquad F_n = \sum_{x \in \Lambda(c_n)} \frac{1}{|C(x) \cap \Lambda(c_n)|} I_{\{|C(x)| < \infty\}},$$

and it is not difficult to see that F_n is the number of finite open clusters of the lattice which intersect $\Lambda(c_n)$. The functions in (9.68) are certainly bounded, as well as being constant on infinite graphs. We require a calculation in order to check (9.66) (see Cox and Grimmett (1984, p. 524)), and the asymptotic normality of F_n follows when $p \neq \frac{1}{2}$. If $p < \frac{1}{2}$ then all

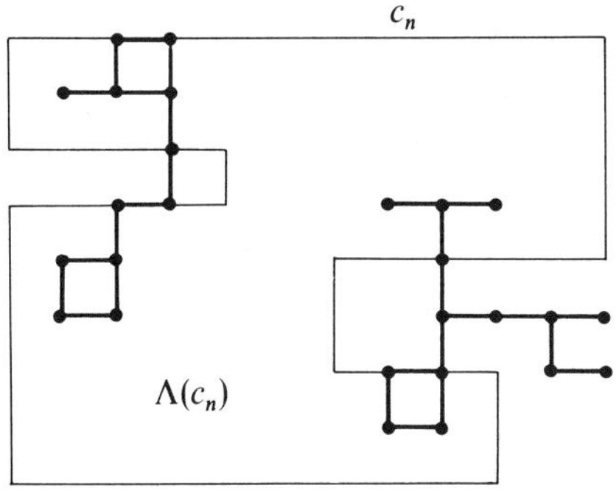

Figure 9.18. All of the vertices of the left-hand cluster belong to $\Lambda(c_n)$ and yet its intersection with $\Lambda(c_n)$ is disconnected. The right-hand cluster has a similar property.

open clusters are finite (almost surely) so that F_n equals (almost surely) the total number of open clusters of $\mathbb{L}^2$ which intersect $\Lambda(c_n)$. If $p > \frac{1}{2}$ then there exists (almost surely) an infinite open cluster, so that the total number of open clusters intersecting $\Lambda(c_n)$ equals either F_n or $F_n + 1$; thus the central limit theorem is valid for this quantity also.

Greater difficulty arises in the study of

$$K_n = \sum_{x \in \Lambda(c_n)} \frac{1}{|C_n(x)|},$$

the number of open clusters of $\Lambda(c_n)$. If $p < \frac{1}{2}$, then Theorem (9.63) is applicable, and we obtain the asymptotic normality of K_n. On the other hand, $f_n(C(x)) = |C_n(x)|^{-1}$ is not constant on infinite graphs, and so the theorem is inapplicable if $p > \frac{1}{2}$. Resorting to a direct application of the method of moments, Cox and Grimmett (1984) have shown the asymptotic normality of K_n when $\frac{1}{2} < p < 1$, so long as the regions $\Lambda(c_n)$ do not have too many bottlenecks. We do not state their theorem here, and neither shall we state precisely the geometry required of **c**. We note, however, that the conclusion is valid whenever c_n is the boundary of the rectangle $[0, k_1(n)] \times [0, k_2(n)]$ for sequences $(k_1(n): n \geq 1)$ and $(k_2(n): n \geq 1)$ satisfying

$$\liminf_{n\to\infty} \left\{ \frac{k_1(n)}{\log(k_1(n)k_2(n))} \right\} = \liminf_{n\to\infty} \left\{ \frac{k_2(n)}{\log(k_1(n)k_2(n))} \right\} = \infty.$$

We leave this topic with a challenge to the reader. Let $a > 0$ and define $k_1(n) = n$, $k_2(n) = \lfloor a \log n \rfloor$. The challenge is to prove the required central limit theorem for the sequence of circuits above, irrespective of the value of a.

9.7 Open Circuits in Annuli

In studying the geometry of the open clusters of percolation on $\mathbb{L}^2$, it is convenient to have simple recipes for constructing certain types of open paths; the probabilities of combinations of such paths may then be estimated with the aid of the FKG and BK inequalities. We have made extensive use already of open crossings of rectangles, and indeed it was by combining such paths in ingenious ways that Russo (1978) and Seymour and Welsh (1978) provided one of the principal foundations of Kesten's proof that $p_c = \frac{1}{2}$ for the square lattice. No longer is the major part of the Russo–Seymour–Welsh technology essential for this result, but such techniques retain their importance for at least two reasons quite apart from historical and intrinsic interest. First, they provide standard tools for approaching certain two-dimensional processes which are similar to percolation but whose detailed behaviours are not understood so

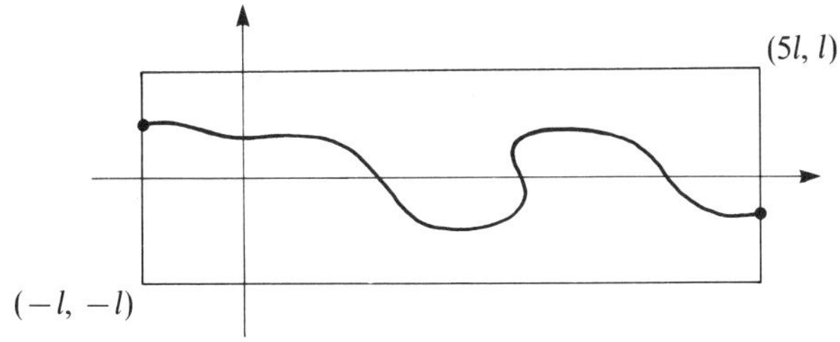

Figure 9.19. A sketch of the event $LR(3l, l)$.

well. Secondly, they enable us to prove certain inequalities such as

$$\theta(p) \le A(p - \tfrac{1}{2})^{\alpha} \quad \text{for} \quad p \ge p_c, \tag{9.69}$$

where A and α are positive constants. Such inequalities have implications for critical exponents, and we shall discuss this matter in greater detail in the next section.

Of the many results dealing with open crossings of rectangles, the Russo–Seymour–Welsh theorem is one of the most fundamental. We begin with some jargon. A *left–right* (respectively *top–bottom*) *crossing* of the rectangle B is an open path in B which joins some vertex on the left (respectively upper) side of B to some vertex on the right (respectively lower) side of B but which uses no edge joining two vertices in the boundary ∂B of B; we require the last condition to overcome a mild technical difficulty later. For positive integers k and l, we write $B(kl, l)$ for the rectangle $[-l, (2k-1)l] \times [-l, l]$; thus $B(l, l) = B(l)$, and $B(kl, l)$ is a rectangle with dimensions $2kl$ by $2l$. We write $LR(kl, l)$ for the event that there is a left–right crossing of $B(kl, l)$, and we abbreviate $LR(l, l)$ to $LR(l)$; see Figure 9.19. Let $A(l)$ be the annulus $B(3l)\backslash B(l)$, and let $O(l)$ be the event that there is an open circuit of $A(l)$ containing the origin of the lattice in its interior; see Figure 9.20. A lasting contribution of Russo, Seymour, and Welsh is a proof that, if there is a significant probability of an open crossing of the box $B(l)$, then there is a significant probability of an open circuit in the annulus $A(l)$.

(9.70) Theorem. RSW theorem. *If* $P_p(LR(l)) = \tau$ *then*

$$P_p(O(l)) \ge \{\tau(1 - \sqrt{1-\tau})^4\}^{12}. \tag{9.71}$$

Our principal application is to the case $p = p_c$ $(=\frac{1}{2})$. We have from Lemma (9.20) that there is probability $\frac{1}{2}$ of an open path in the rectangle $[0, n+1] \times [0, n]$ joining the left side to the right side, when $p = \frac{1}{2}$. Such a path contains a left–right crossing of the rectangle $[0, n+1] \times [-1, n+1]$. Writing x for the lowest right endvertex of all such crossings, we have that there is probability $\frac{1}{2}$ that the edge leading rightwards

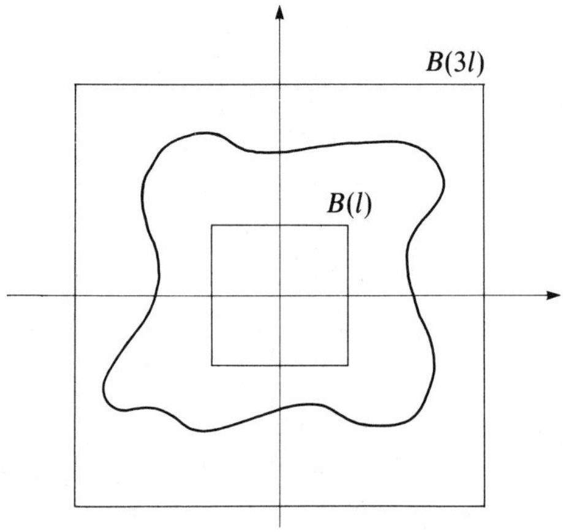

Figure 9.20. An open circuit in the annulus $A(l) = B(3l)\backslash B(l)$, containing the origin in its interior.

out of x is open; therefore, the probability of a left–right crossing of the square $[0, n+2] \times [-1, n+1]$ is at least $\frac{1}{4}$ for $n \geq 0$, and we have proved that $P_{1/2}(LR(l)) \geq \frac{1}{4}$ for all $l \geq 1$. We deduce from the RSW theorem that

$$P_{1/2}(O(l)) \geq 2^{-24}(1 - \tfrac{1}{2}\sqrt{3})^{48} \qquad \text{for all} \quad l \geq 1. \tag{9.72}$$

The actual number on the right-hand side here is immaterial; the important thing is that $P_{1/2}(O(l))$ is bounded away from zero uniformly in l. We shall see applications of (9.72) in the next section.

The RSW theorem is proved in several stages, of which the following is the hardest.

(9.73) Lemma. *If* $P_p(LR(l)) = \tau$ *then*

$$P_p(LR(\tfrac{3}{2}l, l)) \geq (1 - \sqrt{1-\tau})^3. \tag{9.74}$$

The remaining work is contained in the next lemma.

(9.75) Lemma. *It is the case that*

$$P_p(LR(2l, l)) \geq P_p(LR(l))P_p(LR(\tfrac{3}{2}l, l))^2, \tag{9.76}$$

$$P_p(LR(3l, l)) \geq P_p(LR(l))P_p(LR(2l, l))^2, \tag{9.77}$$

$$P_p(O(l)) \geq P_p(LR(3l, l))^4. \tag{9.78}$$

The RSW theorem is an immediate consequence of these two lemmas. We turn next to its proof, and we begin with Lemma (9.73). We shall make repeated use of the square root trick; recall from (9.14) that, if A_1

and A_2 are increasing events with equal probability, then

$$P_p(A_1) \geq 1 - \sqrt{1 - P_p(A_1 \cup A_2)}. \tag{9.79}$$

Proof of Lemma (9.73). We follow Russo (1981). Suppose that $0 \leq p \leq 1$ and consider percolation on $\mathbb{L}^2$ with edge-probability p. The target is to build a left–right crossing of the rectangle $B(\frac{3}{2}l, l)$ using open crossings of $2l \times 2l$ squares. In doing this, we shall encounter certain problems arising from the dependence between certain events; in order to deal with these problems, we shall make use of the idea of the 'lowest' open crossing of a rectangle. Let $\mathscr{C}$ be the collection of all left–right crossings of $B(l)$, and let Π be the lowest such crossing. Some topology is required in order to show that Π exists and is defined uniquely whenever $\mathscr{C}$ is non-empty, but the reader may be prepared to accept the following argument. There is a natural partial order on the set of left–right crossings of $B(l)$; for two such crossings π_1 and π_2 we write $\pi_1 \leq \pi_2$ if π_1 is contained in the closed bounded region of $B(l)$ lying beneath π_2. By geometrical considerations (see Figure 9.21), the union of the edge sets of any two crossings π_1 and π_2 contains a crossing π_3 such that $\pi_3 \leq \pi_1$ and $\pi_3 \leq \pi_2$. It follows that the union of the edge sets of all open crossings of $B(l)$ contains an open crossing Π which satisfies $\Pi \leq \pi$ for all open crossings π, and Π is called the *lowest* left–right crossing. We shall use these ideas in the following way: first, there is a *unique* lowest left–right crossing whenever there is a left–right crossing, and

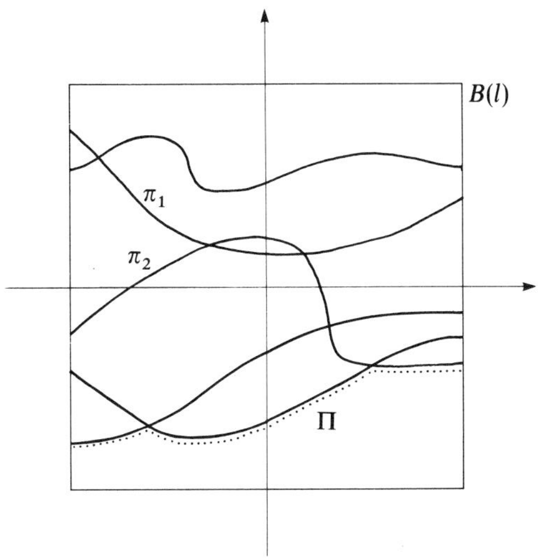

Figure 9.21. Note that the union of any two left–right crossings π_1 and π_2 contains a crossing π_3 satisfying $\pi_3 \leq \pi_1$ and $\pi_3 \leq \pi_2$. The dotted crossing is the lowest left–right crossing of the box.

secondly, the value taken by the lowest left–right crossing Π is independent of the states of edges of $B(l)$ having at least one endvertex in the subset of $B(l)$ lying strictly above Π. This latter property of Π is crucial and requires some elucidation. Let h be a (possibly non-open) path which traverses $B(l)$ from left to right, and consider the event $\{\Pi = h\}$ that h is the lowest left–right crossing. This event depends only on the states of edges in h and below h, since $\Pi = h$ if and only if h is open and there exists no other left–right crossing Π' such that $\Pi' \leq \Pi$. Thus Π enjoys a property similar to that of a stopping time for a Markov process.

Let $\mathscr{T}$ be the set of paths of $B(l)$ which traverse $B(l)$ from left to right but which use no edge contained in the boundary of $B(l)$. Each path $\pi \in \mathscr{T}$ cuts $[-l, l] \times [-l, l]$ into two parts, being the simply connected subsets of $\mathbb{R}^2$ lying above and beneath π. For $\pi \in \mathscr{T}$, we denote by $(0, y_\pi)$ the last vertex encountered on the vertical axis of $\mathbb{L}^2$ when π is traversed from left to right, and we write $\mathscr{T}^-$ (respectively $\mathscr{T}^+$) for the subset of $\mathscr{T}$ containing paths π for which $y_\pi \leq 0$ (respectively $y_\pi \geq 0$). For $\pi \in \mathscr{T}$, we denote by π_r the sub-path of π joining $(0, y_\pi)$ to the right side of $B(l)$. Let π_r' be the reflection of π_r in the line $\{(l, k): -\infty < k < \infty\}$, and let $\pi_r \cup \pi_r'$ be the path obtained by prolonging π_r with π_r'; the path $\pi_r \cup \pi_r'$ traverses the square $B(l)' = [0, 2l] \times [-l, l]$ from left to right. See Figure 9.22 for an illustration of this notation. For any $\pi \in \mathscr{T}$, we write $U(\pi)$ for the set of vertices of $B(l)'$ lying in or above the path $\pi_r \cup \pi_r'$, together with all edges of $B(l)'$ joining pairs of such vertices, at least one of which lies strictly above $\pi_r \cup \pi_r'$.

We are interested in the following events. Let L^- (respectively L^+) be the event that there exists an open path π in $\mathscr{T}^-$ (respectively $\mathscr{T}^+$), so that $L^- = \{y_\Pi \leq 0\}$. For $\pi \in \mathscr{T}$, we denote by L_π the event that π is the lowest left–right crossing Π of $B(l)$. For $\pi \in \mathscr{T}$, we write M_π for the event

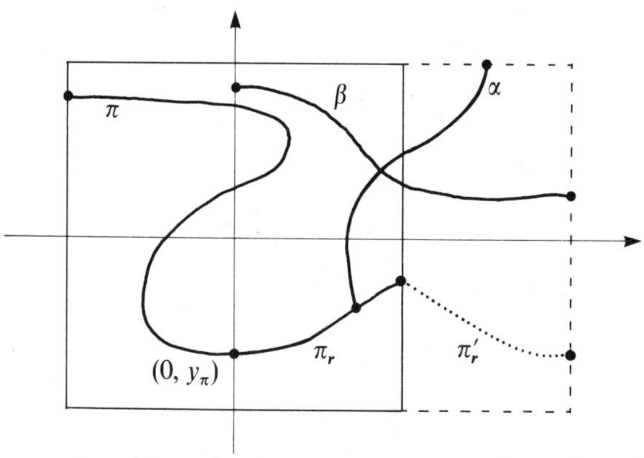

Figure 9.22. The path π crosses $B(l)$ from left to right. The path α guarantees the occurrence of the event M_π^-, and the path β guarantees the occurrence of N^+.

that there exists an open path in $U(\pi)$ joining some vertex on the upper side of $B(l)'$ to some vertex of $\pi_r \cup \pi_r'$, and we write M_π^- (respectively M_π^+) for the event that such a path exists having its second endvertex in π_r (respectively π_r'). Finally, we write N for the event that there exists a left–right crossing of $B(l)'$, and N^- (respectively N^+) for the event that such a path exists having its first endvertex on the non-positive (respectively non-negative) part of the vertical axis.

We shall use these events in the following way. Suppose that, for some $\pi \in \mathscr{T}^-$, the event $L_\pi \cap M_\pi^- \cap N^+$ occurs, and consult Figure 9.22 for a sketch of this event. It is clear from the figure that this event necessarily entails the existence of a left–right crossing of $B(\frac{3}{2}l, l)$, so that

$$P_p(LR(\tfrac{3}{2}l, l)) \geq P_p\left(N^+ \cap \left\{\bigcup_{\pi \in \mathscr{T}^-} (L_\pi \cap M_\pi^-)\right\}\right); \tag{9.80}$$

thus we seek a lower bound for the right-hand side of this inequality. We note first that N^+ is an increasing event. Furthermore,

$$G = \bigcup_{\pi \in \mathscr{T}^-} (L_\pi \cap M_\pi^-)$$

is the event that there exists an open path π in $\mathscr{T}^-$ for which M_π^- occurs; this is an increasing event also, giving by the FKG inequality that

$$P_p(LR(\tfrac{3}{2}l, l)) \geq P_p(N^+)P_p(G). \tag{9.81}$$

We apply the square root trick (9.79) to the events N^+ and N^- to find that

$$\begin{aligned} P_p(N^+) &\geq 1 - \sqrt{1 - P_p(N^+ \cup N^-)} \\ &= 1 - \sqrt{1 - \tau}, \end{aligned} \tag{9.82}$$

where $\tau = P_p(LR(l))$. Now,

$$P_p(G) = \sum_{\pi \in \mathscr{T}^-} P_p(M_\pi^- \mid L_\pi) P_p(L_\pi). \tag{9.83}$$

We claim that

$$P_p(M_\pi^- \mid L_\pi) \geq 1 - \sqrt{1 - \tau} \qquad \text{for all} \quad \pi \in \mathscr{T}; \tag{9.84}$$

once this claim has been proved, the result will follow fairly quickly as follows. We substitute (9.82)–(9.84) into (9.80) to obtain

$$P_p(LR(\tfrac{3}{2}l, l)) \geq \{1 - \sqrt{1 - \tau}\}^2 \sum_{\pi \in \mathscr{T}^-} P_p(L_\pi). \tag{9.85}$$

However,

$$\sum_{\pi \in \mathscr{T}^-} P_p(L_\pi) = P_p(L^-);$$

furthermore, we may apply the square root trick (9.79) to the events L^-

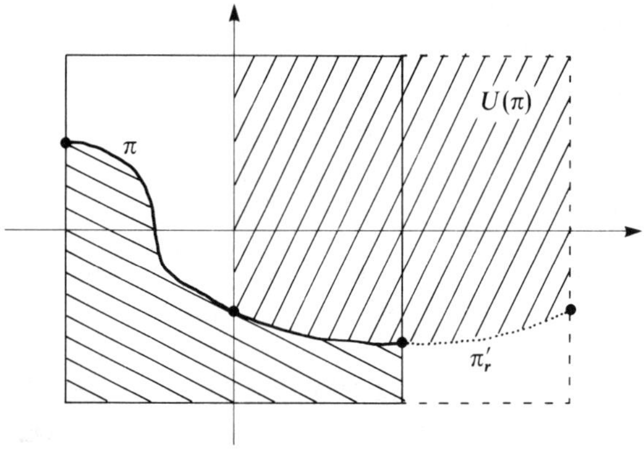

Figure 9.23. The two hatched regions indicate the disjoint sets of edges whose states determine the occurrence or not of M_π^- and L_π.

and L^+ to obtain

$$\begin{aligned} P_p(L^-) &\geq 1 - \sqrt{1 - P_p(L^- \cup L^+)} \\ &= 1 - \sqrt{1 - \tau}, \end{aligned}$$

which, when substituted into (9.85), yields the conclusion of the lemma. It remains to prove (9.84), for which we shall use the square root trick once more. Suppose that $\pi \in \mathscr{T}$. As an illustration of the argument, we consider first the case when π intersects the vertical axis of $\mathbb{L}^2$ once only. In this case, the events M_π^- and L_π are independent, since M_π^- is defined in terms of the states of edges of $B(l)'$ with at least one endvertex strictly above $\pi_r \cup \pi_r'$, whereas L_π is defined in terms of the states of edges in and below π in $B(l)$; see Figure 9.23. Thus

$$P_p(M_\pi^- | L_\pi) = P_p(M_\pi^-) \tag{9.86}$$

in this case; we now apply the square root trick to the events M_π^- and M_π^+ to deduce that

$$\begin{aligned} P_p(M_\pi^-) &\geq 1 - \sqrt{1 - P_p(M_\pi^- \cup M_\pi^+)} \\ &= 1 - \sqrt{1 - P_p(M_\pi)} \\ &\geq 1 - \sqrt{1 - \tau} \end{aligned} \tag{9.87}$$

as required, since M_π occurs whenever $B(l)'$ contains a top–bottom crossing. We now lift the restriction that π intersects the vertical axis once only. Let H be the set of edges of π which lie in $U(\pi)$, and let J be the set of edges of $U(\pi)$ which have at least one endvertex in the simply connected subset of $[-l, l] \times [-l, l]$ ($\subseteq \mathbb{R}^2$) lying strictly beneath π (see Figure 9.24). We claim that

$$P_p(M_\pi^- | L_\pi) = P_p(M_\pi^- | J \cup H \text{ open}),$$

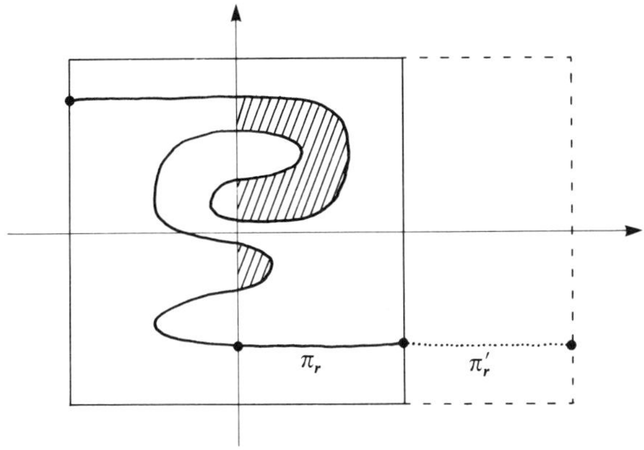

Figure 9.24. The set J is the set of edges in the hatched region.

where $\{J \cup H \text{ open}\}$ is the event that every edge in $J \cup H$ is open. To see this, note from Figure 9.24 that, if π is an open crossing of $B(l)$, we do not have to know the states of edges in J in determining whether or not M_π^- occurs. Now M_π^- and $\{J \cup H \text{ open}\}$ are increasing events, and we apply the FKG inequality to find that

$$P_p(M_\pi^- \mid L_\pi) \geq P_p(M_\pi^-) \qquad \text{for all} \quad \pi \in \mathscr{T}.$$

Inequality (9.87) is valid for all $\pi \in \mathscr{T}$, and (9.84) follows as before. ■

Proof of Lemma (9.75). We follow Russo (1981) again, beginning with (9.76). Let LR_1 (respectively LR_2) be the event that there exists a left–right crossing of the rectangle $[0, 3l] \times [-l, l]$ (respectively $[l, 4l] \times [-l, l]$) and let TB_1 be the event that there exists a top–bottom crossing of the square $[l, 3l] \times [-l, l]$; see Figure 9.25. If all three such paths

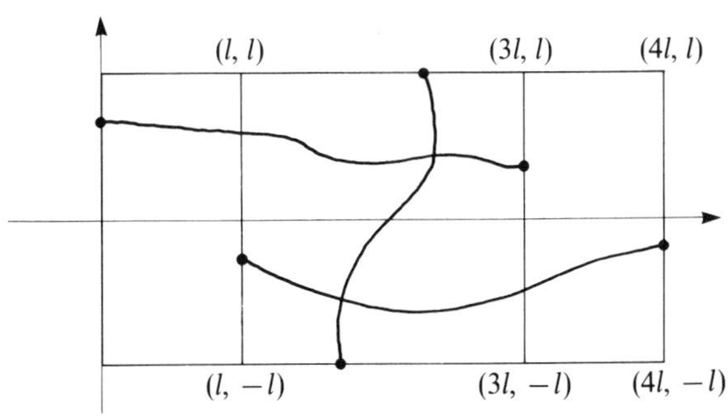

Figure 9.25. Three open paths which guarantee the occurrence of the events LR_1, LR_2, and TB_1. Their union contains a left–right crossing of $[0, 4l] \times [-l, l]$.

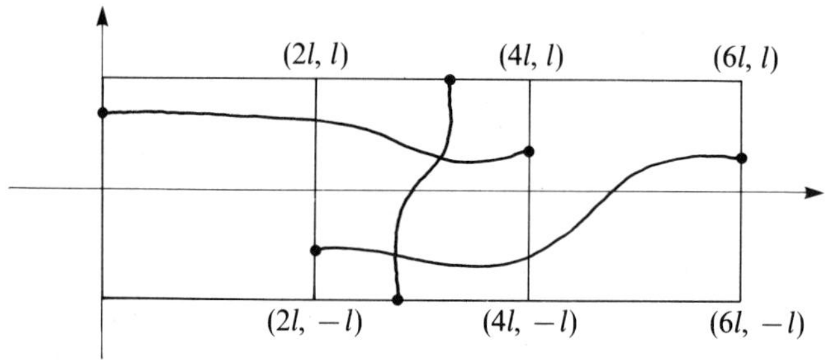

Figure 9.26. Three open paths which guarantee the occurrence of LR_3, LR_4, and TB_2. Their union contains a left–right crossing of $[0, 6l] \times [-l, l]$.

exist, then there is a left–right crossing of $[0, 4l] \times [-l, l]$, so that

$$\begin{aligned} P_p(LR(2l, l)) &\geq P_p(LR_1 \cap LR_2 \cap TB_1) \\ &\geq P_p(LR_1)P_p(LR_2)P_p(TB_1) \end{aligned}$$

by the FKG inequality, and we have proved (9.76).

Inequality (9.77) follows similarly, by considering the events

$$LR_3 = \{\text{there exists a left–right crossing of } [0, 4l] \times [-l, l]\},$$

$$LR_4 = \{\text{there exists a left–right crossing of } [2l, 6l] \times [-l, l]\},$$

$$TB_2 = \{\text{there exists a top–bottom crossing of } [2l, 4l] \times [-l, l]\};$$

see Figure 9.26.

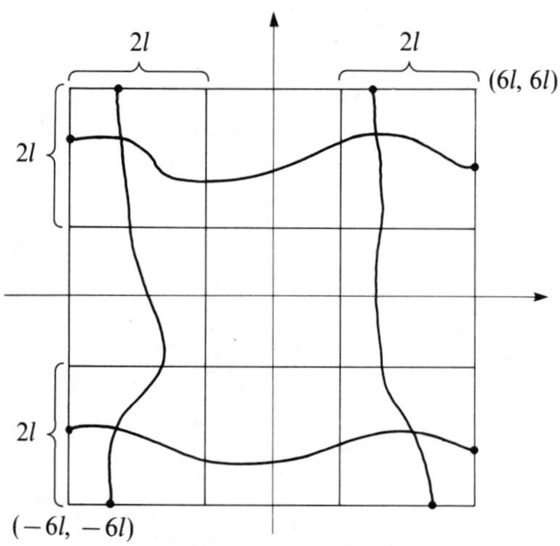

Figure 9.27. If the four $2l \times 6l$ rectangles are traversed by open paths as indicated then there is an open circuit in the annulus $A(l)$ containing the origin in its interior.

It is equally easy to prove (9.78). We express $A(l)$ as the union of the four rectangles with dimensions $2l \times 6l$ illustrated in Figure 9.27. Each of these is traversed by an open path joining its two shorter sides with probability at least $P_p(LR(3l, l))$, and the result follows by the FKG inequality. Note that we have used the convention that 'crossings' of rectangles use no edges in the boundary. ∎

9.8 Power Law Inequalities

The greatest current mystery in percolation is the behaviour of such quantities as θ and χ at and near the critical point. We recall such conjectures as

$$\theta(p) \approx (p - p_c)^{\beta} \qquad \text{as} \quad p \downarrow p_c,$$

$$\chi(p) \approx (p_c - p)^{-\gamma} \qquad \text{as} \quad p \uparrow p_c,$$

and

$$P_{p_c}(|C| = n) \approx n^{-1-1/\delta} \qquad \text{as} \quad n \to \infty,$$

for percolation on $\mathbb{L}^d$ when $d \geq 2$. No rigorous proofs of such relations are known at present. On the other hand, the methods of Chapters 7 and 8 may be used to obtain certain inequalities of the right form: for example, we saw in Section 8.2 that if δ exists then $\delta \geq 2$, which fact suggests the possible validity of an inequality of the form

(9.88) $$P_{p_c}(|C| \geq n) \geq An^{-1/2}$$

for some positive constant $A = A(d)$. In the absence of rigorous proof of power law behaviour, we may seek power law *inequalities* of the form of (9.88). As usual, the case when $d = 2$ is the most tractable, since it is in this case that the technique of duality is most easily available. We describe such results in this section.

We consider percolation on $\mathbb{L}^2$, for which the critical probability is $p_c = \frac{1}{2}$, and $\theta(\frac{1}{2}) = 0$. The following theorem contains our principal estimates at the critical point.

(9.89) Theorem. Power law inequalities at the critical point. *There exist positive finite constants A_i, α_i, for $i \geq 1$, such that the following statements are valid:*

(9.90) $$\tfrac{1}{2}n^{-1/2} \leq P_{1/2}(0 \leftrightarrow \partial B(n)) \leq A_1 n^{-\alpha_1} \qquad \text{for all} \quad n \geq 1,$$

(9.91) $$\tfrac{1}{2}n^{-1/2} \leq P_{1/2}(|C| \geq n) \leq A_2 n^{-\alpha_2} \qquad \text{for all} \quad n \geq 1,$$

(9.92) $$E_{1/2}(|C|^{\alpha_3}) < \infty.$$

We have as consequences of (9.90) and (9.91) that the critical exponents ρ and δ satisfy $2 \leq \rho < \infty$ and $2 \leq \delta < \infty$ in two dimensions, if

ρ and δ exist. Better inequalities than these are available: see Kesten (1987b) for proofs that $\rho \geq 3$ and $\delta \geq 5$. The inequality $\delta \geq 2$ is valid in all dimensions (see Proposition (8.29)), but we present here an alternative proof for two dimensions.

The principal components of Theorem (9.89) are the inequalities (9.90) concerning the probability of an open path from the origin to the surface of the box $B(n)$; the left-hand inequality is proved by the BK inequality, and the right-hand inequality by an application of the RSW theorem. We shall prove Theorem (9.89) later in this section. We turn next to power law inequalities for θ, χ, and χ^f when p is near to its critical value $\frac{1}{2}$.

(9.93) Theorem. Power law inequalities for θ, χ, and χ^f. *There exist positive finite constants A_i, α_i, for $i \geq 1$, such that the following statements are valid:*

(9.94) $$\theta(p) \leq A_4(p - \tfrac{1}{2})^{\alpha_4} \qquad \text{if} \quad p > \tfrac{1}{2},$$

(9.95) $$\chi(p) \leq A_5(\tfrac{1}{2} - p)^{-\alpha_5} \qquad \text{if} \quad p < \tfrac{1}{2},$$

(9.96) $$A_6(p - \tfrac{1}{2})^{-1/4} \leq \chi^f(p) \leq A_7(p - \tfrac{1}{2})^{-\alpha_7} \qquad \text{if} \quad p > \tfrac{1}{2}.$$

There are corresponding lower bounds for θ and χ which are valid in all dimensions: we have from Theorem (3.8) and equation (8.41) that there exists a (>0) such that

(9.97) $$\theta(p) \geq a(p - \tfrac{1}{2}) \qquad \text{if} \quad p > \tfrac{1}{2}$$

and

$$\chi(p) \geq \tfrac{1}{4}(\tfrac{1}{2} - p)^{-1} \qquad \text{if} \quad p < \tfrac{1}{2}.$$

Likewise, the lower bound in (9.96) for χ^f may be improved using arguments of Newman (1986, 1987b, c), subject possibly to a minor assumption concerning the asymptotic behaviour of the cluster size distribution. It is possible to use the arguments in the proofs of the above two theorems to show the existence of positive constants A_8 and α_8 such that

(9.98) $$P_p(n \leq |C| < \infty) \leq A_8 n^{-\alpha_8} \qquad \text{for all} \quad n \geq 1 \quad \text{and} \quad 0 \leq p \leq 1;$$

see Kesten (1981). This inequality has an immediate application to the differentiability of the number $\kappa(p)$ of open clusters per vertex. We know from Theorem (5.108) that κ is analytic on $[0, \frac{1}{2})$, and it follows from Theorem (9.4) that κ is analytic on $(\frac{1}{2}, 1]$ also. The only lingering challenge is to understand its behaviour at $p_c = \frac{1}{2}$. Certainly κ is continuously differentiable at $\frac{1}{2}$, from Theorem (4.31). Expressing $\kappa(p)$ as an infinite series (as in (4.18)) and differentiating twice, term by term, we may use (9.98) to show that the resulting series is uniformly convergent on a neighbourhood of $\frac{1}{2}$; thus κ is twice continuously differentiable at $\frac{1}{2}$. The argument fails to show thrice differentiability, and it is conjectured that the third derivative of κ does not exist at the critical probability (see also the discussion around equation (7.6)).

Proof of Theorem (9.89). We begin with (9.90), of which (9.91) and (9.92) will be easy corollaries. For the left-hand inequality of (9.90) we use the BK inequality, and we follow van den Berg and Kesten (1985). Let $LR(n)$ be the event that there is an open path in the rectangle $R_n = [0, 2n] \times [0, 2n - 1]$ joining some vertex on its left side to some vertex on its right side, and recall from Lemma (9.20) that

(9.99) $$P_{1/2}(LR(n)) = \tfrac{1}{2} \qquad \text{for} \quad n \geq 1.$$

Any such crossing of R_n uses at least one vertex on the centre line $\{(n, k): 0 \leq k \leq 2n - 1\}$ of R_n, and any such vertex is connected by two disjoint paths to the left side and to the right side of R_n; see Figure 9.28. Hence

$$P_{1/2}(LR(n)) \leq \sum_{k=0}^{2n-1} P_{1/2}(A_n(k) \circ A_n(k)),$$

where $A_n(k)$ is the event that the vertex (n, k) is joined by an open path to the surface of the box $B(n, (n, k))$ with side-length $2n$ and centre at (n, k). We use the BK inequality to find that

$$\begin{aligned} P_{1/2}(LR(n)) &\leq \sum_{k=0}^{2n-1} P_{1/2}(A_n(k))^2 \\ &= 2nP_{1/2}(0 \leftrightarrow \partial B(n))^2, \end{aligned}$$

and hence

$$P_{1/2}(0 \leftrightarrow \partial B(n)) \geq \tfrac{1}{2} n^{-1/2}$$

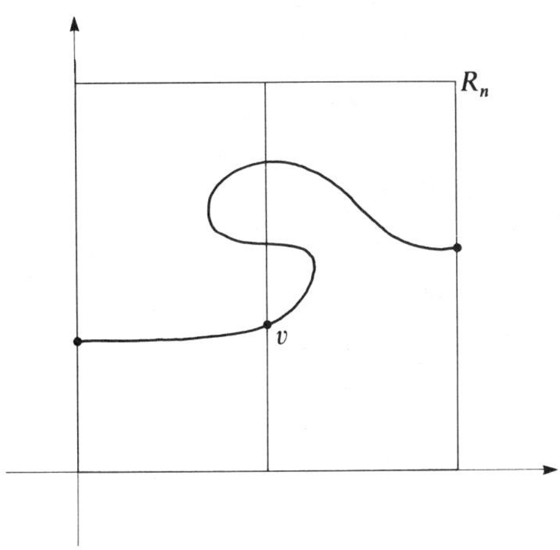

Figure 9.28. Any open path traversing R_n contains some vertex v on the centre line, and this vertex is joined by edge-disjoint open paths to the left and right sides of R_n.

by (9.99), as required. For the upper bound on $P_{1/2}(0 \leftrightarrow \partial B(n))$ we follow Kesten (1981), using the RSW theorem and particularly (9.72): there exists $\zeta > 0$ such that

(9.100) $$P_{1/2}(O(l)) \geq \zeta \qquad \text{for all} \quad l \geq 1,$$

where $O(l)$ is the event that the annulus $A(l) = B(3l) \backslash B(l)$ contains an open circuit having the origin in its interior. We use the dual lattice technique here, writing $\mathbb{L}^2_{\text{d}}$ for the dual of $\mathbb{L}^2$ as before. Let $B(k)_{\text{d}}$ be the square $B(k) + (\frac{1}{2}, \frac{1}{2})$ in the dual lattice, and let $A(l)_{\text{d}} = B(3l)_{\text{d}} \backslash B(l)_{\text{d}}$ be the corresponding annulus in the dual. Let $O(l)_{\text{d}}$ be the event that $A(l)_{\text{d}}$ contains a closed circuit of the dual lattice having the dual origin $O_{\text{d}} = (\frac{1}{2}, \frac{1}{2})$ in its interior. We have from (9.100) that

(9.101) $$P_{1/2}(O(l)_{\text{d}}) \geq \zeta \qquad \text{for all} \quad l \geq 1,$$

since each dual edge is closed with probability $\frac{1}{2}$. Suppose that there is an open path from the origin of $\mathbb{L}^2$ to the surface of the box $B(3^k + 1)$. Such a path traverses each of the dual annuli $A(3^r)_{\text{d}}$, for $0 \leq r < k$, and therefore none of the events $O(3^r)_{\text{d}}$ occurs for $0 \leq r < k$. However, the events $\{O(3^r)_{\text{d}}: 0 \leq r < k\}$ are defined in terms of disjoint sets of dual edges, and they are therefore independent. Thus

$$\begin{aligned} P_{1/2}(0 \leftrightarrow \partial B(3^k + 1)) &\leq P_{1/2}(O(3^r)_{\text{d}} \text{ does not occur for } 0 \leq r < k) \\ &\leq (1 - \zeta)^k \end{aligned}$$

by (9.101). If $3^k + 1 \leq n < 3^{k+1} + 1$ then

(9.102) $$\begin{aligned} P_{1/2}(0 \leftrightarrow \partial B(n)) &\leq P_{1/2}(0 \leftrightarrow \partial B(3^k + 1)) \\ &\leq (1 - \zeta)^k \\ &\leq A_1 n^{-\alpha_1} \end{aligned}$$

for some positive finite constants A_1 and α_1 depending on ζ only. This completes the proof of (9.90).

It is not difficult to deduce (9.91) from (9.90). Clearly

$$P_{1/2}(0 \leftrightarrow \partial B(n)) \leq P_{1/2}(|C| \geq n) \leq P_{1/2}(0 \leftrightarrow \partial B(\lfloor \tfrac{1}{2}\sqrt{n} - 1 \rfloor))$$

since $|C| \geq n$ if $0 \leftrightarrow \partial B(n)$, and a cluster of size n cannot be accommodated in $B(m)$ if $n > |B(m)| = (2m + 1)^2$; (9.91) follows immediately. Similarly, if $0 < a < 1$ then

$$\begin{aligned} E_{1/2}(|C|^a) &= \sum_{n=1}^{\infty} \{n^a - (n - 1)^a\} P_{1/2}(|C| \geq n) \\ &\leq A \sum_{n=1}^{\infty} n^{a-1} n^{-\alpha_2} \end{aligned}$$

for some finite constant A, by (9.91). This summation is finite if $a < \alpha_2$. ■

Proof of Theorem (9.93). We follow largely Kesten (1981, 1982), and begin with the upper bound for θ. We observe first that

(9.103) $$\theta(p) \le P_p(0 \leftrightarrow \partial B(n)) \qquad \text{for all } n,$$

since any infinite open path from the origin contains vertices of $\partial B(n)$. We use inequality (2.31) to relate $P_p(0 \leftrightarrow \partial B(n))$ to $P_{1/2}(0 \leftrightarrow \partial B(n))$. The number of edges in $B(n)$ is no more than $2(2n+1)^2 \le 18n^2$, and therefore

(9.104) $$P_{p_1}(0 \leftrightarrow \partial B(n)) \ge (p_1/p_2)^{18n^2} P_{p_2}(0 \leftrightarrow \partial B(n)) \qquad \text{if} \quad 0 \le p_1 < p_2 \le 1.$$

Setting $p_1 = \frac{1}{2}$ and $p_2 = p > \frac{1}{2}$, we obtain

(9.105) $$P_{1/2}(0 \leftrightarrow \partial B(n)) \ge (2p)^{-18n^2} P_p(0 \leftrightarrow \partial B(n)).$$

When combined with (9.103) and (9.90) this implies that, for $p > \frac{1}{2}$,

$$\theta(p) \le A_1 (2p)^{18n^2} n^{-\alpha_1} \qquad \text{for all} \quad n \ge 1.$$

We now choose n by

(9.106) $$n = \lfloor \{\log(2p)\}^{-1/2} \rfloor,$$

noting that $n = n(p) \sim \{2(p - \frac{1}{2})\}^{-1/2}$ as $p \downarrow \frac{1}{2}$, to obtain

(9.107) $$\begin{aligned} \theta(p) &\le A_1 e^{18} n^{-\alpha_1} \\ &\sim A_4 (p - \tfrac{1}{2})^{\alpha_1/2} \qquad \text{as} \quad p \downarrow \tfrac{1}{2}, \end{aligned}$$

for some constant A_4, and the result follows.

We turn next to the lower bound for χ^f, and we suppose that $p > \frac{1}{2}$. It is the case that $n \le |C| < \infty$ if there is an open path from the origin to the surface of $B(n)$ and there is a closed circuit of the dual in the annulus $A(n)_d$ containing the origin of $\mathbb{L}^2$ in its interior; see Figure 9.29. These two events are defined in terms of disjoint sets of edges, and they are therefore independent. Hence

(9.108) $$\begin{aligned} \chi^f(p) &\ge nP_p(n \le |C| < \infty) \\ &\ge nP_p(0 \leftrightarrow \partial B(n), O(n)_d) \\ &= nP_p(0 \leftrightarrow \partial B(n)) P_p(O(n)_d) \\ &\ge \tfrac{1}{2} n^{1/2} P_p(O(n)_d) \end{aligned}$$

by (9.90), since $P_p(0 \leftrightarrow \partial B(n))$ is a non-decreasing function of p. Also

(9.109) $$P_p(O(n)_d) = P_{1-p}(O(n))$$

since each edge of the dual lattice is closed with probability $1 - p$. We use inequality (2.31) again to find that

$$P_{1-p}(O(n)) \ge \{2(1-p)\}^{98n^2} P_{1/2}(O(n)),$$

where $98n^2$ is an upper bound for the number of edges in the annulus $A(n)$. However $P_{1/2}(O(n)) > \zeta$ (>0) by (9.100), giving from (9.108) and

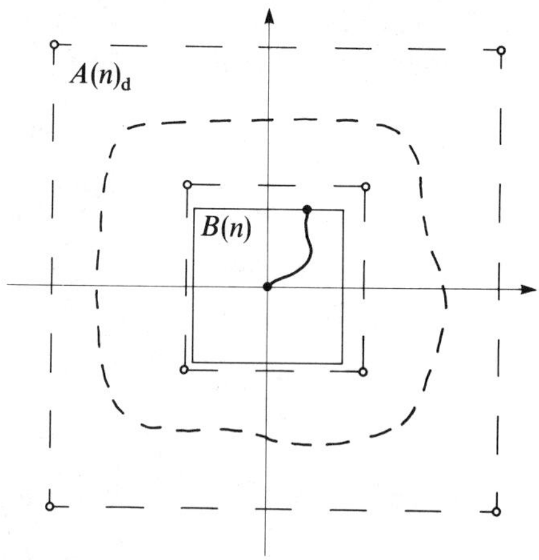

Figure 9.29. If there is an open path from the origin to $\partial B(n)$ together with a closed circuit of $A(n)_d$ surrounding the origin, then the open cluster at the origin is finite with size no less than n.

(9.109) that

$$\chi^f(p) \geq \tfrac{1}{2}\zeta\{2(1-p)\}^{98n^2}n^{1/2}.$$

We choose n by

$$n = \lfloor\{-\log[2(1-p)]\}^{-1/2}\rfloor,$$

noting that $n = n(p) \sim \{2(p-\frac{1}{2})\}^{-1/2}$ as $p \downarrow \frac{1}{2}$, to obtain

$$\begin{aligned}\chi^f(p) &\geq \tfrac{1}{2}\zeta e^{-98}n^{1/2}\\ &\sim \tfrac{1}{2}\zeta e^{-98}\{2(p-\tfrac{1}{2})\}^{-1/4} \qquad \text{as} \quad p \downarrow \tfrac{1}{2}.\end{aligned}$$

It follows that the left-hand inequality in (9.96) is valid for all $p > \frac{1}{2}$ and some constant A_6.

In demonstrating the upper bounds for χ and χ^f we deviate somewhat from the original argument of Kesten (1981), making use instead of material in Section 3.2. Let $S(n)$ be the ball $\{x \in \mathbb{Z}^d : \delta(0, x) \leq n\}$ with radius n and centre at the origin, and let

$$g_p(k) = P_p(0 \leftrightarrow \partial S(k)),$$

as in Section 3.2. We have from (3.22) that

$$(9.110) \quad g_\alpha(n) \leq g_\beta(n) \exp\left(-(\beta-\alpha)\left[\frac{n}{\sum_{i=0}^{n} g_\beta(i)} - 1\right]\right) \qquad \text{if} \quad 0 \leq \alpha < \beta \leq 1.$$

We set $\alpha = p$ $(<\frac{1}{2})$ and $\beta = \frac{1}{2}$ to obtain

$$(9.111) \quad g_p(n) \leq \exp\left(-(\tfrac{1}{2}-p)\left[\frac{n}{\sum_{i=0}^{n} g_{1/2}(i)} - 1\right]\right) \qquad \text{for} \quad 0 \leq p < \tfrac{1}{2}.$$

However

$$g_{1/2}(i) \le P_{1/2}(0 \leftrightarrow \partial B(\lfloor \tfrac{1}{2}i \rfloor))$$

since $B(\lfloor \frac{1}{2}i \rfloor)$ is a subset of $S(i)$. Now, we have from (9.90) that $g_{1/2}(i) \le A_1 i^{-\alpha_1}$ for $i \ge 1$, where $0 < \alpha_1 \le \frac{1}{2}$. We insert this into (9.111) to obtain

(9.112) $$g_p(n) \le A_9 \exp\{-A_{10} n^{\alpha_1}(\tfrac{1}{2} - p)\} \qquad \text{if} \quad 0 \le p < \tfrac{1}{2},$$

for all n and for appropriate positive constants A_9, A_{10}. It is a simple matter to extract the required bounds for χ and χ^f from this inequality. First we recall the obvious inequality (see (3.7))

(9.113) $$P_p(|C| > n) \le g_p(\lfloor \sqrt{n/\pi} \rfloor).$$

Hence, for $0 \le p < \frac{1}{2}$,

$$\begin{aligned} \chi(p) &= \sum_{n=0}^{\infty} P_p(|C| > n) \\ &\le \sum_{n=0}^{\infty} A_9 \exp\{-A_{11} n^{\alpha_1/2}(\tfrac{1}{2} - p)\} \end{aligned}$$

for some positive constant A_{11}. We replace the summation by an integral and make the change of variables $y = n^{\alpha_1/2}(\frac{1}{2} - p)$ to obtain

$$\chi(p) \le A_{12}(\tfrac{1}{2} - p)^{-2/\alpha_1} \int_0^\infty y^{-1+(2/\alpha_1)} e^{-A_{11}y}\, dy$$

for some positive constant A_{12}. The integral converges, and we have shown (9.95).

Finally we prove the upper bound for χ^f, and we assume therefore that $\frac{1}{2} < p < 1$. We recall from the discussion prior to (9.3) that there exists a positive constant λ with the following property: if $|C| = n$ then the origin is in the interior of some closed circuit of the dual lattice having length at least $\lambda\sqrt{n}$, and there exists such a circuit containing at least one of the dual vertices $(k + \frac{1}{2}, \frac{1}{2})$ for $0 \le k < n$. Such a dual vertex is in a closed cluster of the dual with size at least $\lambda\sqrt{n}$, and it follows that

$$\begin{aligned} P_p(|C| = n) &\le \sum_{k=0}^{n-1} P_{1-p}(|C| \ge \lambda\sqrt{n}) \\ &\le n g_{1-p}(\lfloor A_{13} n^{1/4} \rfloor) \end{aligned}$$

by (9.113), for some positive constant A_{13}. Now $1 - p < \frac{1}{2}$, and so

$$g_{1-p}(k) \le A_9 \exp\{-A_{10} k^{\alpha_1}(p - \tfrac{1}{2})\}$$

by (9.112). We now argue as before to deduce that there exists a positive constant A_{14} such that

$$\begin{aligned} \chi^f(p) &= \sum_{n=1}^{\infty} n P_p(|C| = n) \\ &\le A_{14}(p - \tfrac{1}{2})^{-12/\alpha_1} \end{aligned}$$

as required. ■

9.9 Notes

Section 9.2. Duality is used in independent work of Hammersley (1959) and Harris (1960). See Bondy and Murty (1976, p. 140) for a graph-theoretic reference, and Kesten (1982, p. 386) for a more careful treatment than that presented here. Theorem (9.4) appears in Sykes and Essam (1964) together with related material. Their 'exact calculation' contains a hypothesis which remains unjustified to this day, that κ has a singularity at p_c; see the discussion after Theorem (4.31). When applied to other pairs $(\mathscr{L}, \mathscr{L}_d)$ of dual lattices, the Sykes–Essam argument indicates that $p_c(\mathscr{L}) + p_c(\mathscr{L}_d) = 1$. Taken in conjunction with the so-called star/triangle transformation this implies that the critical probability of bond percolation on the triangular (respectively hexagonal) lattice is $2\sin(\pi/18)$ (respectively $1 - 2\sin(\pi/18)$). These exact values have been justified by rigorous arguments of Wierman (1981).

Section 9.3. Kesten (1980a) proved that $p_c = \frac{1}{2}$, building on work of Harris (1960), Russo (1978), and Seymour and Welsh (1978). Lemma (9.12) was first proved using quite different techniques by Harris (1960) in a remarkable paper which contained many of the techniques used in more recent developments. Such results were extended to other dual pairs of two-dimensional lattices by Fisher (1961) and Fisher and Essam (1961), who observed also that similar techniques work for site percolation on so called 'matching pairs' of lattices (see Kesten (1982, Section 2.2)). Harris's proof of Lemma (9.13) has been adapted with considerable originality by Gandolfi, Keane, and Russo (1988) to prove the uniqueness of the infinite open cluster for a certain class of 'non-Bernoulli' percolation processes; see the notes for Section 6.2.

The simple proof given here that $\theta(\frac{1}{2}) = 0$ is due to Zhang (1988, unpublished). The geometrical ingredients of our proofs that $p_c \leq \frac{1}{2}$ are contained in Russo (1978) and Seymour and Welsh (1978).

The arguments of this section are easily adapted to show with the aid of the star/triangle transformation that the critical probabilities of the triangular and hexagonal lattices are as given above. These arguments may be applied also to certain anisotropic processes (see Section 10.1).

Section 9.4. Theorem (9.23) appears in Chayes, Chayes, Grimmett, Kesten, and Schonmann (1988). The argument of the proof of Theorem (9.24) is implicit in Russo (1978). Lemma (9.27) is implicit in the above paper of Chayes et al. (1988) and appears explicitly in Aizenman, Chayes, Chayes, and Newman (1988) and Durrett and Schonmann (1988b).

Section 9.5. Most of this material appears in Grimmett (1981a, 1983); we owe to Chayes and Chayes (1986b) the observation that the function f may be picked in such a way that there exists an infinite open cluster *at* the critical point. The latter authors study the corresponding problem in three and more dimensions also, obtaining improvements to results of Hammersley and Whittington (1985) and Grimmett (1985a). For corresponding results for models of ferromagnetism, see Chayes and Chayes (1986b) and Aizenman, Chayes, Chayes, and Newman (1988).

In related work, Grimmett (1981a) studied the quantity $M_n = \max\{m$: there exists a left-right crossing of the rectangle $[0, m] \times [0, n]\}$, showing that

$$\text{if } p < \tfrac{1}{2} \text{ then } M_n/\log n \to \xi(p) \text{ in probability,}$$

$$\text{if } p > \tfrac{1}{2} \text{ then } n^{-1} \log M_n \to \xi(1-p)^{-1} \text{ in probability,}$$

as $n \to \infty$. Durrett and Schonmann (1988b) have proved that $M_n/E_p(M_n)$ has a limit distribution which is the exponential distribution with mean 1 when $\frac{1}{2} < p < 1$.

Section 9.6. The basic references for central limit theorems are Brånvall (1980), Cox and Grimmett (1981, 1984), and Herrndorf (1985). As for results about large deviations, Durrett and Schonmann (1988a) have studied the probabilities of large deviations in the number of vertices inside $B(n)$ which are contained in the infinite open cluster.

Section 9.7. The RSW theorem appears in Russo (1978, 1981) and Seymour and Welsh (1978). Chayes and Chayes (1986a) describe interesting and useful constructions in the spirit of Lemma (9.75).

Section 9.8. Theorems (9.89) and (9.93) appear with related material in Kesten (1981, 1982), where one may find also the proof that κ is twice differentiable at $\frac{1}{2}$. The exception is the lower bound for $P_{1/2}(0 \leftrightarrow \partial B(n))$, and the consequences thereof; this is due to van den Berg and Kesten (1985). Our proofs of the upper bounds for χ and χ^f are based on Menshikov's proof of the uniqueness of the critical point (Theorem (3.4)) rather than Kesten's original argument; the majority of our argument is valid for any number of dimensions, but we lack the starting point '$\rho < \infty$' if $d \geq 3$; see the relevant remarks in the notes for Section 8.1. (*Added in proof.*) Recent progress of T. Hara and G. Slade (see the notes for Section 8.2) implies certain power law bounds in high dimensions.

10

A Miscellany of Random Processes

10.1 Mixed Percolation on a General Lattice

Bond percolation on $\mathbb{L}^d$ is an attractive representative of the class of all percolation processes. It is simple to formulate, and its analysis contains a minimum of geometrical and analytical detail whilst retaining all general features of substance. We may generalize in many directions, of which we mention a few in this section.

A. *Other lattices.* The hypercubic lattice $\mathbb{L}^d$ is a simple and convenient structure but, apart from self-duality in two dimensions, it has few magical properties. With the exception of the results for bond percolation on $\mathbb{L}^2$, all the results of Chapters 1–8 may be reformulated for bond percolation on a general lattice $\mathscr{L}$. Most people would agree on the meaning to be given to the term 'lattice', and we recall two formal definitions. Kesten (1982, pp. 10–12) defines a *periodic graph in d dimensions* to be a connected, loopless graph $\mathscr{L}$, with bounded vertex degrees, which is embedded in $\mathbb{R}^d$ in such a way that:

(a) the translations $x \to x + f$ are automorphisms of $\mathscr{L}$ for each unit vector f parallel to a coordinate axis;
(b) all edges are of finite length; and
(c) every compact subset of $\mathbb{R}^d$ intersects only finitely many edges.

Such a definition stresses the geometrical aspects of a lattice. Graph-theoretic aspects are stressed more by Grimmett (1978): a *d-dimensional lattice* is a locally finite graph which admits as a group of automorphisms the product of d cyclic groups which act like $\mathbb{Z}^d$, under which the vertex set of $\mathscr{L}$ has only finitely many orbits. See also Rogers (1964), Biggs

(1976), and Godsil and McKay (1980). Needless to say, most percolation theorists avoid such definitions. We defer until Section 10.3 a discussion of 'long-range' percolation.

B. *Site percolation.* Suppose that we wish to block the *vertices* of a lattice rather than its edges. We may declare each vertex to be *open* with probability p and *closed* otherwise, independently of all other vertices. The open clusters of this 'site percolation' model are the connected components of the subgraph of the lattice induced by the set of open vertices. In broad terms, all the results of Chapters 1–9 have analogues for site percolation processes. It is well known that site percolation is more general than bond percolation in the following sense: every bond process may be reformulated as a site process on a different lattice, but there exist site processes which do not arise thus from bond processes. This was observed by Fisher (1961) and Fisher and Essam (1961), and the construction is as follows. Consider a bond percolation process on a lattice $\mathcal{L}$. We construct the *covering* (or *line*) *lattice* $\mathcal{L}_c$ of $\mathcal{L}$ as follows. To each edge of $\mathcal{L}$ there corresponds a vertex of $\mathcal{L}_c$, and two such vertices are adjacent in $\mathcal{L}_c$ if and only if the corresponding edges of $\mathcal{L}$ have an endvertex in common. Figure 10.1 contains a picture of part of the hexagonal lattice and its covering lattice (the so called Kagomé lattice). It should be clear that there is no effective difference between studying bond percolation on $\mathcal{L}$ and studying site percolation on $\mathcal{L}_c$. See Kesten (1982, Chapter 3) for more details.

C. *Mixed percolation.* In mixed percolation models, both edges and vertices may be open or closed, possibly with different probabilities. See McDiarmid (1980, 1981) and Hammersley (1980). One may even block the faces, in two dimensions at least; see Wierman (1984).

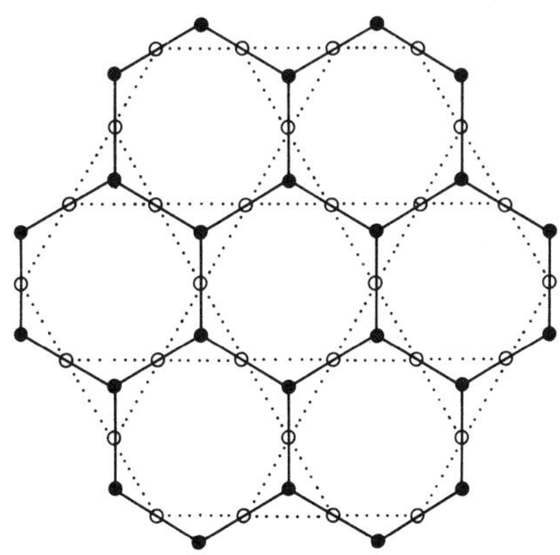

Figure 10.1. The dark circles are the vertices of part of the hexagonal lattice, and the open circles are the vertices of its covering lattice.

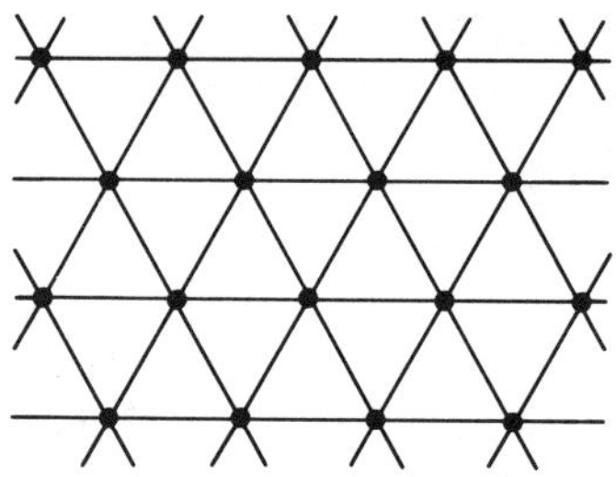

Figure 10.2. Part of the triangular lattice.

D. *Anisotropic percolation.* Here are two examples of anisotropic bond percolation processes:

(i) bond percolation on $\mathbb{L}^2$ in which each horizontal edge is open with probability p_h, and each vertical edge is open with probability p_v;
(ii) bond percolation on the triangular lattice (see Figure 10.2) in which an edge is open with probability p_1, p_2, or p_3, depending on whether its inclination to the horizontal is 0, $\pi/3$, or $2\pi/3$.

We refer the reader to Kesten (1982, Chapter 3) for a formal definition. Anisotropic processes have broadly similar properties to those of isotropic processes, although the critical probability p_c of the latter becomes the 'critical surface' $\varphi(p_1, p_2, \ldots) = 0$ of the former, where p_1, p_2, ... are the edge-probabilities of the process in question. We make this remark more transparent by reference to the two examples above.

(i) For the square lattice we define $\varphi(p_h, p_v) = p_h + p_v - 1$. It turns out that all open clusters are almost surely finite if $\varphi(p_h, p_v) < 0$, and there exists (almost surely) a unique infinite cluster if $\varphi(p_h, p_v) > 0$.
(ii) The same conclusions are valid in the case of the triangular lattice with the function φ defined by $\varphi(p_1, p_2, p_3) = p_1 + p_2 + p_3 - p_1 p_2 p_3 - 1$.

These (and related) results are proved or implied by Kesten (1982), Menshikov, Molchanov, and Sidorenko (1986), Aizenman and Barsky (1987), and Aizenman, Kesten, and Newman (1987).

10.2 *AB* Percolation

We are provided with a lattice $\mathscr{L}$ and two types of atom, type A and type B. Each vertex of $\mathscr{L}$ is occupied by an atom of one or other type, and we suppose that there is probability p that any given vertex is occupied by an atom of type A. We assume further that different vertices are occupied independently of each other. Each edge of $\mathscr{L}$ is declared to be *open* if its endvertices are occupied by atoms of *different* types; the idea is that dissimilar atoms bond together, whereas similar atoms repel each other. Of principal interest is the probability $\theta(p; \mathscr{L})$ that the origin of the lattice is in an infinite cluster of open edges; we may think of $\theta(p; \mathscr{L})$ as being the probability that the origin is an endvertex of an infinite '*AB* path'.

As a model for gelation, this process has been studied somewhat intuitively by Halley (1983) and, under the title 'antipercolation', by Sevsek, Debierre, and Turban (1983). Here is a pencil sketch of recent results of Appel and Wierman (1987), Scheinerman and Wierman (1987), and Wierman and Appel (1987). In the negative direction, it is known that $\theta(p; \mathscr{L}) = 0$ for all values of p if $\mathscr{L}$ belongs to a certain class of bipartite lattices; the square lattice does not lie in this general class, but separate arguments may be used to establish $\theta(p; \mathbb{L}^2) = 0$ for all p. In the positive direction, we have that $\theta(p; \mathscr{L}) > 0$ if $\mathscr{L}$ is the triangular lattice, for values of p lying in a certain non-empty interval containing $\frac{1}{2}$. We note that currently known results have most power when applied to two-dimensional lattices.

We mention an open problem which is similar to *AB* percolation in spirit if not in practice. We start with the square lattice $\mathbb{L}^2$, a bucket of small double-sided mirrors, and a parameter p taking values in $[0, 1]$. For each vertex v we perform the following experiment, independently of all other vertices. With probability p, we pick a mirror from the bucket and place it at the vertex v, in such a way that the plane of the mirror is perpendicular to that of the lattice, and the angle between the mirror and the horizontal axis of the lattice is equally likely to be $\pi/4$ or $3\pi/4$. The remaining probability $1 - p$ is the chance that we do nothing. Thus each vertex is occupied by a mirror with probability p, and vacant otherwise; see Figure 10.3. We think of the mirrors as being random scatterers of light, since rays of light passing along edges of the lattice are deflected through an angle of $\frac{1}{2}\pi$ whenever they meet a mirror, but

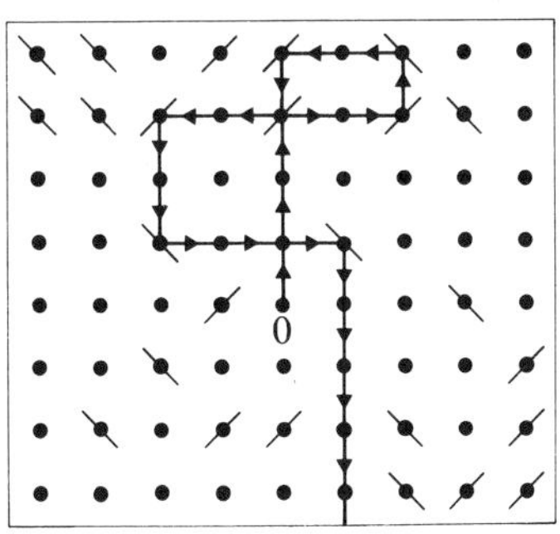

Figure 10.3. Double-faced mirrors are placed at randomly chosen vertices and are inclined at random. The marked path is the route of a light ray which leaves the origin heading northwards.

the direction of the deflection is random. How many vertices can see the origin? More precisely, we suppose that four rays of light are emitted along the coordinate axes from a light source placed at the origin. Let C be the set of vertices of the lattice which are illuminated by one or more of these light rays, and let $\theta(p)$ be the probability that C has infinite cardinality. Clearly $\theta(0) = 1$, and it is not too difficult to show that $\theta(1) = 0$ using the fact that, for bond percolation on the square lattice with edge-probability $\frac{1}{2}$, the origin is almost surely in the interior of some open circuit. We conjecture that $\theta(p) > 0$ whenever $p < 1$. This problem is a version of Ehrenfest's 'wind-tree' model (see Ehrenfest (1959, p. 229) and the more recent papers of Hauge and Cohen (1967) and van Beijeren (1982)).

10.3 Long-Range Percolation in One Dimension

Bond percolation on $\mathbb{L}^d$ is a process which involves 'nearest neighbour interaction' only, in the sense that only pairs of neighbours may be joined by open edges. Greater generality is obtained by considering 'long-range' processes, constructed as follows. Let $\mathbf{p} = (p(x): x \in \mathbb{Z}^d)$ be a collection of numbers from the interval $[0, 1)$, indexed by the set $\mathbb{Z}^d$ of d-vectors of integers; we suppose that $p(x) = p(-x)$ for all $x \in \mathbb{Z}^d$. We examine in turn each unordered pair $\{u, v\}$ of distinct points in $\mathbb{Z}^d$; we join u and v by an open edge with probability $p(v - u)$, independently of all other pairs. The resulting process is a translation-invariant long-range bond percolation process, and we may recover ordinary bond percolation with edge-probability p by setting

$$p(x) = \begin{cases} p & \text{if } x \text{ or } -x \text{ is a unit vector,} \\ 0 & \text{otherwise.} \end{cases}$$

We have ruled out the possibility that $p(x) = 1$ to avoid certain cases of little intrinsic interest.

We define the *range* of a long-range process to be the integer M given by $M = \inf\{m: p(x) = 0 \text{ if } |x| > m\}$, with the convention that the infimum of the empty set is ∞. We call the process *finite-range* if $M < \infty$ and *infinite-range* otherwise. A finite-range process with range M may be reformulated as a nearest-neighbour process on the lattice obtained by decorating $\mathbb{Z}^d$ with edges between all pairs of points which are within distance M of each other. The ensuing process will generally be 'anisotropic', since different edges may have different probabilities of being open, but such anisotropy is unlikely to affect substantially the qualitative features of the process over long distances. On the other hand, the consequences of *infinite-range* interactions are somewhat less predictable, and it is upon such processes that we concentrate here.

It is perhaps in one dimension that the effect of infinite-range interactions is most startling. Whereas nearest-neighbour percolation on the line $\mathbb{L}$ has few (or no) properties of interest to mathematical physicists, infinite-range processes have rich and beautiful structure. This will not surprise those familiar with the analogous one-dimensional Ising model, of which we summarize briefly the properties. Let us consider the Ising model of ferromagnetism on the line $\mathbb{Z}$, in which each pair $\{u, v\}$ of points has an interaction with strength $J(|v-u|)$. It is elementary that there is no critical phenomenon if $J(n) = 0$ for all large n. The picture is much more interesting if the interaction has infinite range. For the case when $J(n) \sim n^{-\alpha}$ as $n \to \infty$, it turns out that the process is totally ordered at all temperatures if $\alpha \leq 1$, and disordered at all temperatures if $\alpha > 2$. There is a phase transition if $1 < \alpha \leq 2$, in the sense that there is a critical temperature above which there is no magnetization, and below which the spontaneous magnetization is non-zero (see Ruelle (1968), Dyson (1969a, b, 1971), and Fröhlich and Spencer (1982)). Furthermore, the phase transition is of the first order when $J(n) \sim n^{-2}$, which is to say that the spontaneous magnetization is a discontinuous function of temperature at the critical point (see Aizenman, Chayes, Chayes, and Newman (1988)).

What are the corresponding results for long-range percolation in one dimension? We may think of the probability vector $\mathbf{p}$ as being a vector $(p(n): n \geq 1)$ of numbers lying in the interval $[0, 1)$, so that each pair $\{u, v\}$ of points in $\mathbb{Z}$ is joined by an (open) edge with probability $p(|v-u|)$. The resulting graph G is a random graph with vertex set $\mathbb{Z}$, and the problem is to relate the existence (or not) of infinite components to the particular choice of probability vector $\mathbf{p}$.

It is not surprising that all the components of G are finite if the $p(n)$'s are sufficiently small (such conclusions should be interpreted with the caveat 'almost surely'). We describe two sufficient conditions for G to contain no infinite component. Let A_k be the event that no vertex u $(\leq k)$ is joined to any vertex v $(\geq k+1)$. The sequence $(I_{A_k}: -\infty < k < \infty)$ of indicator functions is stationary with trivial tail σ-field, and the common mean value of these random variables is $P(A_0)$. There are exactly n edges of the form $\langle u, u+n \rangle$ where $u \leq 0$ and $u + n \geq 1$, and therefore

$$\begin{aligned} P(A_0) &= 1 - \prod_{n=1}^{\infty} \{1 - p(n)\}^n \\ &\geq 1 - \exp\left(- \sum_{n=1}^{\infty} np(n) \right). \end{aligned}$$

Suppose now that $\sum_n np(n) < \infty$, so that $P(A_0) > 0$. We apply the ergodic theorem to the stationary sequence of indicator functions to deduce that infinitely many of the A_k's occur, with probability 1, and thus all components of G are finite (see Figure 10.4). We have proved

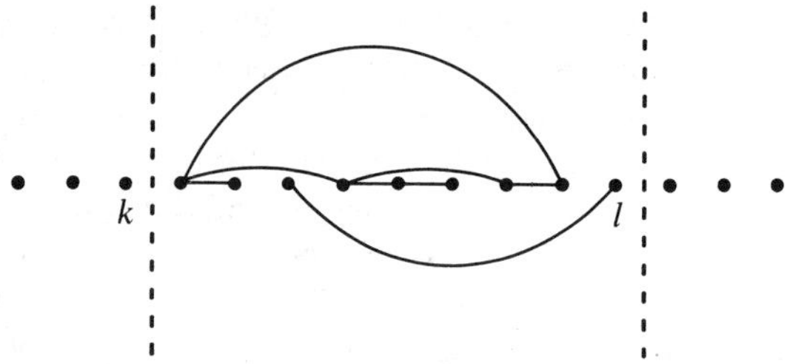

Figure 10.4. If A_k and A_l occur then no vertex in $\{k+1, k+2, \ldots, l\}$ is joined to any vertex outside this set.

that G contains no infinite component if

(10.1) $$\sum_{n=1}^{\infty} np(n) < \infty.$$

For the second such condition, we compare the component at the origin with a branching process; it is not difficult to make the following argument rigorous. The mean number of vertices which are adjacent to the origin is $\mu = 2\sum_n p(n)$. Each such vertex v is adjacent to a certain number of new vertices (that is, vertices which are not adjacent to the origin), and the mean number of such new vertices is at most μ. We may build the component at the origin by generations; in the resulting branching tree, each point has a family with mean size not exceeding μ. Therefore the size of the component at the origin is no larger in distribution than the size of a certain branching process with mean family-size μ. If $\mu \le 1$, such a branching process is (almost surely) finite, and the component at the origin follows suit; the special case when $\mu = 1$ poses no serious difficulty. It is a small step to deduce that all components of G are finite if

(10.2) $$2\sum_{n=1}^{\infty} p(n) \le 1.$$

The above discussion appeared in Schulman (1983).

We turn next to conditions which are sufficient for G to contain an infinite component. The next result is not surprising when we remember that the mean degree of each point is $2\sum_n p(n)$.

(10.3) Theorem. *The graph G is almost surely connected if and only if*

(10.4) *the greatest common divisor of $\{n: p(n) > 0\}$ is* 1, *and*

(10.5) $$\sum_{n=1}^{\infty} p(n) = \infty.$$

Condition (10.4) requires that the probability vector $\mathbf{p}$ be 'aperiodic' in the sense that, for all pairs u, v of vertices, there is strictly positive probability that u and v are joined by a path of G; it is clear that this condition is necessary for G to be (almost surely) connected. Similarly, the necessity of (10.5) is obvious. To see this, we note that the probability that any given vertex is isolated equals

$$\prod_{n=1}^{\infty} \{1 - p(n)\}^2,$$

which is strictly positive if $\sum_n p(n) < \infty$; it follows that, under this condition, G contains (almost surely) infinitely many isolated vertices. The core of Theorem (10.3) is the assertion that, subject to the aperiodicity of $\mathbf{p}$, the graph G is almost surely connected if $\sum_n p(n) = \infty$. That is to say, if $\sum_n p(n) = \infty$ then not only does G contain an infinite component, but *every vertex belongs to this component.*

Theorem (10.3) was proved first by Grimmett, Keane, and Marstrand (1984) in the context of long-range models in d dimensions where $d \geq 1$. There is a considerably simpler proof due to Kalikow (see Kalikow and Weiss (1988)) which is valid (apparently) in the one-dimensional case only, and we present this proof at the end of this section. Kalikow has reached a somewhat stronger conclusion than that of the above theorem: his argument implies that, subject to (10.4) and (10.5), the random graph *G restricted to the half-line* $\{0, 1, 2, \ldots\}$ is almost surely connected. Theorem (10.3) is a corollary also of the uniqueness theorem for the infinite open cluster of long-range percolation on $\mathbb{L}^d$ where $d \geq 1$; see Aizenman, Kesten, and Newman (1987), and the simplified proof of Gandolfi, Grimmett, and Russo (1988) described in Section 6.2. The line of reasoning is the following. If $\sum_n p(n) = \infty$ then, by the second Borel–Cantelli lemma, each vertex has (almost surely) infinite degree; thus each vertex is in an infinite component. However there is a unique infinite component, and thus all vertices belong (almost surely) to the same component.

Here is an open problem. Grimmett, Keane, and Marstrand (1984) have proved the d-dimensional version of Theorem (10.3) for general d. Prove the corresponding result for the restricted space $\{0, 1, 2, \ldots\}^d$. The best result to date in this direction is that of Kesten (1988, unpublished) who has dealt with the half-space $H = \{0, 1, 2, \ldots\} \times \mathbb{Z}^{d-1}$.

We return to long-range percolation in one dimension, and specifically to the twilight zone containing probability vectors $\mathbf{p}$ satisfying

$$\sum_{n=1}^{\infty} np(n) = \infty, \qquad \tfrac{1}{2} < \sum_{n=1}^{\infty} p(n) < \infty.$$

We specialize to the class of vectors for which

$$p(n) \sim \beta n^{-\alpha} \quad \text{as} \quad n \to \infty \tag{10.6}$$

for some positive constants α and β. We have from the foregoing remarks that (almost surely)

$$\text{for} \quad \alpha \le 1, \qquad G \text{ is connected,}$$

$$\text{for} \quad \alpha > 2, \qquad G \text{ has no infinite component.}$$

As in the case of the one-dimensional Ising model, there is a critical phenomenon if $1 < \alpha \le 2$. In order to formulate the result correctly, we note that both short-range and long-range interactions are relevant to the occurrence of a critical phenomenon, and we parametrize the process accordingly in the following somewhat arbitrary manner. Let α and β be positive constants and let p satisfy $0 \le p < 1$. We consider a probability vector $\mathbf{p}$ satisfying

(10.7) $$p(1) = p, \qquad p(n) \sim \beta n^{-\alpha} \quad \text{as} \quad n \to \infty,$$

so that p measures the strength of interaction between nearest neighbours, and α and β measure the strengths of interactions over long distances. We write $\theta(p, \alpha, \beta)$ for the probability that the origin is in an infinite component of G, and we are concerned with the case $1 < \alpha \le 2$.

(10.8) Theorem. *Suppose that* $1 < \alpha < 2$ *and* $\beta > 0$. *There exists a critical value* $p_c(\alpha, \beta)$ *satisfying* $0 < p_c(\alpha, \beta) < 1$ *such that*

$$\theta(p, \alpha, \beta)\begin{cases} = 0 & \textit{if} \quad p < p_c(\alpha, \beta), \\ > 0 & \textit{if} \quad p > p_c(\alpha, \beta). \end{cases}$$

(10.9) Theorem. *Suppose that* $\alpha = 2$.

(a) *If* $\beta \le 1$ *then* $\theta(p, 2, \beta) = 0$ *for all* $p < 1$.
(b) *If* $\beta > 1$, *there exists a critical value* $p_c(2, \beta)$ *satisfying* $0 < p_c(2, \beta) < 1$ *such that*

$$\theta(p, 2, \beta)\begin{cases} = 0 & \textit{if} \quad p < p_c(2, \beta), \\ \ge \beta^{-1/2} & \textit{if} \quad p > p_c(2, \beta). \end{cases}$$

Thus there is a critical phenomenon if $1 < \alpha < 2$ or $\alpha = 2$, $\beta > 1$. In the latter case, the phenomenon is of the first order in the sense that $\theta(p, 2, \beta)$ is a discontinuous function of p at the critical point $p_c(\beta)$. These results are due to Newman and Schulman (1986) and Aizenman and Newman (1986), from the latter of which we reproduce in Figure 10.5 a picture of the phase diagram when $p(n) \sim \beta n^{-2}$.

One of the principal purposes of this section is to make available Kalikow's contribution to the proof of Theorem (10.3).

Partial proof of Theorem (10.3). We restrict ourselves to proving that G is almost surely connected if

(10.10) $$p(1) > 0$$

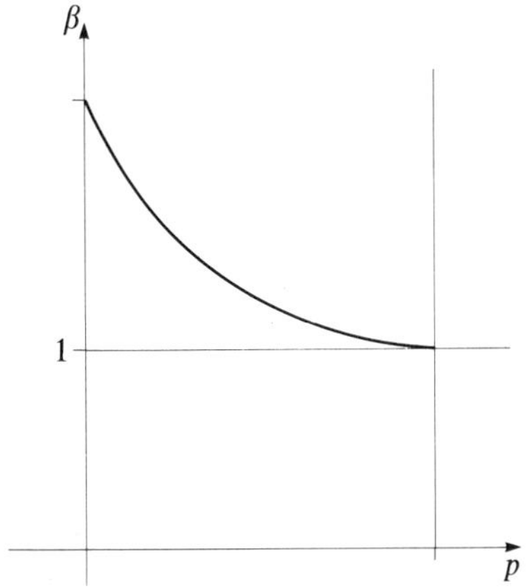

Figure 10.5. The curve $p = p_c(2, \beta)$. We have that $\theta(p, 2, \beta) = 0$ beneath the curve, and $\theta(p, 2, \beta) \geq \beta^{-1/2}$ above the curve. Thus θ is discontinuous as we cross the critical curve.

and

$$\sum_{n=1}^{\infty} p(n) = \infty. \tag{10.11}$$

Inequality (10.10) is the simplest condition which implies the aperiodicity of the probability vector $\mathbf{p}$; we refer the reader to Grimmett, Keane, and Marstrand (1984) for the extra details in the general case. It is not hard to make rigorous the remarks after the statement of the theorem, and for that reason we omit also the proof that (10.11) is necessary for the connectedness of G. In partial compensation for these omissions we shall prove a stronger conclusion than is required, namely that (10.10) and (10.11) imply the almost-sure connectedness of the random graph G^+ on the restricted vertex set $\{0, 1, 2, \ldots\}$ with edges distributed at random in the usual way. The result for the doubly infinite vertex set $\mathbb{Z}$ will be an immediate consequence, by the following argument. If the two random graphs on the vertex sets $\{0, 1, 2, \ldots\}$ and $\{\ldots, -3, -2, -1\}$ are connected, then G is disconnected if and only if there is no edge of the form $\langle u, v \rangle$ with $u < 0$ and $v \geq 0$. This event has probability

$$\prod_{n=1}^{\infty} \{1 - p(n)\}^n,$$

which equals 0, by assumption (10.11).

Let G_n be the graph G restricted to the vertex set $\{0, 1, 2, \ldots, n\}$, and let A_n be the event that the vertices 0 and 1 are in the same component of G_n; that is to say, $A_n = \{0 \leftrightarrow 1 \text{ in } G_n\}$. We note that $A_n \subseteq A_{n+1}$ for all n, so that $P(A_n) \to P(\lim_{n\to\infty} A_n) = P(0 \leftrightarrow 1 \text{ in } G^+)$ as $n \to \infty$. We shall prove that

$$P(A_n) \to 1 \quad \text{as} \quad n \to \infty, \tag{10.12}$$

implying that there exists almost surely a path from 0 to 1 in G^+; the required conclusion will follow quickly from this. We write B_n for the event that there exists in G_n some edge of the form $\langle 0, i \rangle$ for $2 \le i \le n$; the probability that B_n does not occur is

$$P(B_n^c) = \prod_{i=2}^{n} \{1 - p(i)\}, \tag{10.13}$$

which satisfies

$$P(B_n^c) \to 0 \quad \text{as} \quad n \to \infty, \tag{10.14}$$

by (10.11).

Now,

$$P(A_n^c) \le P(0 \not\leftrightarrow 1 \text{ in } G_n'), \tag{10.15}$$

where G_n' is the random graph obtained from G_n by removing the edge $\langle 0, 1 \rangle$ if it exists; we note that G_n' does not depend on the existence or non-existence of the edge $\langle 0, 1 \rangle$. It is the case that

$$P(A_n^c) \le P(0 \not\leftrightarrow 1 \text{ in } G_n', B_n) + P(B_n^c), \tag{10.16}$$

of which the last term is controlled in (10.14). To control the first term, we argue as follows. If the event $\{0 \not\leftrightarrow 1 \text{ in } G_n', B_n\}$ occurs, and in addition the edge $\langle 0, 1 \rangle$ exists, then G_n has strictly fewer components than the number C_{n-1} of components of the random graph on the smaller vertex set $\{1, 2, 3, \dots, n\}$. Writing K_n for the number of components of G_n, we have that

$$\begin{aligned} p(1)P(0 \not\leftrightarrow 1 \text{ in } G_n', B_n) &\le P(K_n < C_{n-1}) \\ &= P(K_n < K_{n-1}), \end{aligned}$$

since the pair (K_n, C_{n-1}) has the same joint distribution as the pair (K_n, K_{n-1}). We substitute this into (10.16), remembering that the K_n's are integer-valued, to obtain

$$P(A_n^c) \le \frac{1}{p(1)} P(K_{n-1} - K_n \ge 1) + P(B_n^c). \tag{10.17}$$

Now $K_{n-1} - K_n \ge -1$, since the inclusion of one new vertex increases the number of components by 1 at most. Therefore

$$\begin{aligned} P(K_{n-1} - K_n \ge 1) &= \sum_{i=1}^{\infty} P(K_{n-1} - K_n = i) \\ &\le E(K_{n-1} - K_n) + P(K_{n-1} - K_n = -1) \\ &\le E(K_{n-1} - K_n) + P(B_n^c), \end{aligned} \tag{10.18}$$

since $K_{n-1} - K_n = -1$ if and only if vertex n is isolated in G_n, an event with probability $\{1 - p(1)\}P(B_n^c)$.

We wish to take the limit in (10.17) and (10.18) as $n \to \infty$. It is the case that

$$\liminf_{n\to\infty} E(K_{n-1} - K_n) \le 0 \tag{10.19}$$

since, if $\liminf_{n\to\infty} E(K_{n-1} - K_n) > \varepsilon$ where $\varepsilon > 0$, then $E(K_{n-1} - K_n) > \frac{1}{2}\varepsilon$ for all but finitely many values of n, which implies that

$$E(K_n) < E(K_N) - \tfrac{1}{2}\varepsilon(n - N)$$

for some N and all $n \ge N$; this is impossible since $E(K_n) \ge 0$ for all n. We choose a subsequence along which $E(K_{n-1} - K_n)$ approaches $\liminf_{n\to\infty} E(K_{n-1} - K_n)$, and we take the limit along this subsequence to obtain from (10.17) and (10.18) that

$$\begin{aligned}\liminf_{n\to\infty} P(A_n^c) &\le \frac{1}{p(1)} \liminf_{n\to\infty} E(K_{n-1} - K_n) \qquad \text{by (10.14)}\\ &\le 0.\end{aligned}$$

However $\lim_{n\to\infty} P(A_n^c)$ exists, since $A_{n+1}^c \subseteq A_n^c$ for all n, and thus (10.12) is proved.

Only a detail remains. If G^+ is disconnected, then there exists a pair x, $x + 1$ of non-negative integers such that x and $x + 1$ are in different components of G^+. Such a pair x, $x + 1$ has the property that they are not joined by a path in the subgraph of G^+ on the vertex set $\{x, x + 1, x + 2, \ldots\}$, and therefore

$$\begin{aligned}P(G^+ \text{ is disconnected}) &\le \sum_{x=0}^{\infty} P(x \nleftrightarrow x + 1 \text{ in } \{x, x + 1, \ldots\})\\ &= \sum_{x=0}^{\infty} P(0 \nleftrightarrow 1 \text{ in } \{0, 1, \ldots\})\end{aligned}$$

by translation invariance. The final summand equals 0, and the proof is complete. ∎

10.4 Surfaces in Three Dimensions

We return to bond percolation on $\mathbb{L}^d$. The main weapon in two dimensions is planar duality. Using duality, we have seen that the critical point $p_c(2)$ marks the crossover between the existence of infinite closed clusters on the dual lattice and the existence of infinite open clusters on the primal lattice. The picture is not so clear in three dimensions.

Whereas in two dimensions every finite open cluster is surrounded by a closed *circuit* of the dual, such clusters in three dimensions are sur-

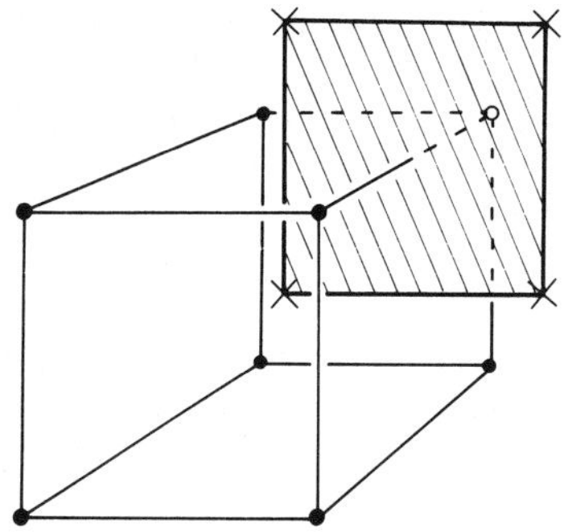

Figure 10.6. An elementary cube in $\mathbb{L}^3$ together with a dual plaquette. The crosses are vertices of the dual lattice, placed at the centres of the elementary cubes. Each edge of $\mathbb{L}^3$ passes through the centre of a face of the dual lattice.

rounded by closed *surfaces* on an appropriately defined lattice. We construct the dual lattice $\mathbb{L}^3_{\mathrm{d}}$ of $\mathbb{L}^3$ as follows. We take as vertices of $\mathbb{L}^3_{\mathrm{d}}$ the set $\mathbb{Z}^3 + (\frac{1}{2}, \frac{1}{2}, \frac{1}{2}) = \{(x_1 + \frac{1}{2}, x_2 + \frac{1}{2}, x_3 + \frac{1}{2}): x \in \mathbb{Z}^3\}$ of centres of the fundamental unit cubes of $\mathbb{L}^3$. We join two such vertices u and v by an edge whenever $\sum_{i=1}^3 |u_i - v_i| = 1$. Thus $\mathbb{L}^3_{\mathrm{d}}$ is isomorphic to $\mathbb{L}^3$, and each edge of $\mathbb{L}^3$ passes through the centre of a unit face of $\mathbb{L}^3_{\mathrm{d}}$; following Aizenman, Chayes, Chayes, Fröhlich, and Russo (1983), we call such faces *plaquettes*. Figure 10.6 is a sketch of a small part of $\mathbb{L}^3$ and its dual.

Let us consider bond percolation on $\mathbb{L}^3$ with edge-probability p. We declare each plaquette of the dual lattice $\mathbb{L}^3_{\mathrm{d}}$ to be *open* (respectively *closed*) if the corresponding edge of $\mathbb{L}^3$ is open (respectively closed). Whereas in two dimensions we were interested in the existence of closed circuits in the dual, in three dimensions we study instead the existence of surfaces of closed plaquettes.

Each plaquette π has four bounding edges whose union we call the *boundary* of π, denoted by $\partial\pi$. Every collection $\Pi = (\pi_i: 1 \le i \le m)$ of distinct plaquettes determines a *surface*

$$S_\Pi = \bigcup_{i=1}^{m} \pi_i$$

together with a *boundary* ∂S_Π, defined as the set of edges of $\mathbb{L}^3_{\mathrm{d}}$ which belong to an odd number of plaquettes in Π. For the sake of brevity and concreteness, we shall study here the existence of surfaces of the following form only. Let c_n be the square boundary of the set $\{(x_1 + \frac{1}{2}, x_2 + \frac{1}{2}, \frac{1}{2}): 0 \le x_1, x_2 \le n\}$ of vertices of $\mathbb{L}^3_{\mathrm{d}}$; c_n is a circuit of $\mathbb{L}^3_{\mathrm{d}}$ with a particularly simple form. We denote by C_n the event that there exists a collection Π of closed plaquettes whose boundary ∂S_Π is c_n. Thus C_n is the event that the circuit c_n is 'spanned' by a surface of closed plaquettes.

Certain topological questions raise their heads at this point: do we want merely 'surfaces', or might we be more interested in orientable surfaces, or perhaps surfaces homeomorphic to the unit disc? We shall not discuss such questions here, but refer the reader to Aizenman, Chayes, Chayes, Fröhlich, and Russo (1983).

How does the probability of C_n behave as a function of the parameter p? It is easy to see that

$$P_p(C_n) \geq (1-p)^{n^2}, \tag{10.20}$$

since C_n occurs whenever all the n^2 plaquettes in the smallest surface spanning c_n are closed. On the other hand

$$P_p(C_n) \leq (1-p^4)^{4(n-1)}, \tag{10.21}$$

since, if C_n occurs, then at least one of the four plaquettes abutting each edge of c_n is closed. The bounds in (10.20) and (10.21) have qualitatively different asymptotic behaviours as $n \to \infty$, the lower bound decaying exponentially at a rate having the order of the *area* n^2 of c_n, and the upper bound at a rate having the order of its *perimeter length* $4n$. What is the true asymptotic behaviour of $P_p(C_n)$ for a given value of p? Closed surfaces of $\mathbb{L}_\text{d}^3$ cut off open paths of $\mathbb{L}^3$, and we are led to conjecture that the answer to the question depends on the existence (or not) of an infinite open cluster in the primal lattice $\mathbb{L}^3$.

(10.22) Theorem.

(a) Perimeter law. *If* $0 < p < p_c$, *there exists* $\alpha(p)$ *satisfying* $0 < \alpha(p) < \infty$ *such that*

$$P_p(C_n) \approx e^{-n\alpha(p)} \qquad \textit{as} \quad n \to \infty.$$

(b) Area law. *If* $p_c^+ < p < 1$, *there exists* $\beta(p)$ *satisfying* $0 < \beta(p) < \infty$ *such that*

$$P_p(C_n) \approx e^{-n^2\beta(p)} \qquad \textit{as} \quad n \to \infty.$$

In this theorem, p_c and p_c^+ are the critical probabilities of the whole lattice $\mathbb{L}^3$ and the half-space $\{0, 1, 2, \dots\} \times \mathbb{Z}^2$, respectively. It is conjectured that $p_c^+ = p_c$. We recall that '$a_n \approx b_n$' means that $\log a_n / \log b_n \to 1$.

Thus $\log P_p(C_n)$ is roughly proportional to the area of c_n when the density $1-p$ of closed plaquettes is small, and to the length of its perimeter when the density is large. Subject to the usual conjecture concerning the critical probability of the half-space, the critical density of closed plaquettes is therefore $1 - p_c$.

There is another percolation-type phenomenon associated with plaquettes: the formation of infinite connected clusters of closed plaquettes (with the obvious notion of adjacency). It is not difficult to see that this phenomenon occurs at a critical plaquette-density which is quite different from $1 - p_c$. By taking a horizontal slice through the dual lattice we may satisfy ourselves that the critical density for the formation of infinite clusters of plaquettes is no more than $\frac{1}{2}$, whereas $p_c < \frac{1}{2}$ and therefore $1 - p_c > \frac{1}{2}$.

The results of this section appeared in Aizenman, Chayes, Chayes, Fröhlich, and Russo (1983), who considered also their generalizations to higher dimensions. They proved the area law for values of p exceeding the limit of the critical probabilities of slabs; Barsky, Grimmett, and Newman (1988) have proved that this limit equals p_c^+. See Kesten (1986a, 1987a) for related work.

10.5 Snails on a Lily Pond

Young lilies inhabit a beautiful large square pond. They grow at a uniform rate. How long need a snail wait beside the pond before it becomes possible for him to traverse the pond without getting his feet wet? This question may be rephrased in terms of the following classical problem of stochastic geometry: ascertain the minimal density of unit discs in the plane which guarantees the existence of an infinite cluster. It leads to a percolation-type process which has been dubbed the 'Poisson blob model' by the less poetically minded. A picture of the snail's route appears in Figure 10.7.

Let $(X_i: i \in I)$ be the points of a Poisson process with intensity λ in $\mathbb{R}^d$, where $d \geq 2$. For each $i \in I$, we position a closed d-dimensional sphere of

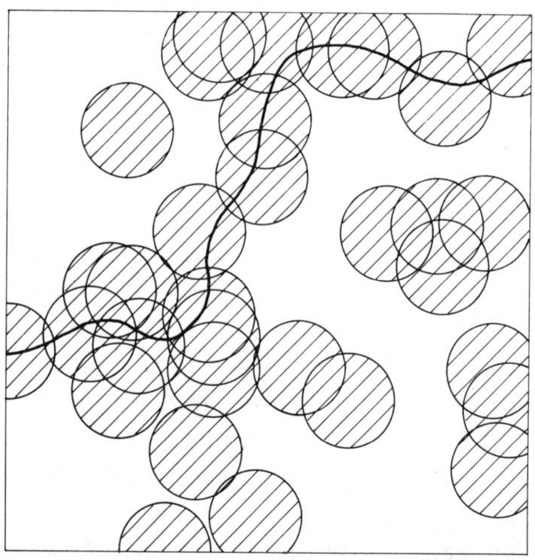

Figure 10.7. The thick line indicates a possible dry crossing of the lily pond. The centres of the circles are the points of a Poisson process. When the circles are sufficiently large, there exists a crossing of the box which is contained in the union of the circles.

radius ρ with its centre at X_i. Writing S_i for the sphere with centre X_i, we have that $S_i = S + X_i$, where

$$S = \left\{x \in \mathbb{R}^d: \sum_{i=1}^{d} x_i^2 \leq \rho^2\right\}.$$

We are interested in the shapes of clusters of overlapping spheres, and particularly in the possible existence of infinite clusters.

We call two spheres S_i and S_j *adjacent* if $S_i \cap S_j \neq \varnothing$. We write $S_i \leftrightarrow S_j$ if there exists a sequence $S_{i_1}, S_{i_2}, \ldots, S_{i_k}$ of spheres such that $S_{i_1} = S_i$, $S_{i_k} = S_j$, and S_{i_j} is adjacent to $S_{i_{j+1}}$ for $1 \leq j < k$. A *cluster* of spheres is a set $(S_i: i \in J)$ of spheres which is maximal with the property that $S_i \leftrightarrow S_j$ for all $i, j \in J$. We identify a cluster with the set of centres of the spheres in the cluster. The *size* of a cluster is thus the number of spheres belonging to it.

We wish to study the sizes and shapes of clusters for specified values of λ and ρ. From the re-scaling property of the Poisson process, the cluster shapes have the same distributions (except for a change in the length scale) whenever the parameters have the form $\alpha^{-d}\lambda$ and $\alpha\rho$, as α ranges over $(0, \infty)$. We may therefore assume without loss of generality that $\rho = 1$, so that the unique parameter of the process is the intensity λ of points.

From our experience of bond percolation on $\mathbb{L}^d$, we can be fairly sure that there exists a non-trivial critical intensity λ_c such that (almost surely) there is an infinite cluster when $\lambda > \lambda_c$ and not when $\lambda < \lambda_c$. We shall prove this and rather more, although we do not attempt a complete description of the process here.

As for notation, we write P^λ for the probability measure associated with a Poisson process with intensity λ, and E^λ for the corresponding expectation operator. We denote by $W(x)$ the cluster containing the point x of $\mathbb{R}^d$; more formally, $W(x)$ is defined to be the set of points in the Poisson process which are centres of spheres in the cluster containing x. The *percolation probability* $\gamma(\lambda)$ is defined by

$$\gamma(\lambda) = P^\lambda(|W| = \infty), \tag{10.23}$$

where $W = W(0)$ is the cluster at the origin. It is clear from the argument leading to Theorem (1.11) that

$$P^\lambda(\text{there exists } x \text{ with } |W(x)| = \infty) = \begin{cases} 0 & \text{if } \gamma(\lambda) = 0, \\ 1 & \text{if } \gamma(\lambda) > 0. \end{cases}$$

(10.24) Theorem. *There exists λ_c satisfying $0 < \lambda_c < \infty$ such that*

$$\gamma(\lambda) \begin{cases} = 0 & \textit{if } \lambda < \lambda_c, \\ > 0 & \textit{if } \lambda > \lambda_c. \end{cases} \tag{10.25}$$

Furthermore, the mean cluster size $E^\lambda|W|$ is finite whenever $\lambda < \lambda_c$.

The proof of this theorem is due to Zuev and Sidorenko (1985a, b); see also Menshikov, Molchanov, and Sidorenko (1986) and Menshikov and Sidorenko (1987). The main idea is to approximate to the 'continuous' problem by site percolation problems on special lattices constructed by partitioning $\mathbb{R}^d$ into small cubes. This approximation technique may be used to explore many other properties of the continuous problem. For example, it is not difficult to show that

$$\zeta(\lambda) = \lim_{n\to\infty} \left\{ -\frac{1}{n} \log P^{\lambda}(|W| = n) \right\}$$

exists and is strictly positive when $0 < \lambda < \lambda_c$. Furthermore, there is almost surely a *unique* infinite cluster of overlapping spheres whenever $\gamma(\lambda) > 0$.

Numerous generalizations of the problem spring readily to mind, and some are discussed by Zuev and Sidorenko (1985a, b) and Menshikov and Sidorenko (1987). Interesting phenomena may occur if the random sets are permitted to be large, and Hall (1985) has shown that there exist such processes for which the mean cluster size and the percolation probability have quite different critical behaviours. Suppose that a sphere of random radius is positioned at each point of a Poisson process with intensity λ; we assume that the radii of the spheres are independent and identically distributed random variables. Hall has shown that there exists a non-trivial critical intensity λ_T for the divergence of the mean cluster size if and only if a typical sphere-radius R satisfies $E(R^{2d}) < \infty$; if $E(R^{2d}) = \infty$ then the mean cluster size is infinite for all $\lambda > 0$. On the other hand, if $E(R^{2d-1}) < \infty$ then all clusters are almost surely finite for all positive but sufficiently small values of λ. If $E(R^{2d-1}) < \infty$ but $E(R^{2d}) = \infty$ then the mean cluster size is infinite for all $\lambda > 0$ but there exists $\lambda_c > 0$ such that all clusters are almost surely finite when $\lambda < \lambda_c$.

Further references are given by Kesten (1987e), who discusses other applications to polymerization, cluster analysis, and communication systems. See Roy (1987) for more details of the two-dimensional case, and Hall (1988) for an account of the general theory of 'coverage' processes.

Proof of Theorem (10.24). There is a natural way of relating the continuous-space process to an ordinary site percolation process. Let n be a positive integer and let $\mathbb{Z}_n^d$ be the set $n^{-1}\mathbb{Z}^d$; that is to say

$$\mathbb{Z}_n^d = \{(x_1/n, x_2/n, \dots, x_d/n) : x \in \mathbb{Z}^d\}.$$

Later we shall take the limit as $n \to \infty$. We partition $\mathbb{R}^d$ into cubes whose centres are at the points of $\mathbb{Z}_n^d$, defining

$$B_n(x) = \prod_{i=1}^{d} \left[x_i - \frac{1}{2n}, x_i + \frac{1}{2n} \right) \qquad \text{for} \quad x \in \mathbb{Z}_n^d.$$

We turn $\mathbb{Z}_n^d$ into a lattice $\mathscr{L}_n$ by defining the adjacency relation $\sim$ on $\mathbb{Z}_n^d$ with the rule that $x \sim y$ if and only if there exist points $u \in B_n(x)$ and $v \in B_n(y)$ such that

$$\left(\sum_{i=1}^{d} (u_i - v_i)^2\right)^{1/2} \leq 2. \tag{10.26}$$

We shall consider site percolation on the ensuing lattice $\mathscr{L}_n$. We declare a vertex x of $\mathscr{L}_n$ to be *open* if there exist one or more points of the Poisson process in the cube $B_n(x)$, and *closed* otherwise. It is elementary that the states of different vertices are independent random variables, and that the probability $p_n(\lambda)$ that any given vertex is open is given by

$$\begin{aligned} p_n(\lambda) &= 1 - P^{\lambda}(\text{no point in } B_n(0)) \\ &= 1 - e^{-\lambda n^{-d}}. \end{aligned} \tag{10.27}$$

We shall make liberal use of the properties of site percolation on $\mathscr{L}_n$ with no greater proof than the imprecation to adapt the arguments given earlier for bond percolation on $\mathbb{L}^d$. Thus we shall assume for example that this process has a non-trivial critical probability, and that the mean cluster size is finite throughout the subcritical phase.

Each point X of the Poisson process lies in some unique elementary cube $B_n(x)$ of $\mathscr{L}_n$ and we write $x = \pi_n(X)$. The function π_n is a many–one mapping from the points of the Poisson process into the set $\mathbb{Z}_n^d$ of vertices of $\mathscr{L}_n$. If S_1 and S_2 are two intersecting spheres with centres X_1 and X_2, then the euclidean distance between X_1 and X_2 is no greater than 2, giving by the definition of the adjacency relation of $\mathscr{L}_n$ that $\pi_n(X_1) \sim \pi_n(X_2)$; see Figure 10.8. If follows that, if the origin is in an infinite cluster of spheres, then there exists an infinite open cluster in $\mathscr{L}_n$. In terms of the critical intensity

$$\lambda_c = \sup\{\lambda \colon \gamma(\lambda) = 0\} \tag{10.28}$$

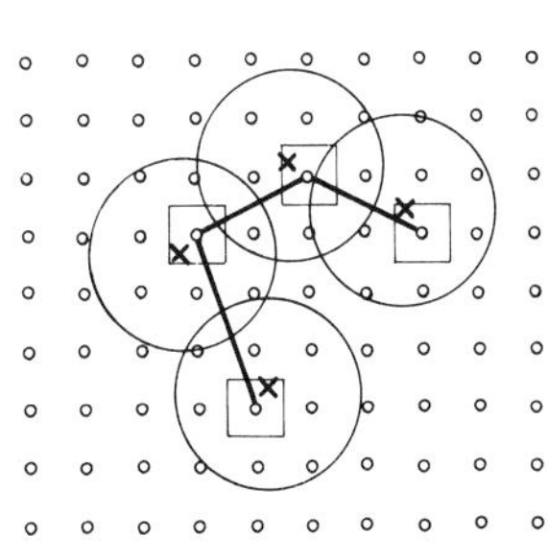

Figure 10.8. The crosses mark the centres of four overlapping circles. Each cross is associated with the vertex closest to it, and the adjacency relation of the lattice $\mathscr{L}_n$ has been chosen so that two such vertices are adjacent whenever the corresponding circles overlap.

and the critical probability $p_c(\mathscr{L}_n)$ of site percolation on $\mathscr{L}_n$, we have proved that $p_n(\lambda) \geq p_c(\mathscr{L}_n)$ whenever $\lambda > \lambda_c$. We rewrite this implication with the aid of (10.27) as

(10.29) $$\text{`}\lambda > \lambda_c\text{'} \quad \text{implies} \quad \text{`}\lambda \geq -n^d \log\{1 - p_c(\mathscr{L}_n)\}\text{'},$$

which is to say that

(10.30) $$\lambda_c \geq -n^d \log\{1 - p_c(\mathscr{L}_n)\}.$$

However, $0 < p_c(\mathscr{L}_n) < 1$, and therefore $\lambda_c > 0$. Actually, the foregoing argument was largely for practice only, since inequality (10.30) is not itself of great value in studying the mean cluster size $E^\lambda|W|$. Of greater value would be the corresponding inequality for

(10.31) $$\lambda_T = \sup\{\lambda\colon E^\lambda|W| < \infty\}.$$

Clearly

(10.32) $$\lambda_T \leq \lambda_c$$

and our principal object is to prove that equality holds here. In order to obtain the inequality for λ_T corresponding to (10.30), we need to relate the mean number of spheres in W to the mean size of open clusters of $\mathscr{L}_n$. Let D_n be the set of open vertices of $\mathscr{L}_n$ which are joined by open paths to either the origin or some vertex in the set $\{x \in \mathbb{Z}_n^d\colon x \sim 0\}$ of neighbours of the origin, and suppose that $|D_n| = m$. Each vertex x of D_n is the centre of a cube $B_n(x)$ which contains at least one point of the Poisson process. The mean number of such points in any such cube is therefore $E(N|N \geq 1)$, where N is a random variable having the Poisson distribution with parameter λn^{-d}; hence

$$E^\lambda|\pi_n^{-1}(D_n)| = \sum_{m=0}^{\infty} mE(N|N \geq 1)P^\lambda(|D_n| = m)$$

$$= \frac{\lambda n^{-d}}{1 - e^{-\lambda n^{-d}}} E^\lambda|D_n|$$

by an elementary calculation. The origin of $\mathscr{L}_n$ has at most $(4n)^d$ neighbours, and therefore

$$E^\lambda|D_n| \leq \{1 + (4n)^d\} E^\lambda|C_n|,$$

where C_n is the open cluster of $\mathscr{L}_n$ at the origin. It is the case that $W \subseteq \pi_n^{-1}(D_n)$, by the remarks prior to (10.28), and therefore

$$E^\lambda|W| \leq \frac{\lambda n^{-d}\{1 + (4n)^d\}}{\{1 - e^{-\lambda n^{-d}}\}} E^\lambda|C_n|.$$

It follows that $E^\lambda|C_n| = \infty$ whenever $E^\lambda|W| = \infty$, which is to say that

(10.33) $$\lambda_T \geq -n^d \log\{1 - p_c(\mathscr{L}_n)\}$$

as in (10.30); we have used the fact that $E^{\lambda}|C_n| = \infty$ only when $p \geq p_c(\mathcal{L}_n)$.

We seek next an upper bound for λ_c, and intend to prove that

$$\lambda_c \leq -l_n^d n^d \log\{1 - p_c(\mathcal{L}_n)\}, \tag{10.34}$$

where $l_n = 1 + n^{-1}\sqrt{d}$. It is an immediate consequence of (10.32)–(10.34) that

$$\lambda_T = \lambda_c = \lim_{n\to\infty} [-n^d \log\{1 - p_c(\mathcal{L}_n)\}], \tag{10.35}$$

using the fact that $l_n \to 1$ as $n \to \infty$. Incidentally, (10.34) implies that $0 < \lambda_c < \infty$. We now prove (10.34). We have from the definition of the adjacency relation of $\mathcal{L}_n$ that

$$\left(\sum_{i=1}^{d} (u_i - v_i)^2\right)^{1/2} \leq 2l_n$$

whenever $u \in B_n(x)$, $v \in B_n(y)$, and $x \sim y$; just note that the diagonal of the typical elementary cube $B_n(0)$ has length $n^{-1}\sqrt{d}$. Suppose that we replace each sphere of radius 1 by a sphere of radius l_n. If x and y are adjacent open vertices of $\mathcal{L}_n$, then *all* such enlarged spheres with centres in $B_n(x)$ intersect *all* such enlarged spheres with centres in $B_n(y)$; see Figure 10.9. Thus the cluster of spheres (with radius l_n) at the origin is infinite whenever the open cluster of $\mathcal{L}_n$ at the origin is infinite. However, the process of spheres with radius l_n is equivalent to a process of unit-radius spheres

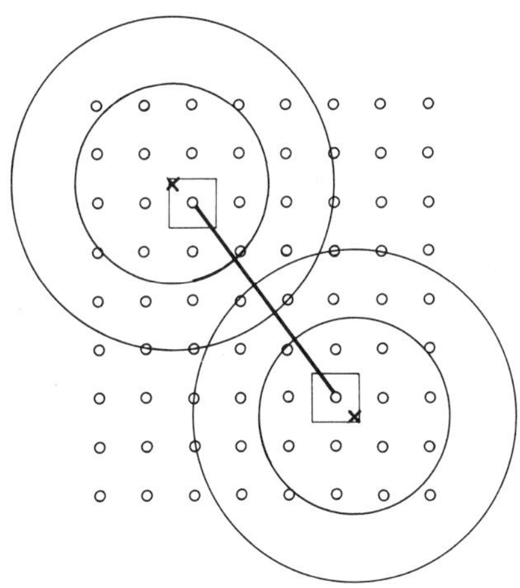

Figure 10.9. Here are two vertices of the lattice $\mathcal{L}_n$ which are adjacent even though there exist corresponding points (marked with crosses) which are separated by a distance exceeding 2. Unit circles centred at these points do not overlap, but concentric circles with increased radius $l_n = 1 + n^{-1}\sqrt{d}$ do overlap.

with an amended intensity λl_n^d. We now see as in (10.29) that

$$\text{`}p_n(\lambda) > p_c(\mathscr{L}_n)\text{'} \qquad \text{implies} \qquad \text{`}\lambda l_n^d \geq \lambda_c\text{'},$$

which is to say that

$$-n^d \log\{1 - p_c(\mathscr{L}_n)\} \geq l_n^{-d}\lambda_c$$

as required in (10.34). ■

10.6 Invasion Percolation

Suppose that fluid is pumped under pressure into a random medium, and that each unit of the fluid displaces the least resistant unit of neighbouring background material. The following dynamic invasion process has received some attention as a model for such a situation. Let $(X(e): e \in \mathbb{E}^d)$ be independent random variables indexed by the edge set of $\mathbb{L}^d$, each having the uniform distribution on [0, 1]. We construct a sequence $\mathbf{C} = (C_i: i \geq 0)$ of random connected subgraphs of the lattice in the following way. The graph C_0 contains the origin and no edges. Having defined C_i, we obtain C_{i+1} by adding to C_i the edge e_{i+1} for which $X(e_{i+1}) = \min\{X(e): e \in \Delta C_i\}$; that is to say, we examine the edges in the edge boundary of C_i, and add to C_i the edge with the smallest X-value, together with any endvertex of this edge not already belonging to C_i.

A principal object of study is the empirical distribution function of the set $(X(e_j): 1 \leq j \leq n)$ of X-values of edges in C_n, and particularly the limiting behaviour of this function as $n \to \infty$. Thus we define

$$Q_n(y) = \frac{1}{n} \sum_{j=1}^{n} I_{\{X(e_j) \leq y\}}$$

to be the proportion of edges e in C_n for which $X(e) \leq y$. The following intuitive argument is useful in understanding the behaviour of Q_n for large n. We call an edge e *p-open* if $X(e) < p$. Let π be a little larger than p_c, so that the lattice $\mathbb{L}^d$ contains (almost surely) an infinite π-open cluster. Once the invading fluid has reached some vertex in this cluster, then it need never subsequently invade an edge with X-value exceeding π, since it is provided with an infinite supply of edges with X-values no greater than π. This motivates the guess that

(10.36) $$Q_n(y) \to 1 \qquad \text{if} \quad y > p_c.$$

On the other hand, as the fluid devours this infinite π-open cluster, the X-values of the edges therein have a conditional distribution which is uniform on $[0, \pi]$, suggesting that

(10.37) $$Q_n(y) \to y/p_c \qquad \text{if} \quad y < p_c.$$

Chayes, Chayes, and Newman (1985) have proved these results, subject to one of the standard conjectures of percolation. We state part of their conclusions.

(10.38) Theorem.

(a) *It is the case that, as* $n \to \infty$,

$$Q_n(y) \to 1 \qquad \textit{for} \quad p_c^+ \leq y \leq 1,$$

where p_c^+ *is the critical probability of bond percolation on the half-space.*

(b) *If* $p_c = p_c^+$ *then, as* $n \to \infty$,

$$Q_n(y) \to y/p_c \qquad \textit{for} \quad 0 \leq y \leq p_c.$$

In practice, background material cannot be forced out of an edge by the invading fluid if the edge in question is already 'trapped' in a cavity of the invading cluster. 'Invasion percolation with trapping' takes this fact into account; see Chayes, Chayes, and Newman (1988).

For more motivation and information see de Gennes and Guyon (1978), Lenormand and Bories (1980), Chandler, Koplick, Lerman, and Willemsen (1982), Wilkinson and Willemsen (1983), Chayes, Chayes, and Newman (1985, 1987), and Kesten (1987e).

10.7 First-Passage Percolation

First-passage percolation is the half-brother of ordinary percolation. It was formulated by Hammersley and Welsh (1965) as a time-dependent model for the passage of fluid through a porous medium, and has since been a source of probabilistic problems endowed with both excellent motivation and mathematical appeal; some of the best of these questions remain unanswered. In addition, its birth witnessed the conception of the idea of a subadditive stochastic process, of which the ergodic theorem is now a major weapon of standard issue to graduate students in probability. Only inadequate justice can be done to the subject in a few pages only, and we refer the reader to the monograph of Smythe and Wierman (1978) and the recent work of Kesten (1986a, 1987a, e) for more detailed information and references.

To each edge e of $\mathbb{L}^d$ we allocate a random *time coordinate* $T(e)$, which we think of as being the time required for fluid to flow along e; we assume that the family $(T(e): e \in \mathbb{E}^d)$ of time coordinates contains independent non-negative random variables with common distribution function F. For any path π we define the *passage time* $T(\pi)$ of π by

$$T(\pi) = \sum_{e \in \pi} T(e),$$

the sum of the time coordinates of the edges of π. The *first-passage time* $a(x, y)$ between vertices x and y is given by

$$a(x, y) = \inf\{T(\pi): \pi \text{ a path from } x \text{ to } y\};$$

thus $a(x, y)$ is the time taken by fluid to wet y, having been supplied at x. If we supply an unlimited quantity of fluid at the origin, the set of wet vertices at time t is

$$W(t) = \{x \in \mathbb{Z}^d: a(0, x) \le t\},$$

and the main questions pertain to the rate of growth of $W(t)$ as $t \to \infty$. The easiest way of describing the asymptotic behaviour of $W(t)$ involves re-scaling the lattice as $t \to \infty$, and the result is easiest stated for the set $\tilde{W}(t)$ obtained from $W(t)$ by filling in the holes: we define

$$\tilde{W}(t) = \{y + B: y \in W(t)\},$$

where

$$B = \{x \in \mathbb{R}^d: |x_i| \le \tfrac{1}{2} \text{ for } 1 \le i \le d\}$$

is the unit cube with centre at the origin.

It turns out that $\tilde{W}(t)$ grows approximately linearly as time passes, in the sense that, subject to a suitable moment condition on the time coordinate distribution, there exists a non-random limit set L for $t^{-1}\tilde{W}(t)$ having non-empty interior such that either

(a) L is compact with

$$(1 - \varepsilon)L \subseteq \frac{1}{t}\tilde{W}(t) \subseteq (1 + \varepsilon)L \qquad \text{eventually, almost surely,}$$

for all $\varepsilon > 0$, or

(b) $L = \mathbb{R}^d$ with

$$\frac{1}{t}\tilde{W}(t) \supseteq \{x \in \mathbb{R}^d: |x| \le M\} \qquad \text{eventually, almost surely,}$$

for all $M > 0$.

The earliest such 'shape' result was proved by Richardson (1973), and the last words were due to Cox and Durrett (1981) and Kesten (1986a). Few non-trivial facts are known about the shape of the limit set L; worthy of note in this regard is the finding of Durrett and Liggett (1981) that, for a closely related problem, the boundary of L may contain straight line segments when $d = 2$.

Specializing to the spread of fluid in an axial direction, we define

$$a_{mn} = a(e_m, e_n) \qquad \text{for} \quad -\infty < m \le n < \infty,$$

the first-passage time from $e_m = (m, 0, 0, \dots, 0)$ to $e_n = (n, 0, 0, \dots, 0)$. Two of the most useful properties of the family $\mathbf{a} = (a_{mn}: m \le n)$ are *stationarity*, in that $(a_{mn}: m \le n)$ and $(a_{m+1,n+1}: m \le n)$ have the same finite-dimensional distributions, and *subadditivity*, in that

$$a_{mn} \le a_{mr} + a_{rn} \qquad \text{whenever} \quad m \le r \le n.$$

The subadditive inequality is a consequence of the fact that a_{mn} is no greater than the infimum of the passage times of paths from e_m to e_n which pass through e_r. Subject to a moment condition, we have by the subadditive ergodic theorem that

$$\lim_{n\to\infty}\left\{\frac{1}{n}a_{0n}\right\} = \mu \qquad \text{almost surely}$$

for some constant $\mu = \mu(F, d)$ called the *time constant* of the process. This result is one of the main steps towards establishing the 'shape' theorem above. Little is known about the way in which $\mu(F, d)$ depends on the choice of the distribution function F.

The sequence **a** satisfies a large-deviation theorem (Grimmett and Kesten (1984a)), but it is an open problem to find sequences $(\alpha(n), \beta(n): n \geq 0)$ such that $(a_{0n} - \alpha(n))/\beta(n)$ has a non-trivial limit distribution as $n \to \infty$. See Kesten (1986a, 1987e).

10.8 Capacitated Networks

The emphasis of first-passage percolation is upon flow rate subject to constraints of time. An alternative model involves constraints of capacity, and the classical model of this form is the 'capacitated network' of Ford and Fulkerson (1962). Suppose that we are provided with a finite graph G, of which each edge e has a non-negative *capacity* $c(e)$; we may think of $c(e)$ as an upper bound for the quantity of fluid which may flow along e in unit time. Let S and T be disjoint sets of vertices, called *source* and *sink* vertices, respectively. A *flow* from S to T is an assignment of a non-negative number $f(e)$ and an orientation to each edge e of G such that

$$I(v) = \sum_{w:w\sim v}^{-} f(\langle v, w\rangle) - \sum_{w:w\sim v}^{+} f(\langle v, w\rangle) \tag{10.39}$$

satisfies $I(v) = 0$ for all vertices $v \notin S \cup T$, where the first (respectively second) summation is calculated over all neighbours w of v for which $\langle v, w\rangle$ is oriented away from (respectively towards) v; thus fluid is conserved at all vertices except possibly at sources and sinks. A flow is *admissible* if

$$f(e) \leq c(e) \qquad \text{for all edges } e,$$

and the *value* of such a flow is defined to be $\sum_{v\in S} I(v)$, the aggregate amount of fluid entering G at the source vertices. The *maximum flow* is the largest value of all admissible flows.

The max–flow min–cut theorem characterizes the maximal flow through the network in terms of the sizes of the cutsets. A set E of edges is called a (S, T)-*cutset* if all paths from S to T use at least one edge of E. The *size* of a (S, T)-cutset E is defined to be the sum of the capacities of the edges in E. It is clear that the maximum flow cannot exceed the size of any cutset, and the max–flow min–cut theorem asserts that the maximum flow is actually equal to the minimum size of all (S, T)-cutsets.

We now specialize to the lattice $\mathbb{L}^d$. Suppose that the capacities $(c(e): e \in \mathbb{E}^d)$ are independent random variables having the common distribution function F. We are interested in flows through large boxes, and denote by U_n the cube $[0, n]^d$ of $\mathbb{L}^d$. As sources we take the 'bottom' of U_n,

$$S_n = [0, n]^{d-1} \times \{0\},$$

and as sinks we take the 'top',

$$T_n = [0, n]^{d-1} \times \{n\}.$$

What can be said about the maximum flow φ_n from S_n to T_n through U_n? It is illustrative to consider in some detail the special case of two dimensions. By the max–flow min–cut theorem, φ_n equals the minimum size of (S_n, T_n)-cutsets. It is not difficult to see from Figure 10.10 that each such cutset corresponds to a path in the dual graph which traverses the rectangle U_n from left to right. If to each edge of the dual lattice we allocate a time coordinate equal in value to the capacity of the edge which it crosses, then the size of such a cutset equals the passage time of the corresponding dual path. That is to say, the maximum flow from bottom to top in U_n equals the minimum passage time from left to right in the dual. We now appeal to the theory of first-passage percolation to deduce (after some work: see Grimmett and Kesten (1984a)) that, as $n \to \infty$,

$$\frac{1}{n}\varphi_n \to \mu(F, 2) \qquad \text{almost surely,}$$

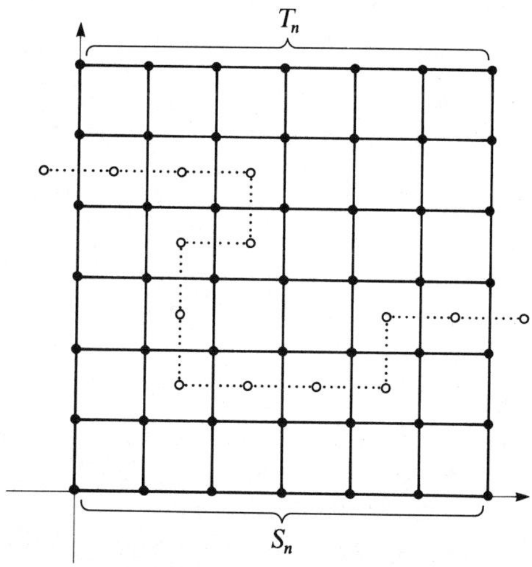

Figure 10.10. Every left–right crossing in the dual traverses edges in U_n which form a cutset. It is not difficult to see that every minimal cutset arises in such a way.

where $\mu(F, 2)$ is the time constant of first-passage percolation on $\mathbb{L}^2$ with time coordinate distribution F.

The picture in higher dimensions is less clear, although Kesten (1985, 1986a, 1987a) has made considerable progress towards establishing results analogous to those of first-passage percolation. Consider, for example, the case $d = 3$. The asymptotic behaviour of φ_n depends qualitatively upon the size of the atom of F at zero, since if a large number of edges have capacity 0 then it is likely that there exists no path from S_n to T_n with non-zero capacity. The critical value of $F(0)$ is probably $1 - p_c(3)$, where $p_c(3)$ is the critical probability of bond percolation on $\mathbb{L}^3$. Certainly if $F(0) > 1 - p_c(3)$ then $\varphi_n = 0$ for all large n (almost surely). It remains an open problem to establish the asymptotic behaviour of φ_n when $F(0) < 1 - p_c(3)$. One expects that φ_n is proportional to the cross-sectional area n^2 of U_n, in that $n^{-2}\varphi_n$ converges to a positive limit as $n \to \infty$, but this has been established only for values of $F(0)$ not exceeding $\frac{1}{27}$ (see Kesten (1987a)).

Further discussions may be found in Grimmett and Kesten (1984a), Grimmett (1985c), and Kesten (1987e).

10.9 Random Electrical Networks

Percolation theory is one of the principal tools for analysing disordered media, and there are few better examples of its use than in the study of disordered electrical networks. It may not be too difficult to calculate the effective electrical resistance of a block of either material A or material B, but what is the effective resistance of a mixture of the two materials? The answer depends, of course, on the way in which the composite block is constructed from the materials in question. If the composition is disordered, then the most practicable assumption may be that each individual component is made of a randomly chosen material. It is not known how to calculate the effective resistance of such large blocks even if we are prepared to assume the maximum amount of independence between components.

We make the discussion more concrete as follows. Let $d \geq 2$. Instead of the capacities of the last section, we allocate to each edge e of $\mathbb{L}^d$ an electrical resistance $r(e)$ satisfying $0 \leq r(e) \leq \infty$, and we assume that the collection $(r(e): e \in \mathbb{E}^d)$ is a set of independent and identically distributed random variables. The value $r(e) = 0$ corresponds to a superconductor, and $r(e) = \infty$ corresponds to an insulator. We denote by U_n the cube $[0, n]^d$ of the lattice $\mathbb{L}^d$. We turn U_n into an electrical network by applying a potential difference between the bottom and top (hyper)faces of U_n,

$$S_n = [0, n]^{d-1} \times \{0\}, \qquad T_n = [0, n]^{d-1} \times \{n\};$$

see Figure 1.3 for a two-dimensional illustration. We write R_n for the effective resistance of the network, and we seek to understand how the asymptotic properties of R_n depend upon the underlying distribution of the edge-resistances. It is reasonable to guess that R_n is roughly proportional to the depth of U_n and inversely proportional to its cross-sectional area, and this leads to the conjecture that the limit

$$\lim_{n\to\infty} \{n^{d-2}R_n\} \tag{10.40}$$

exists almost surely, for a large class of distributions at least. Similar results involving L^2 convergence have been obtained by Papanicolaou and Varadhan (1979) and Golden and Papanicolaou (1982) for the related problem of conduction in the continuous sheet $\mathbb{R}^2$, and by Künnemann (1983a, b) for the two-dimensional lattice $\mathbb{L}^2$ and edge-resistances which are uniformly bounded away from 0 and ∞. Although electrical resistance is not a subadditive function of the underlying volume of material, Hammersley (1988) has shown that the dissipated energy is subadditive when the problem is suitably reformulated; using this observation, he has proved the existence of the limit corresponding to (10.40) for thin slabs of (non-superconducting) material.

The simplest special case in which percolation comes to the forefront is when each edge-resistance is either an insulator or a standard conductor, which is to say that there exists a number p such that

$$P(r(e) = 1) = p, \qquad P(r(e) = \infty) = 1 - p,$$

for all edges e. If p is beneath the critical probability $p_c(d)$ of bond percolation on $\mathbb{L}^d$, then there are insufficiently many conducting edges to guarantee non-zero current, which is to say that

$$P(R_n = \infty \text{ for all large } n) = 1 \qquad \text{if} \quad p < p_c(d).$$

The case $p > p_c(d)$ is of much greater interest, for it is in this case that we conjecture the existence of a constant $\rho = \rho(p, d)$ satisfying $0 < \rho < \infty$ such that

$$n^{d-2}R_n \to \rho \qquad \text{almost surely as} \quad n \to \infty. \tag{10.41}$$

Although this is unproven, various upper and lower bounds are known for $n^{d-2}R_n$, and the best are those of Chayes and Chayes (1986d). In seeking a lower bound for R_n, we observe that there is a dominating influence from those edges of U_n which are conducting and whose end-vertices are contained in disjoint infinite conducting paths of $\mathbb{L}^d$. This leads by way of the BK inequality to the bound

$$\liminf_{n\to\infty} \{n^{d-2}R_n\} \geq \{dp\theta(p)^2\}^{-1} \qquad \text{almost surely,}$$

where $\theta(p)$ is the percolation probability for bond percolation on $\mathbb{L}^d$. Thus

$$\liminf_{n\to\infty} \{n^{d-2}R_n\} > 0 \qquad \text{almost surely, if} \quad p > p_c(d).$$

For a decent upper bound for R_n, we need to find a large number of edge-disjoint paths of conducting edges joining S_n to T_n, and it is here that Theorem (2.45) is useful. If the probability that there is no such path decays sufficiently quickly as $n \to \infty$, then we may apply Theorem (2.45) to find that the number of such paths has (with overwhelming probability) order n^{d-1}. This would be adequate for our purposes, implying as it does that

$$\text{(10.42)} \qquad \limsup_{n\to\infty} \{n^{d-2} R_n\} < \infty \qquad \text{almost surely,}$$

whenever the above hypothesis is true. This argument should demonstrate (10.42) whenever $p > p_c(d)$, but the hypothesis in question has been proved only under the condition that $p > p_c^+$, the critical probability of the half-space of $\mathbb{L}^d$ (see Aizenman, Chayes, Chayes, Fröhlich, and Russo (1983), Chayes and Chayes (1986d), and Barsky, Grimmett, and Newman (1988)).

Random media are discussed in some detail in papers contained in Hughes and Ninham (1983). Other references to the above work include Kesten (1982, 1987e). Stinchcombe (1974), Grimmett and Kesten (1984b), and Grimmett (1985c) discuss random electrical networks on trees and complete graphs.

10.10 Oriented Percolation

Modellers of galaxies, semiconductors, and elementary particles have had reason to study percolation on oriented lattices, and in particular on the 'north-east' lattice $\vec{\mathbb{L}}^d$ obtained by orienting each edge of $\mathbb{L}^d$ in the direction of increasing coordinate-value (see Figure 10.11 for a two-dimensional illustration). There are many similarities between results for oriented percolation and those for ordinary percolation; on the other hand the corresponding proofs often differ greatly, largely because the existence of one-way streets restricts the degree of spatial freedom of the traffic.

Suppose that $0 \le p \le 1$. We examine each edge of $\vec{\mathbb{L}}^d$ in turn, declaring it to be *open* with probability p and *closed* otherwise; as usual, the edge-states are independent of each other. We now supply fluid at the origin, but allow it to travel along open edges in the directions of their orientations only. We denote by C the set of vertices which may be reached from the origin along open directed paths; C contains exactly those vertices which are wetted by fluid supplied at the origin. The *percolation probability* is given by

$$\vec{\theta}(p) = P_p(|C| = \infty),$$

and the critical probability $\vec{p}_c(d)$ by

$$\vec{p}_c(d) = \sup\{p : \vec{\theta}(p) = 0\}.$$

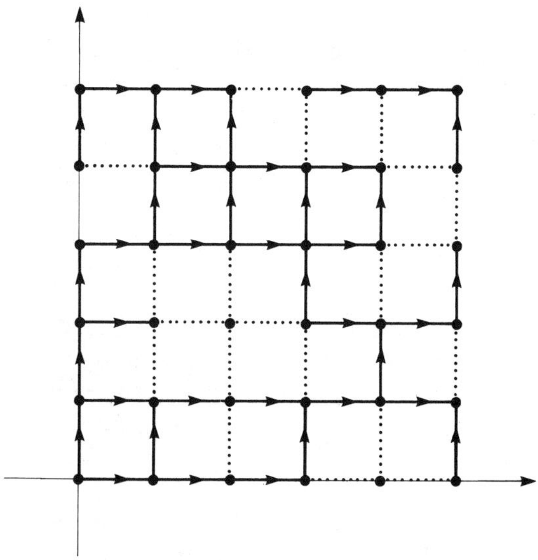

Figure 10.11. Part of the 'north-east' lattice in which each edge has been deleted with probability $1 - p$, independently of all other edges.

It is an interesting exercise to show that $0 < \vec{p}_c(d) < 1$ if $d \geq 2$. The exact numerical value of $\vec{p}_c(d)$ is of course unknown; as for inequalities, we know that $\vec{p}_c(2) \geq 0.6298$ (see Durrett (1984)) and D. Williams (1983, unpublished) has had good reason to conjecture the provability of $\vec{p}_c(2) \leq \frac{2}{3}$.

The process is understood quite well when $d = 2$, for which case Durrett (1984) has written a rather comprehensive survey. By looking at the set A_n of wet vertices on the diagonal $\{x \in \mathbb{Z}^2 : x_1 + x_2 = n\}$ of $\vec{\mathbb{L}}^2$, one may reformulate two-dimensional oriented percolation as a one-dimensional contact process in discrete time (see Liggett (1985, Chapter 6)). It turns out that $\vec{p}_c(2)$ may be characterized in terms of the velocity of the rightwards edge of a contact process on $\mathbb{Z}$ whose initial distribution places infectives to the left of the origin and susceptibles to the right. With the support of arguments from branching processes and ordinary percolation, one may prove such results as the exponential decay of the cluster size distribution when $p < \vec{p}_c(2)$, and its sub-exponential decay when $p > \vec{p}_c(2)$: there exist $\alpha(p), \beta(p) > 0$ such that

$$e^{-\alpha(p)\sqrt{n}} \leq P_p(n \leq |C| < \infty) \leq e^{-\beta(p)\sqrt{n}} \qquad \text{if } \vec{p}_c(2) < p < 1.$$

Progress in higher dimensions has been more limited. Cox and Durrett (1983) have explored the asymptotic behaviour of $p_c(d)$ as $d \to \infty$, showing in particular that $p_c(d) \sim d^{-1}$. More recent work is contained in Wierman (1985a), and in Durrett and Schonmann (1987, 1988a).

We close with an open problem of a different sort. Suppose that each edge of $\mathbb{L}^2$ is oriented in a random direction, horizontal edges being oriented eastwards with probability p and westwards otherwise, and vertical edges being oriented northwards with probability p and southwards otherwise. Let $\eta(p)$ be the probability that there exists an infinite oriented path from the origin. It is not hard to use Theorem (9.12) to show that $\eta(\frac{1}{2}) = 0$. We conjecture that $\eta(p) > 0$ if $p \neq \frac{1}{2}$.

Appendix I

The Infinite-Volume Limit for Percolation

Certain steps were omitted from the Aizenman–Barsky proof of the uniqueness of critical points, as presented in Section 3.3. The calculations necessary for these minor steps are rather standard, and we indicate here how they may be performed.

We shall use the notation of Section 3.3 without further elaboration. The principal quantities are

$$\theta(p,\gamma) = 1 - \sum_{n=1}^{\infty} (1-\gamma)^n P_p(|C| = n), \tag{I.1}$$

together with the corresponding quantity defined on the box $B(N)$ with periodic boundary conditions,

$$\theta_N(p,\gamma) = 1 - \sum_{n=1}^{\infty} (1-\gamma)^n P_p(|C_N| = n), \tag{I.2}$$

where C_N is the open cluster of $B(N)$ containing the origin.

We are required to prove the following:

(I.3) $\theta(p,\gamma)$ is continuously differentiable in p, for $\gamma > 0$,

and

$$\theta_N(p,\gamma) \to \theta(p,\gamma) \quad \text{as} \quad N \to \infty, \tag{I.4}$$

$$\frac{\partial \theta_N}{\partial p} \to \frac{\partial \theta}{\partial p} \quad \text{as} \quad N \to \infty, \tag{I.5}$$

$$\frac{\partial \theta_N}{\partial \gamma} \to \frac{\partial \theta}{\partial \gamma} \quad \text{as} \quad N \to \infty, \tag{I.6}$$

if $0 < p, \gamma < 1$. In demonstrating these limits we use techniques related to those of Chapter 4 and Section 6.8. Thus we write

$$P_p(|C| = n) = \sum_{m,b} a_{nmb} p^m (1-p)^b \tag{I.7}$$

and

$$P_p(|C_N| = n) = \sum_{m,b} a_{nmb}(N) p^m (1-p)^b, \tag{I.8}$$

where a_{nmb} (respectively $a_{nmb}(N)$) is the number of animals of $\mathbb{L}^d$ (respectively $B(N)$ with periodic boundary conditions) with n vertices, m edges, and b boundary edges. We note as in (4.14) and (4.15) that $a_{nmb} = a_{nmb}(N) = 0$ if either $b > 2dn$ or $m > dn$, and we may therefore assume that $b \le 2dn$ and $m \le dn$.

We prove (I.3) first, restricting ourselves to the case when $0 < p < 1$; the special cases $p = 0, 1$ are of no importance in Section 3.3 and may be dealt with somewhat similarly. Suppose then that $0 < p < 1$ and $\gamma > 0$. It suffices to show that the term-by-term derivative of the infinite series in (I.1) is uniformly convergent on some neighbourhood of p. The tail of the term-by-term derivative is

$$\sum_{n=N}^{\infty} (1-\gamma)^n \sum_{m,b} a_{nmb} p^m (1-p)^b \left(\frac{m}{p} - \frac{b}{1-p}\right),$$

where we have used (I.7). This is no larger in absolute value than

$$\sum_{n=N}^{\infty} (1-\gamma)^n \frac{2dn}{p(1-p)} P_p(|C| = n)$$

since $m, b \le 2dn$; the last summation converges uniformly as $N \to \infty$ for values of p in any strict sub-interval of $[0, 1]$. It follows that $\theta(\cdot, \gamma)$ is continuously differentiable at p.

We move on to (I.4)–(I.6) of which we shall prove (I.4) only, the other two limits having analogous proofs. We have from (I.1) and (I.2) that

$$|\theta(p, \gamma) - \theta_N(p, \gamma)| \le \sum_{n=1}^{\infty} (1-\gamma)^n \big|P_p(|C| = n) - P_p(|C_N| = n)\big|.$$

Now $P_p(|C| = n) = P_p(|C_N| = n)$ if $n < N$, since no cluster of $\mathbb{L}^d$ containing the origin intersects the boundary of $B(N)$ unless it has more than N vertices. Thus

$$\begin{aligned} |\theta(p, \gamma) - \theta_N(p, \gamma)| &\le \sum_{n=N}^{\infty} 2(1-\gamma)^n \\ &= 2(1-\gamma)^N/\gamma \qquad \text{if} \quad \gamma > 0, \end{aligned}$$

and hence $\theta_N(p, \gamma) \to \theta(p, \gamma)$ as $N \to \infty$ whenever $\gamma > 0$. The remaining limits (I.5) and (I.6) may be proved by using the expansion (I.8) and combining the arguments leading to (I.3) and (I.4).

Appendix II

The Subadditive Inequality

Let $(x_r: r \geq 1)$ be a sequence of real numbers. If the x_r's satisfy the 'subadditive inequality'

(II.1) $$x_{m+n} \leq x_m + x_n \qquad \text{for all } m, n,$$

then we say that the sequence is *subadditive*. Subadditive sequences are nearly additive, in the sense of the following theorem.

(II.2) Theorem. Subadditive limit theorem. *If* $(x_r: r \geq 1)$ *is subadditive then*

(II.3) $$\lambda = \lim_{r\to\infty} \left\{\frac{1}{r} x_r\right\}$$

exists and satisfies $-\infty \leq \lambda < \infty$. *Furthermore*

(II.4) $$\lambda = \inf\left\{\frac{1}{m} x_m : m \geq 1\right\},$$

and thus $x_m \geq m\lambda$ *for all* m.

See Hille (1948, Theorem 6.6.1) for a proof of this standard result. In many potential applications of the subadditive limit theorem, we encounter sequences $(x_r: r \geq 1)$ which are nearly subadditive, in that they satisfy a 'generalized subadditive inequality'

(II.5) $$x_{m+n} \leq x_m + x_n + g_{m+n} \qquad \text{for all } m, n,$$

for some given sequence $(g_r: r \geq 1)$. Such a sequence $(x_r: r \geq 1)$ is nearly additive, so long as the g's do not grow too fast.

(II.6) Theorem. Generalized subadditive limit theorem. *Suppose that* $(x_r: r \geq 1)$ *satisfies the generalized subadditive inequality (II.5), where*

$(g_r\colon r \geq 1)$ is a non-decreasing sequence satisfying

$$\sum_{r=1}^{\infty} \frac{g(r)}{r(r+1)} < \infty. \tag{II.7}$$

Then

$$\lambda = \lim_{r \to \infty} \left\{ \frac{1}{r} x_r \right\} \tag{II.8}$$

exists and satisfies $-\infty \leq \lambda < \infty$. *Furthermore*

$$x_m \geq m\lambda + g_m - 4m \sum_{r=2m}^{\infty} \frac{g(r)}{r(r+1)} \qquad \textit{for all } m. \tag{II.9}$$

This was proved by Hammersley (1962). A similar result was found by M.E. Fisher (unpublished) in independent and roughly contemporaneous work.

It is not unusual to encounter sequences which satisfy inequalities somewhat similar to the generalized subadditive inequality; for example, in Chapter 6 we came across a real sequence $(x_r\colon r \geq 1)$ satisfying

$$x_{m+n+2} \leq x_m + x_n + g_n \qquad \text{for all } m, n. \tag{II.10}$$

It is not difficult to adapt the proof of the subadditive limit theorem to deal with such inequalities. It turns out that (II.10), together with the condition $n^{-1} g_n \to 0$ as $n \to \infty$, is sufficient to ensure the existence of $\lambda = \lim_{r \to \infty} \{r^{-1} x_r\}$, and implies also that

$$x_r \geq (r+2)\lambda - g_r \qquad \text{for all } r. \tag{II.11}$$

It is easy to obtain more refined versions of such results.

List of Notation

Against each entry appears the number of the section in which the notation first appears.

Graphs and sets

G_{d}	9.2	Dual graph of G
$\langle a, b\rangle$	1.3	Edge joining a and b
$\mathbb{Z}$	1.3	The set $\{\dots, -1, 0, 1, \dots\}$
$\mathbb{Z}_+$	6.4	The non-negative integers $\{0, 1, 2, \dots\}$
$\mathbb{L}^d$	1.3	The d-dimensional cubic lattice
$\mathbb{E}^d$	1.3	The set of edges of $\mathbb{L}^d$
$\mathbb{L}^2_{\mathrm{d}}$	9.2	The dual lattice of $\mathbb{L}^2$
x_i	1.3	The value of the ith component of the vertex x of $\mathbb{L}^d$, unless otherwise specified
$B(n, x)$	1.3	Box with side-length $2n$ and centre at x
$B(n)$	1.3	Box with side-length $2n$ and centre at 0
$B(a, b)$	1.3	Box with vertices x satisfying $a_i \le x_i \le b_i$ for $i = 1, 2, \dots, d$
$A(k)$	9.7	The annulus $B(3k)\backslash B(k)$
H	6.4	The half-space $\mathbb{Z}_+ \times \mathbb{Z}^{d-1}$
S_k	6.4	The slab $\{0, 1, 2, \dots, k-1\} \times \mathbb{Z}^{d-1}$
T_k	6.4	The slice $\{0, 1, 2, \dots, k-1\}^{d-2} \times \mathbb{Z}^2_+$
∂A	1.3	The surface of the set A of vertices, being those members of A which are adjacent to vertices not in A
D_e	4.2	The set of edges of a graph D
ΔD	4.2	The edge boundary of D, being the set of edges not in D but incident to at least one vertex of D
$\lambda(d)$	1.4	The connective constant of $\mathbb{L}^d$
$\mathscr{R}_b(\mathbb{L}^d)$	6.7, 7.3	The lattice $\mathbb{L}^d$ renormalized by a linear factor b

$\delta(x, y)$	1.3	The number of edges in the shortest path from x to y
$\|x\|$	1.3	$\delta(0, x)$, for $x \in \mathbb{Z}^d$
$\|\|x\|\|$	1.3	$\max\{\|x_i\|: 1 \leq i \leq d\}$, for $x \in \mathbb{Z}^d$
$x \sim y$	1.3	x is adjacent to y
$\mathrm{rad}(D)$	5.1	The radius $\max\{\|x\|: x \in D\}$ of a subgraph D of $\mathbb{L}^d$ containing 0
$\mathrm{diam}(D)$	6.5	$\max\{\|x_i - y_i\|: x, y \in D, 1 \leq i \leq d\}$, the maximum of the differences of the coordinates of vertices in the graph D
$S(n, x)$	3.2	The sphere $\{y \in \mathbb{Z}^d: \|y - x\| \leq n\}$ with radius n and centre at x
$S(n)$	3.2	The sphere $S(n, 0)$
	9.3	The rectangle $[0, n + 1] \times [0, n]$ in $\mathbb{L}^2$
$\|A\|$	1.3	The cardinality of a set A, or the number of vertices of a graph A
a_{nmb}	4.2	Number of animals of $\mathbb{L}^d$ with n vertices, m edges, and b edges in the edge boundary

Probability notation

p	1.3	The probability that an edge is open
q	1.3	$1 - p$
P_p	1.3	Product measure with density p
E_p	1.3	Expectation operator corresponding to P_p
P_p^e	1.3	Product measure on $\mathbb{L}^d$ with the edge e deleted
I_A	1.3	Indicator function of an event A
$E_p(X; A)$	1.3	Expectation $E_p(XI_A)$ of X on the event A
cov_p	2.5	Covariance corresponding to P_p
var_p	2.5	Variance corresponding to P_p
ω	1.3	Typical realization of open and closed edges
$K(\omega)$	1.3	The set of edges which are open in ω
A^c	1.3	Complement of event A
$A \circ B$	2.3	Event that A and B occur 'disjointly', for increasing events A, B
$A \,\square\, B$	2.3	Event that A and B occur 'disjointly'

Percolation notation

$C(x)$	1.3	Open cluster at x
C	1.3	Open cluster at 0
$p_c = p_c(d)$	1.4	Critical probability of bond percolation on $\mathbb{L}^d$
p_c^+	6.4	The critical probability of the half-space H
$p_c(G)$	6.4	The critical probability of a connected graph G
$p_c(S)$	6.4	The limit of the critical probabilities of slabs

$p_c(T)$	6.4	The limit of the critical probabilities of slices
$\theta(p)$	1.4	Percolation probability $P_p(\lvert C\rvert = \infty)$
$\chi(p)$	1.5	Mean cluster size $E_p\lvert C\rvert$
$\chi^f(p)$	1.5	Mean size $E_p(\lvert C\rvert; \lvert C\rvert < \infty)$ of a finite open cluster
$\kappa(p)$	1.5	Number $E_p(\lvert C\rvert^{-1})$ of open clusters per vertex
$\xi(p)$	5.2, 6.6	Correlation length
$\tau_p(x, y)$	5.1	Connectivity function $P_p(x \leftrightarrow y)$
$\tau_p^f(x, y)$	6.6	Truncated connectivity function $P_p(x \leftrightarrow y, \lvert C(x)\rvert < \infty)$
$\alpha, \beta, \gamma, \delta$; Δ, η, ρ, ν	7.1	Critical exponents, unless otherwise specified
$\{A \leftrightarrow B\}$	1.3	Event that some vertex of A is joined by an open path to some vertex of B
$\{A \not\leftrightarrow B\}$	1.3	Complement of $\{A \leftrightarrow B\}$

Asymptotics

$a_n \sim b_n$	1.3	$a_n/b_n \to 1$ as $n \to \infty$
$a_n \approx b_n$	1.3	$\log a_n/\log b_n \to 1$ as $n \to \infty$
$a_n \simeq b_n$	7.1	Asymptotic properties of a_n are comparable to those of b_n

Similar asymptotic relations are used for functions of p, in the limit as p approaches p_c.

Finally

$\lceil c \rceil$	1.3	Greatest integer not greater than c
$\lfloor c \rfloor$	1.3	Least integer not less than c

References

Ahlswede, R. and Daykin, D. E.
1979 Inequalities for a pair of maps $S \times S \to S$ with S a finite set, *Mathematische Zeitschrift* 165, 267–289.

Aizenman, M.
1987 General results in percolation theory, in *Proceedings of Taniguchi Workshop and Symposium on Probabilistic Methods in Mathematical Physics*, ed. N. Ikeda, 1–22, Kataka and Kyoto.

Aizenman, M. and Barsky, D. J.
1987 Sharpness of the phase transition in percolation models, *Communications in Mathematical Physics* 108, 489–526.
1988 Percolation critical exponents under the triangle condition, preprint.

Aizenman, M., Barsky, D. J., and Fernández, R.
1987 The phase transition in a general class of Ising-type models is sharp, *Journal of Statistical Physics* 47, 343–374.

Aizenman, M., Chayes, J. T., Chayes, L., Fröhlich, J., and Russo, L.
1983 On a sharp transition from area law to perimeter law in a system of random surfaces, *Communications in Mathematical Physics* 92, 19–69.

Aizenman, M., Chayes, J. T., Chayes, L., and Newman, C. M.
1988 Discontinuity of the magnetization in one-dimensional $1/|x - y|^2$ Ising and Potts models, *Journal of Statistical Physics* 50, 1–40.

Aizenman, M., Delyon, F., and Souillard, B.
1980 Lower bounds on the cluster size distributions, *Journal of Statistical Physics* 23, 267–280.

Aizenman, M., Kesten, H., and Newman, C. M.
1987 Uniqueness of the infinite cluster and continuity of connectivity functions for short- and long-range percolation, *Communications in Mathematical Physics* 111, 505–532.

Aizenman, M. and Newman, C. M.
1984 Tree graph inequalities and critical behavior in percolation models, *Journal of Statistical Physics* 36, 107–143.

1986 Discontinuity of the percolation density in one-dimensional $1/|x-y|^2$ percolation models, *Communications in Mathematical Physics* 107, 611–647.

Appel, M. J. and Wierman, J. C.
1987 On the absence of infinite *AB* percolation clusters in bipartite graphs, *Journal of Physics A: Mathematical and General* 20, 2527–2531.

Asmussen, S. and Hering, H.
1983 *Branching Processes*, Birkhäuser, Boston.

Athreya, K. B. and Ney, P. E.
1972 *Branching Processes*, Springer, Berlin.

Barlow, R. N. and Proschan, F.
1965 *Mathematical Theory of Reliability*, Wiley, New York.
1975 *Statistical Theory of Reliability and Life Testing*, Holt, Rinehart and Winston, New York.

Barsky, D. J., Grimmett, G. R., and Newman, C. M.
1988 Percolation in half-spaces: equality of critical probabilities and continuity of the percolation probability, in preparation.

Batty, C. J. K. and Bollman, H. W.
1980 Generalized Holley–Preston inequalities on measure spaces and their products, *Zeitschrift für Wahrscheinlichkeitstheorie und Verwandte Gebiete* 53, 157–173.

Beijeren, H. van
1982 Transport properties of stochastic Lorentz models, *Reviews in Modern Physics* 54, 195–234.

Ben-Avraham, D. and Havlin, S.
1982 Diffusion on percolation clusters at criticality, *Journal of Physics A: Mathematical and General* 15, L691–L697.

Berg, J. van den
1981 Percolation theory on pairs of matching lattices, *Journal of Mathematical Physics* 22, 152–157.
1982 A note on percolation theory, *Journal of Physics A: Mathematical and General* 15, 605–610.
1985 Disjoint occurrences of events: results and conjectures, in *Particle Systems, Random Media and Large Deviations*, ed. R. Durrett, 357–361, Contemporary Mathematics 41, American Mathematical Society, Providence, Rhode Island.

Berg, J. van den and Fiebig, U.
1987 On a combinatorial conjecture concerning disjoint occurrences of events, *Annals of Probability* 15, 354–374.

Berg, J. van den and Keane, M.
1984 On the continuity of the percolation probability function, in *Conference on Modern Analysis and Probability*, ed. R. Beals et al., 61–65, Contemporary Mathematics 26, American Mathematical Society, Providence, Rhode Island.

Berg, J. van den and Kesten, H.
1985 Inequalities with applications to percolation and reliability, *Journal of Applied Probability* 22, 556–569.

Biggs, N. L.
1976 *Interaction Models*, London Mathematical Society Lecture Notes no. 30, Cambridge University Press, Cambridge.

Bolthausen, E.
1982 On the central limit theorem for stationary mixing random fields, *Annals of Probability* 10, 1047–1050.

Bondy, J. A. and Murty, U. S. R.
1976 *Graph Theory with Applications*, Macmillan, New York.

Brånvall, G.
1980 A note on limit theorems in percolation, *Zeitschrift für Wahrscheinlichkeitstheorie und Verwandte Gebiete* 53, 317–328.

Broadbent, S. R.
1954 In discussion of Symposium on Monte Carlo Methods, *Journal of the Royal Statistical Society* (*B*) 16, 68.

Broadbent, S. R. and Hammersley, J. M.
1957 Percolation processes I. Crystals and mazes, *Proceedings of the Cambridge Philosophical Society* 53, 629–641.

Burton, R. M. and Keane, M.
1988 Density and uniqueness in percolation, preprint.

Campanino, M. and Russo, L.
1985 An upper bound for the critical probability for the three-dimensional cubic lattice, *Annals of Probability* 13, 478–491.

Chandler, R., Koplick, J., Lerman, K., and Willemsen, J. F.
1982 Capillary displacement and percolation in porous media, *Journal of Fluid Mechanics* 119, 249–267.

Chayes, J. T. and Chayes, L.
1986a Percolation and random media, in *Critical Phenomena, Random Systems and Gauge Theories*, Les Houches, Session XLIII, 1984, eds. K. Osterwalder and R. Stora, 1001–1142, Elsevier, Amsterdam.
1986b Critical points and intermediate phases on wedges of $\mathbb{Z}^d$, *Journal of Physics A: Mathematical and General* 19, 3033–3048.
1986c An inequality for the infinite cluster density in percolation, *Physical Review Letters* 56, 1619–1622.
1986d Bulk transport properties and exponent inequalities for random resistor and flow networks, *Communications in Mathematical Physics* 105, 133–152.
1987 The mean field bound for the order parameter of Bernoulli percolation, in *Percolation Theory and Ergodic Theory of Infinite Particle Systems*, ed. H. Kesten, 49–71, Springer, Berlin.
1988 On the upper critical dimension of Bernoulli percolation, *Communications in Mathematical Physics*, to appear.

Chayes, J. T., Chayes, L., and Durrett, R.
1987 Inhomogeneous percolation problems and incipient infinite clusters, *Journal of Physics A: Mathematical and General* 20, 1521–1530.
1988 Connectivity properties of Mandelbrot's percolation process, *Probability Theory and Related Fields* 77, 307–324.

Chayes, J. T., Chayes, L., Fisher, D., and Spencer, T.
1986 Finite-size scaling and correlation lengths for disordered systems, *Physical Review Letters* 57, 2999–3002.

Chayes, J. T., Chayes, L., and Fröhlich, J.

1985 The low-temperature behavior of disordered magnets, *Communications in Mathematical Physics* 100, 399–437.

Chayes, J. T., Chayes, L., Grimmett, G. R., Kesten, H., and Schonmann, R. H.

1988 The correlation length for the high density phase of Bernoulli percolation, *Annals of Probability*, to appear.

Chayes, J. T., Chayes, L., and Newman, C. M.

1985 The stochastic geometry of invasion percolation, *Communications in Mathematical Physics* 101, 383–407.

1987 Bernoulli percolation above threshold: an invasion percolation analysis, *Annals of Probability* 15, 1272–1287.

Chow, Y. S. and Teicher, H.

1978 *Probability Theory*, Springer, New York.

Coniglio, A.

1985 Shapes, surfaces and interfaces in percolation clusters, in *Physics of Finely Divided Matter*, Les Houches, Session XLIV, 1985, eds. N. Boccara and M. Daoud, 84–101, Springer, Berlin.

Cox, J. T. and Durrett, R.

1981 Some limit theorems for percolation processes with necessary and sufficient conditions, *Annals of Probability* 9, 583–603.

1983 Oriented percolation in dimensions $d \geq 4$: bounds and asymptotic formulas, *Mathematical Proceedings of the Cambridge Philosophical Society* 93, 151–162.

1988 Limit theorems for the spread of epidemics and forest fires, *Stochastic Processes and their Applications*, to appear.

Cox, J. T. and Grimmett, G. R.

1981 Central limit theorems for percolation models, *Journal of Statistical Physics* 25, 237–251.

1984 Central limit theorems for associated random variables and the percolation model, *Annals of Probability* 12, 514–528.

Delyon, F.

1980 Taille, forme et nombre des amas dans les problèmes de percolation, Thèse de 3me cycle, *Université Pierre et Marie Curie*, Paris.

De Masi, A., Ferrari, P. A., Goldstein, S., and Wick, W. D.

1985 Invariance principle for reversible Markov processes with application to diffusion in the percolation regime, in *Particle Systems, Random Media and Large Deviations*, ed. R. Durrett, 71–85, Contemporary Mathematics 41, American Mathematical Society, Providence, Rhode Island.

Deutscher, G., Zallen, R., and Adler, J.

1983 *Percolation Structures and Processes*, editors, Annals of the Israel Physical Society 5.

Doyle, P. and Snell, J. L.

1984 *Random Walk and Electric Networks*, Carus Mathematical Monograph no. 22, American Mathematical Association, Washington, DC.

Dunford, N. and Schwartz, J. T.

1958 *Linear Operators*, volume I, Wiley–Interscience, New York.

Durrett, R.
1984 Oriented percolation in two dimensions, *Annals of Probability* 12, 999–1040.
1985a Some general results concerning the critical exponents of percolation processes, *Zeitschrift für Wahrscheinlichkeitstheorie und Verwandte Gebiete* 69, 421–437.
1985b *Particle Systems, Random Media and Large Deviations*, editor, Contemporary Mathematics 41, American Mathematical Society, Providence, Rhode Island.
1988 *Lecture Notes on Particle Systems and Percolation*, Wadsworth and Brooks/Cole, Monterey, California.

Durrett, R. and Liggett, T. M.
1981 The shape of the limit set in Richardson's growth process, *Annals of Probability* 9, 186–193.

Durrett, R. and Nguyen, B.
1985 Thermodynamic inequalities for percolation, *Communications in Mathematical Physics* 99, 253–269.

Durrett, R. and Schonmann, R. H.
1987 Stochastic growth models, in *Percolation Theory and Ergodic Theory of Infinite Particle Systems*, ed. H. Kesten, 85–119, Springer, Berlin.
1988a Large deviations for the contact process and two-dimensional percolation, *Annals of Probability* 77, 583–603.
1988b The contact process on a finite set II, *Annals of Probability*, to appear.

Dyson, F. J.
1969a Existence of a phase-transition in a one-dimensional Ising ferromagnet, *Communications in Mathematical Physics* 12, 91–107.
1969b Non-existence of spontaneous magnetization in a one-dimensional Ising ferromagnet, *Communications in Mathematical Physics* 12, 212–215.
1971 Phase transitions in ferromagnets, in *Statistical Mechanics at the Turn of the Decade*, ed. E. G. D. Cohen, 129–144, Dekker, New York.

Efros, A. L.
1986 *Physics and Geometry of Disorder*, Mir, Moscow.

Ehrenfest, P.
1959 *Collected Scientific Papers*, ed. M. J. Klein, North-Holland, Amsterdam.

Esary, J. D. and Proschan, F.
1963 Coherent structures of non-identical components, *Technometrics* 5, 191–209.

Essam, J. W.
1971 Graph theory and statistical physics, *Discrete Mathematics* 1, 83–112.
1972 Percolation and cluster size, in *Phase Transitions and Critical Phenomena*, volume II, eds. C. Domb and M. S. Green, 197–270, Academic Press, London.
1980 Percolation theory, *Reports on Progress in Physics* 43, 833–949.

Feller, W.
1968 *An Introduction to Probability Theory*, volume I, Wiley, New York.
1971 *An Introduction to Probability Theory*, volume II, Wiley, New York.

Fisher, M. E.
1961 Critical probabilities for cluster size and percolation problems, *Journal of Mathematical Physics* 2, 620–627.

1983 Scaling, universality and renormalization group, in *Critical Phenomena*, ed. F. J. W. Hahne, 1–139, Lecture Notes in Physics no. 186, Springer, Berlin.

Fisher, M. E. and Essam, J. W.
1961 Some cluster size and percolation problems, *Journal of Mathematical Physics* 2, 609–619.

Flory, P. J.
1941 Molecular size distribution in three dimensional gelation I–III, *Journal of the American Chemical Society* 63, 3083–3100.

Ford, L. R. and Fulkerson, D. R.
1962 *Flow in Networks*, Princeton University Press, Princeton.

Fortuin, C. M.
1972 On the random cluster model. II. The percolation model, *Physica* 58, 393–418.

Fortuin, C. M. and Kasteleyn, P. W.
1972 On the random cluster model. I. Introduction and relation to other models, *Physica* 57, 536–564.

Fortuin, C. M., Kasteleyn, P. W., and Ginibre, J.
1971 Correlation inequalities on some partially ordered sets, *Communications in Mathematical Physics* 22, 89–103.

Frisch, H. L., Gordon, S. B., Vyssotsky, V. A., and Hammersley, J. M.
1962 Monte Carlo solution of bond percolation processes on various crystal lattices, *The Bell System Technical Journal* 41, 909–920.

Frisch, H. L. and Hammersley, J. M.
1963 Percolation processes and related topics, *Journal of the Society for Industrial and Applied Mathematics* 11, 894–918.

Fröhlich, J. and Spencer, T.
1982 The phase transition in the one-dimensional Ising model with $1/r^2$ interaction energy, *Communications in Mathematical Physics* 84, 87–101.

Gandolfi, A.
1988 Uniqueness of the infinite cluster for stationary finite range Gibbs states, *Annals of Probability*, to appear.

Gandolfi, A., Grimmett, G. R., and Russo, L.
1988 On the uniqueness of the infinite open cluster in the percolation model, *Communications in Mathematical Physics* 114, 549–552.

Gandolfi, A., Keane, M., and Russo, L.
1988 On the uniqueness of the infinite occupied cluster in dependent two-dimensional site percolation, *Annals of Probability* 16, 1147–1157

Gennes, P. G. de
1976 La percolation: un concept unificateur, *La Recherche* 7, 919–927.

Gennes, P. G. de and Guyon, E.
1978 Lois générales pour l'injection d'un fluide dans un milieu poreux aléatoire, *Journal de Mécanique* 119, 249–267.

Gilbert, E. N.
1961 Random plane networks, *Journal of the Society for Industrial and Applied Mathematics* 9, 533–543.

Godsil, C. and McKay, B.
1980 The dimension of a graph, *Quarterly Journal of Mathematics, Oxford, Series* (2), 31, 423–427.

Golden, K. and Papanicolaou, G.
1982 Bounds for effective parameters of heterogeneous media by analytic continuation, *Communications in Mathematical Physics* 90, 473–491.

Gordon, D. M.
1988 Percolation in high dimensions, preprint.

Greene, J. W. and El Gamal, A.
1984 Configuration of VLSI arrays in the presence of defects, *Journal of the Association for Computing Machinery* 31, 694–717.

Griffiths, R. B.
1967 Correlations in Ising ferromagnets. II. External magnetic fields, *Journal of Mathematical Physics* 8, 484–489.

Grigorchuk, R. I.
1983 On Milnor's problem of group growth, *Soviet Mathematics Doklady* 28, 23–26.

Grimmett, G. R.
1976 On the number of clusters in the percolation model, *Journal of the London Mathematical Society* 13, 346–350.
1978 Multidimensional lattices and their partition functions, *Quarterly Journal of Mathematics, Oxford, Series* (2), 29, 142–157.
1979 Central limit theorems in percolation theory, unpublished manuscript.
1981a Critical sponge dimensions in percolation theory, *Advances in Applied Probability* 13, 314–324.
1981b On the differentiability of the number of clusters per site in the percolation model, *Journal of the London Mathematical Society* 23, 372–384.
1983 Bond percolation on subsets of the square lattice, and the transition between one-dimensional and two-dimensional behaviour, *Journal of Physics A: Mathematical and General* 16, 599–604.
1985a On a conjecture of Hammersley and Whittington concerning bond percolation on subsets of the simple cubic lattice, *Journal of Physics A: Mathematical and General* 18, L49–L52.
1985b The largest components in a random lattice, *Studia Scientiarum Mathematicarum Hungarica* 20, 325–331.
1985c Random flows: network flows and electrical flows through random media, in *Surveys in Combinatorics* 1985, ed. I. Anderson, 59–96, Cambridge University Press, Cambridge.
1986 Percolation processes and dimensionality, in *Stochastic Spatial Processes*, ed. P. Tautu, 165–174, Lecture Notes in Mathematics no. 1212, Springer, Berlin.
1987a Long paths and cycles in a random lattice, in *Random Graphs* '85, Annals of Discrete Mathematics 33, 69–76, North-Holland, Amsterdam.
1987b Recent progress in percolation, *Proceedings of the First World Congress of the Bernoulli Society*, Tashkent 1986, eds. Yu. V. Prohorov and V. V. Sazonov, VNU Science Press, Utrecht.

Grimmett, G. R., Keane, M., and Marstrand, J. M.
1984 On the connectedness of a random graph, *Mathematical Proceedings of the Cambridge Philosophical Society* 96, 151–166.

Grimmett, G. R. and Kesten, H.

1984a First-passage percolation, network flows and electrical resistances, *Zeitschrift für Wahrscheinlichkeitstheorie und Verwandte Gebiete* 66, 335–366.

1984b Random electrical networks on complete graphs, *Journal of the London Mathematical Society* 30, 171–192.

Grimmett, G. R. and Newman, C. M.

1988 Percolation in $\infty + 1$ dimensions, preprint.

Grimmett, G. R. and Stirzaker, D. R.

1982 *Probability and Random Processes*, Clarendon Press, Oxford.

Hall, P.

1985 On continuum percolation, *Annals of Probability* 13, 1250–1266.

1986 Clump counts in a mosaic, *Annals of Probability* 14, 424–458.

1988 *An Introduction to the Theory of Coverage Processes*, Wiley, New York.

Halley, J. W.

1983 Polychromatic percolation, in *Percolation Structures and Processes*, Annals of the Israel Physical Society 5, 323–351.

Hammersley, J. M.

1957a Percolation processes. Lower bounds for the critical probability, *Annals of Mathematical Statistics* 28, 790–795.

1957b Percolation processes II. The connective constant, *Proceedings of the Cambridge Philosophical Society* 53, 642–645.

1959 Bornes supérieures de la probabilité critique dans un processus de filtration, in *Le Calcul des Probabilités et ses Applications*, 17–37, CNRS, Paris.

1961 Comparison of atom and bond percolation, *Journal of Mathematical Physics* 2, 728–733.

1962 Generalization of the fundamental theorem on subadditive functions, *Proceedings of the Cambridge Philosophical Society* 58, 235–238.

1963 A Monte Carlo solution of percolation in a cubic lattice, in *Methods in Computational Physics*, volume I, eds. B. Alder, S. Fernbach and M. Rotenberg, 281–298, Academic Press, London.

1980 A generalization of McDiarmid's theorem for mixed Bernoulli percolation, *Proceedings of the Cambridge Philosophical Society* 88, 167–170.

1983 Origins of percolation theory, *Annals of the Israel Physical Society* 5, 47–57.

1988 Mesoadditive processes and the specific conductivity of lattices, *Journal of Applied Probability* 25A, 347–358.

Hammersley, J. M. and Welsh, D. J. A.

1965 First-passage percolation, subadditive processes, stochastic networks and generalized renewal theory, in *Bernoulli, Bayes, Laplace Anniversary Volume*, eds. J. Neyman and L. LeCam, 61–110, Springer, Berlin.

1980 Percolation theory and its ramifications, *Contemporary Physics* 56, 13–20.

Hammersley, J. M. and Whittington, S. G.

1985 Self-avoiding walks in wedges, *Journal of Physics A: Mathematical and General* 18, 101–111.

Hankey, A.

1978 Three properties of the infinite cluster in percolation theory, *Journal of Physics A: Mathematical and General* 11, L49–L55.

Hara, T. and Slade, G.

1988 Mean-field critical phenomena for percolation in high dimensions, preprint.

Harris, T. E.
1960 A lower bound for the critical probability in a certain percolation process, *Proceedings of the Cambridge Philosophical Society* 56, 13–20.
1963 *The Theory of Branching Processes*, Springer, Berlin.

Hauge, E. H. and Cohen, E. G. D.
1967 Normal and abnormal effects in Ehrenfest wind-tree model, *Physics Letters* 25A, 78–79.

Herrndorf, N.
1985 Mixing conditions for random fields in percolation theory, preprint.

Higuchi, Y.
1982 Coexistence of the infinite (*) clusters: a remark on the square lattice site percolation, *Zeitschrift für Wahrscheinlichkeitstheorie und Verwandte Gebiete* 61, 75–81.

Hille, E.
1948 *Functional Analysis and Semi-groups*, American Mathematical Society Colloquium Publications no. 31, New York.

Hollander, W. Th. F. den and Keane, M.
1986 Inequalities of FKG type, *Physica* 138A, 167–182.

Holmes, R. A.
1975 Random graphs with the natural numbers as vertices, unpublished manuscript.

Hughes, B. D. and Ninham, B. W.
1983 *The Mathematics and Physics of Disordered Media*, editors, Lecture Notes in Mathematics no. 1035, Springer, Berlin.

Imbrie, J. Z. and Newman, C. M.
1988 An intermediate phase with slow decay of correlations in one-dimensional $1/|x-y|^2$ percolation, Ising and Potts models, *Communications in Mathematical Physics* 118, 303–336.

Kalikow, S. and Weiss, B.
1988 When are random graphs connected?, preprint.

Kapitulnik, A., Aharony, A., Deutscher, G., and Stauffer, D.
1983 Self similarity and correlations in percolation, *Journal of Physics A: Mathematical and General* 16, L269–L274.

Kasteleyn, P. W. and Fortuin, C. M.
1969 Phase transitions in lattice systems with random local properties, *Journal of the Physical Society of Japan* 26, Supplement, 11–14.

Kesten, H.
1980a The critical probability of bond percolation on the square lattice equals $\frac{1}{2}$, *Communications in Mathematical Physics* 74, 41–59.
1980b On the time constant and path length of first passage percolation, *Advances in Applied Probability* 12, 848–863.
1981 Analyticity properties and power law estimates in percolation theory, *Journal of Statistical Physics* 25, 717–756.
1982 *Percolation Theory for Mathematicians*, Birkhäuser, Boston.
1985 First-passage percolation and a higher dimensional generalization, in *Particle Systems, Random Media and Large Deviations*, ed. R. Durrett, 235–251, Contemporary Mathematics 41, American Mathematical Society, Providence, Rhode Island.

1986a Aspects of first-passage percolation, in *Ecole d'Eté de Probabilités de Saint Flour* XIV-1984, ed. P. L. Hennequin, 125–264, Lecture Notes in Mathematics no. 1180, Springer, Berlin.

1986b The incipient infinite cluster in two-dimensional percolation, *Probability Theory and Related Fields* 73, 369–394.

1986c Subdiffusive behavior of random walk on a random cluster, *Annales de l'Institut Henri Poincaré* 22, 425–487.

1987a Surfaces with minimal random weights and maximal flows: A higher dimensional version of first-passage percolation, *Illinois Journal of Mathematics* 31, 99–166.

1987b Scaling relations for 2D-percolation, *Communications in Mathematical Physics* 109, 109–156.

1987c A scaling relation at criticality for 2D-percolation, in *Percolation Theory and Ergodic Theory of Infinite Particle Systems*, ed. H. Kesten, 203–212, Springer, Berlin.

1987d *Percolation Theory and Ergodic Theory of Infinite Particle Systems*, editor, Springer, Berlin.

1987e Percolation theory and first-passage percolation, *Annals of Probability* 15, 1231–1271.

1988a Correlation length and critical probabilities of slabs for percolation, preprint.

1988b Asymptotics in high dimensions for percolation, preprint.

1988c Asymptotics in high dimensions for the Fortuin–Kasteleyn cluster model, preprint.

Kesten, H. and Schonmann, R. H.

1988 Behavior in large dimensions of the Potts and Heisenberg models, preprint.

Kesten, H. and Zhang, Y.

1987 Strict inequalities for some critical exponents in 2D-percolation, *Journal of Statistical Physics* 46, 1031–1055.

1988 The probability of a large finite cluster in supercritical Bernoulli percolation, preprint.

Kirkpatrick, S.

1979 Models of disordered materials, Course 5 in *La Matière Mal Condensée*, Les Houches, Session XXXI, 1978, eds. R. Balian et al., North-Holland, Amsterdam.

Klein, S. T. and Shamir, E.

1982 An algorithmic method for studying percolation clusters, *Stanford University Department of Computer Science Report* no. STAN-CS-82-933.

Künnemann, R.

1983a Effective conductivity on a lattice as a limit of box conductivities, Ph. D. Thesis, University of Heidelberg.

1983b The diffusion limit for reversible jump processes on $\mathbb{Z}^d$ with ergodic random bond connectivities, *Communications in Mathematical Physics* 80, 27–68.

Kunz, H. and Souillard, B.

1978 Essential singularity in percolation problems and asymptotic behavior of cluster size distribution, *Journal of Statistical Physics* 19, 77–106.

Lenormand, R. and Bories, S.

1980 Description d'un mécanisme de connexion de liaison destiné a l'étude du drainage avec piégeage en milieu poreux, *Comptes Rendus des Séances de l'Académie des Sciences* 291, *Série B*, 279–282.

Lieb, E. H.
1980 A refinement of Simon's correlation inequality, *Communications in Mathematical Physics* 77, 127–135.

Liggett, T. M.
1985 *Interacting Particle Systems*, Springer, New York.

Lyons, R.
1988 Random walks and percolation on trees, preprint.

Malyshev, V. A.
1975 The central limit theorem for Gibbsian random fields, *Soviet Mathematics Doklady* 16, 1141–1145.

McDiarmid, C. J. H.
1980 Clutter percolation and random graphs, *Mathematical Programming Study* 13, 17–25.
1981 General percolation and random graphs, *Advances in Applied Probability* 13, 40–60.
1983 General first-passage percolation, *Advances in Applied Probability* 15, 149–161.

Menshikov, M. V.
1985 Estimates of percolation thresholds for lattices in $\mathbb{R}^n$, *Soviet Mathematics Doklady* 32, 368–370.
1986 Coincidence of critical points in percolation problems, *Soviet Mathematics Doklady* 33, 856–859.
1987a Numerical bounds and strict inequalities for critical points of graphs and their subgraphs, *Theory of Probability and its Applications* (in Russian) 32, 599–602 (544–547 in translation).
1987b Critical points in the mathematical theory of percolation, Doctoral Dissertation, University of Moscow.

Menshikov, M. V., Molchanov, S. A., and Sidorenko, A. F.
1986 Percolation theory and some applications, *Itogi Nauki i Techniki*, Series of Probability Theory, Mathematical Statistics, Theoretical Cybernetics 24, 53–110.

Menshikov, M. V. and Sidorenko, A. F.
1987 Coincidence of critical points for Poisson percolation models, *Theory of Probability and its Applications* (in Russian) 32, 603–606 (547–550 in translation).

Moore, E. F. and Shannon, C. E.
1956 Reliable circuits using less reliable components, *Journal of the Franklin Institute* 262, 191–208 and 281–297.

Neaderhouser, C. C.
1978 Limit theorems for multiply indexed mixing random variables with application to Gibbs random fields, *Annals of Probability* 6, 207–215.
1981 Some limit theorems for random fields, *Communications in Mathematical Physics* 61, 293–305.

Newman, C. M.
1980 Normal fluctuations and the FKG inequalities, *Communications in Mathematical Physics* 74, 119–128.
1986 Some critical exponent inequalities for percolation, *Journal of Statistical Physics* 45, 359–368.
1987a Percolation theory: a selective survey of rigorous results, in *Advances in Multiphase Flow and Related Problems*, ed. G. Papanicolaou, 146–167, SIAM, Philadelphia.

1987b Inequalities for γ and related exponents in short- and long-range percolation, in *Percolation Theory and Ergodic Theory of Infinite Particle Systems*, ed. H. Kesten, 229–244, Springer, Berlin.

1987c Another critical exponent inequality for percolation: $\beta \geq 2/\delta$, *Journal of Statistical Physics* 47, 695–699.

Newman, C. M. and Schulman, L. S.

1981a Infinite clusters in percolation models, *Journal of Statistical Physics* 26, 613–628.

1981b Number and density of percolating clusters, *Journal of Physics A: Mathematical and General* 14, 1735–1743.

1985 Infinite chains and clusters in one-dimensional directed and undirected percolation, in *Particle Systems, Random Media and Large Deviations*, ed. R. Durrett, 297–317, Contemporary Mathematics 41, American Mathematical Society, Providence, Rhode Island.

1986 One-dimensional $1/|j-i|^s$ percolation models, *Communications in Mathematical Physics* 104, 547–571.

Newman, C. M. and Wright, A. L.

1981 An invariance principle for certain dependent sequences, *Annals of Probability* 9, 671–675.

Nguyen, B. G.

1985 Correlation lengths for percolation processes, Ph. D. Thesis, University of California at Los Angeles.

1987a Correlation length and its critical exponent for percolation processes, *Journal of Statistical Physics* 46, 517–523.

1987b Gap exponents for percolation processes, *Journal of Statistical Physics* 49, 235–243.

1988 Fourier transform of the percolation free energy, *Probability Theory and Related Fields* 78, 165–168.

Nienhuis, B., Riedel, E. K., and Schick, M.

1980 Magnetic exponents of the two-dimensional q-state Potts model, *Journal of Physics A: Mathematical and General* 13, L189–L192.

Nijs, M. P. M. den

1979 A relation between the temperature exponents of the eight-vertex and q-state Potts model, *Journal of Physics A: Mathematical and General* 12, 1857–1868.

Papanicolaou, G. C. and Varadhan, S. R. S.

1979 Boundary value problems with rapidly oscillating random coefficients, in *Random Fields*, Colloquia of Mathematical Society János Bolyai, Esztergom, 27, 835–873, North-Holland, Amsterdam.

Pearson, R. P.

1980 Conjecture for the extended Potts model magnetic eigenvalue, *Physical Review B* 22, 2579–2580.

Reynolds, P. J., Stanley, H. E., and Klein, W.

1980 Large-cell Monte Carlo renormalization group for percolation, *Physical Review B* 21, 1223–1245.

Richardson, D.

1973 Random growth in a tesselation, *Proceedings of the Cambridge Philosophical Society* 74, 515–528.

Rogers, C. A.

1964 *Packing and Covering*, Cambridge University Press, Cambridge.

Roy, R.
1987 The Russo–Seymour–Welsh theorem for continuum percolation on $\mathbb{R}^2$, preprint.

Ruelle, D.
1968 Statistical mechanics of a one-dimensional lattice gas, *Communications in Mathematical Physics* 9, 267–278.

Rüschendorf, L.
1982 Comparison of percolation probabilities, *Journal of Applied Probability* 19, 864–868.

Russo, L.
1978 A note on percolation, *Zeitschrift für Wahrscheinlichkeitstheorie und Verwandte Gebiete* 43, 39–48.
1981 On the critical percolation probabilities, *Zeitschrift für Wahrscheinlichkeitstheorie und Verwandte Gebiete* 56, 229–237.
1982 An approximate zero–one law, *Zeitschrift für Wahrscheinlichkeitstheorie und Verwandte Gebiete* 61, 129–139.

Scheinerman, E. R. and Wierman, J. C.
1987 Infinite *AB* clusters exist, *Journal of Physics A: Mathematical and General* 20, 1305–1307.

Schulman, L. S.
1983 Long-range percolation in one dimension, *Journal of Physics A: Mathematical and General* 16, L639–L641.

Sevsek, F., Debierre, J. M., and Turban, L.
1983 Antipercolation on Bethe and triangular lattices, *Journal of Physics A: Mathematical and General* 16, 801–810.

Seymour, P. D. and Welsh, D. J. A.
1978 Percolation probabilities on the square lattice, in *Advances in Graph Theory*, ed. B. Bollobás, 227–245, Annals of Discrete Mathematics 3, North-Holland, Amsterdam.

Shante, V. K. S. and Kirkpatrick, S.
1971 An introduction to percolation theory, *Advances in Physics* 20, 325–357.

Simon, B.
1980 Correlation inequalities and the decay of correlations in ferromagnets, *Communications in Mathematical Physics* 77, 111–126.

Smythe, R. T. and Wierman, J. C.
1978 *First-Passage Percolation on the Square Lattice*, Lecture Notes in Mathematics no. 671, Springer, Berlin.

Stanley, H. E.
1971 *Introduction to Phase Transitions and Critical Phenomena*, Clarendon Press, Oxford.

Stanley, H. E. and Ostrowsky, N.
1986 *On Growth and Form*, editors, Nijhoff, Dordrecht.

Stanley, H. E., Reynolds, P. J., Redner, S., and Family, F.
1982 Position-space renormalization group for models of linear polymers, branched polymers, and gels, in *Real-Space Renormalization*, eds. T. W. Burkhardt and J. M. J. van Leeuwen, 169–206, Springer, Berlin.

Stauffer, D.
1979 Scaling theory of percolation clusters, *Physics Reports* 54, 1–74.

1981 Scaling properties of percolation clusters, in *Disordered Systems and Localization*, eds. C. Castellani, C. DiCastro and L. Pettiti, 9–25, Springer, Berlin.
1985 *Introduction to Percolation Theory*, Taylor and Francis, London.
1986 Percolation and cluster size distribution, in *On Growth and Form*, eds. H. E. Stanley and N. Ostrowsky, 79–100, Nijhoff, Dordrecht.

Stinchcombe, R. B.
1974 Conductivity and spin-wave stiffness in disordered systems—an exactly soluble model, *Journal of Physics C: Solid State Physics* 7, 179–203.

Straley, J. P.
1977 Critical exponents for the conductivity of random resistor lattices, *Physical Review B* 15, 5733–5737.

Suzuki, M.
1965 A theorem on the exact solution of a spin system with a finite magnetic field, *Physics Letters* 18, 233–234.

Sykes, M. F. and Essam, J. W.
1964 Exact critical percolation probabilities for site and bond problems in two dimensions, *Journal of Mathematical Physics* 5, 1117–1127.

Tasaki, H.
1987a Geometric critical exponent inequalities for general random cluster models, *Journal of Statistical Physics* 49, 841–847.
1987b Hyperscaling inequalities for percolation, *Communications in Mathematical Physics* 113, 49–65.
1988 On the upper critical dimension of random spin systems, preprint.

Tempel'man, A. A.
1972 Ergodic theorems for general dynamical systems, *Transactions of the Moscow Mathematical Society* 26, 94–132.

Temperley, H. N. V. and Lieb, E. H.
1971 Relations between the 'percolation' and 'colouring' problem and other graph-theoretical problems associated with regular planar lattices; some exact results for the 'percolation' problem, *Proceedings of the Royal Society of London, Series A* 322, 251–280.

Toth, B.
1985 A lower bound for the critical probability of the square lattice site percolation, *Zeitschrift für Wahrscheinlichkeitstheorie und Verwandte Gebiete* 69, 19–22.

Toulouse, G.
1974 Perspectives from the theory of phase transitions, *Il Nuovo Cimento B* 23, 234–240.

Welsh, D. J. A.
1986 Correlated percolation and repulsive particle systems, in *Stochastic Spatial Processes*, ed. P. Tautu, 300–311, Lecture Notes in Mathematics no. 1212, Springer, Berlin.

Wierman, J. C.
1978 On critical probabilities in percolation theory, *Journal of Mathematical Physics* 19, 1979–1982.
1981 Bond percolation on honeycomb and triangular lattices, *Advances in Applied Probability* 13, 293–313.
1982 Percolation theory, *Annals of Probability* 10, 509–524.

1983 A bond percolation critical probability determination based on the star-triangle transformation, *Journal of Physics A: Mathematical and General* 17, 1525–1530.

1984 Mixed percolation on the square lattice, *Journal of Applied Probability* 21, 247–259.

1985a Duality for directed site percolation, in *Particle Systems, Random Media and Large Deviations*, ed. R. Durrett, 363–380, Contemporary Mathematics 41, American Mathematical Society, Providence, Rhode Island.

1985b Critical percolation probabilities, in *Random Graphs* '83, ed. M. Karoński, 349–359, Annals of Discrete Mathematics 28, North-Holland, Amsterdam.

1987 Duality for k-degree percolation on the square lattice, in *Percolation Theory and Ergodic Theory of Infinite Particle Systems*, ed. H. Kesten, 311–323, Springer, Berlin.

Wierman, J. C. and Appel, M. J.

1987 Infinite *AB* percolation clusters exist on the triangular lattice, *Journal of Physics A: Mathematical and General* 20, 2533–2537.

Wilkinson, D. and Willemsen, J. F.

1983 Invasion percolation: a new form of percolation theory, *Journal of Physics A: Mathematical and General* 16, 3365–3376.

Willemsen, J. F.

1984 Investigations on scaling and hyperscaling for invasion percolation, *Physical Review Letters* 52, 2197–2200.

Wilson, R. J.

1979 *Introduction to Graph Theory*, Longman, London.

Zuev, S. A.

1987 Percolation threshold bounds for the square lattice, *Theory of Probability and its Applications* (in Russian) 32, 606–609 (551–553 in translation).

Zuev, S. A. and Sidorenko, A. F.

1985a Continuous models of percolation theory I, *Theoretical and Mathematical Physics* 62, 76–86 (51–58 in translation from Russian).

1985b Continuous models of percolation II, *Theoretical and Mathematical Physics* 62, 253–262 (171–177 in translation from Russian).

Index of Names

Subject Index